Heritage and Memory of War

Every large nation in the world was directly or indirectly affected by the impact of war during the course of the twentieth century, and while the historical narratives of war of these nations are well known, far less is understood about how small islands coped. These islands—often not nations in their own right but small outposts of other kingdoms, countries, and nations—have been relegated to mere footnotes in history and heritage studies as interesting case studies or unimportant curiosities. Yet for many of these small islands, war had an enduring impact on their history, memory, intangible heritage and future cultural practices, leaving a legacy that demanded some form of local response. This is the first comprehensive volume dedicated to what the memories, legacies and heritage of war in small islands can teach those who live outside them, through closely related historical and contemporary case studies covering twentieth- and twenty-first-century conflict across the globe.

This volume investigates a number of important questions. Why and how is war memory so enduring in small islands? Do factors such as population size, island size, isolation or geography have any impact? Do close ties of kinship and group identity enable collective memories to shape identity and its resulting war-related heritage? This book contributes to heritage and memory studies and to conflict and historical archaeology by providing a globally wide-ranging comparative assessment of small islands and their experiences of war. *Heritage and Memory of War: Responses from Small Islands* is of relevance to students, researchers, heritage and tourism professionals, local governments and NGOs.

Gilly Carr is a Senior Lecturer and Academic Director in Archaeology at the University of Cambridge's Institute of Continuing Education, UK. She is also a Fellow and Director of Studies in Archaeology and Anthropology at St. Catharine's College, UK. She is author of *Legacies of Occupation: Archaeology, Heritage and Memory in the Channel Islands* (Routledge, 2014) and co-editor (with Harold Mytum) of *Cultural Heritage and Prisoners of War: Creativity Behind Barbed Wire* (Routledge, 2012) and *Prisoners of War: Archaeology, Memory and Heritage of 19th and 20th-century mass internment* (Springer 2013).

Keir Reeves is Professor of Australian history and is the founding director of the Collaborative Research Centre for Australian History (CRCAH) at Federation University Australia. In 2013, he was a Visiting Fellow at Clare Hall, Cambridge, and a visiting researcher at the McDonald Institute for Archaeological Research at the University of Cambridge, UK, where he worked with the Heritage Research Group in the Department of Archaeology and Anthropology. Keir is co-editor (with Bill Logan) of *Places of Pain and Shame: Dealing with 'Difficult Heritage'* (Routledge, 2009).

Routledge Studies in Heritage

1 **Intangible Natural Heritage**
New Perspectives on Natural Objects
Edited by Eric Dorfman

2 **Cultural Heritage and Prisoners of War: Creativity Behind Barbed Wire**
Edited by Gilly Carr and Harold Mytum

3 **International Heritage and Historic Building Conservation**
Saving the World's Past
Zeynep Aygen

4 **Corporate Responsibility for Cultural Heritage**
Conservation, Sustainable Development, and Corporate Reputation
Fiona Starr

5 **Counterheritage**
Critical Perspectives on Heritage Conservation in Asia
Denis Byrne

6 **Industrial Heritage Sites in Transformation**
Clash of Discourses
Edited by Heike Oevermann and Harald A. Mieg

7 **Conserving Cultural Heritage**
Challenges and New Directions
Edited by Ken Taylor, Archer St Clair and Nora Mitchell

8 **The Making of Heritage**
Seduction and Disenchantment
Edited by Camila del Mármol, Marc Morell and Jasper Chalcraft

9 **Heritage and Memory of War**
Responses from Small Islands
Edited by Gilly Carr and Keir Reeves

Heritage and Memory of War

Responses from Small Islands

Edited by
Gilly Carr and Keir Reeves

Routledge
Taylor & Francis Group
NEW YORK AND LONDON

First published 2015
by Routledge
605 Third Avenue, New York, NY 10017
4 Park Square, Milton Park, Abingdon, Oxon OX14 4RN

First issued in paperback 2017

Routledge is an imprint of the Taylor & Francis Group, an informa business

Library of Congress Cataloging-in-Publication Data

CIP data has been applied for.

ISBN 13: 978-0-8153-4696-8 (pbk)
ISBN 13: 978-1-138-83172-8 (hbk)

Typeset in Sabon
by Apex CoVantage, LLC

Contents

SECTION II
Islands of Tourism, Landscapes of War

SECTION III
Islands of War, Islands of Dark and Difficult Heritage

Illustrations

CHAPTER 1

CHAPTER 2

CHAPTER 3

CHAPTER 4

CHAPTER 5

CHAPTER 6

CHAPTER 7

CHAPTER 8

CHAPTER 9

CHAPTER 10

CHAPTER 11

CHAPTER 12

CHAPTER 13

CHAPTER 14

CHAPTER 15

Introduction
Islands of War, Islands of Memory

Gilly Carr and Keir Reeves

During the twentieth century, every nation in the world was directly or indirectly associated with or affected by the impact of war. While the historical narratives and legacies of war among nations are well known, far less is understood about how small islands coped. These islands, often not nations in their own right but rather small outposts of other kingdoms, countries and nations, have frequently been relegated to mere footnotes in history and heritage studies as interesting case studies (at best) and as unimportant curiosities (at worst). Yet for many of these small islands, war had an enduring impact on their history, memory, tangible and intangible heritage and future cultural practices and left behind a legacy that demanded some form of local response.

While some islands have been perceived to be on the periphery of war (the Channel Islands, the Orkney Islands, Texel, Bermuda and Vanuatu), others have been in the epicentre of where the action has taken place, often because of their strategic position in the theatres of war (for instance Guadalcanal in the Solomon Islands and East Falkland and South Georgia in the Falkland Islands). Islands are often noted for their stereotypical images of tranquillity and rurality, and yet when war encroaches upon their shores, the impacts upon the local population are often disproportionate and continue to affect identity and heritage long after the conflict has passed.

For this book, we seek to discover what the memories, legacies and heritage of war in small islands can teach those who live outside them, on the mainland. What can we learn from these islands of memory? Can islands really act as 'controls' or 'laboratories' or even bounded entities that allow us to understand the macrocosm of war memory/heritage microcosm? Conversely, do we regard them as closed and bounded societies when that isn't the case?

This volume also investigates a number of research themes. Why and how can war memory be so enduring in small islands? Do factors such as population size, island size, isolation, climate or geography have anything to do with it? We argue here that close ties of kinship and group identity enable collective memories to last longer and in turn more strongly shape identity and its resulting war-related heritage, as discussed in this volume. In some islands, the shock of conflict and the arrival of armies as 'strangers', 'outsiders', 'occupiers' and 'invaders' are also enough to ensure their long-term impact. Also explored is how contested war memory manifests itself in present-day island communities and in the many ways that these communities have tackled such contestation. Small islands are also places that contain dark heritage sites associated with war such as captivity and internment, and the perceived darkness of such places can seep out, or 'contaminate', the surrounding area. Given the small size of many islands of conflict, this can be problematic, and we seek to understand the process by which that darkness is remembered with positivity and can be turned around to create a positive self-identity for islanders. In other places, dark heritage has been treated only with silence and neglect in the hope that this will allow healing, but the effects of this can be surprising and can have a negative impact on later generations who live in the same small communities.

That this volume was long overdue, necessary and valuable for those who work in this area was revealed in the international conference Islands of War, Islands of Memory that preceded this volume, held at the McDonald Institute for Archaeological Research at the University of Cambridge in April 2013. The deliberate aim of the meeting was to bring islands away from the periphery, a position that they have occupied for too long. At core, we wanted to ask what (if anything) is special about small islands in the way that they nurture their heritage and war memory. This was the central research question of the conference from which this book emerged and was based upon the observation and hypothesis of the editors that there was something different, still raw, and particularly long-lived about the island response to war that begged analysis. In this volume, our central position is that the answer is to be found in two main areas: first, in the particular way that memory is passed down in small, close-knit communities; and second, in the way that the continued circulation, curation and reuse of war debris, broadly defined, appears to play a key role in the creation of heritage, which itself is also used to construct and maintain—and is constructed and maintained by—island identities.

This volume seeks to situate and integrate the research therein firmly and centrally within the field of critical heritage studies, rather than in the often marginal field of island studies. While islands themselves may, by definition, be insular, there is no reason why the study of them should also be so. Accordingly, this book aims to bring islands from the margins to the centre stage in the analysis of the heritage and memory of war during the twentieth century.

While an early concern for the volume was whether we could compare the many island experiences, memories and heritage of war without typecasting the island war experience as a generic one, the chapters collected here show that there is no generic or single experience. If anything, this volume shows that each island has used the memory and heritage of war to construct its own identity, often in direct and deliberate contrast to that which is nurtured and narrated on the mainland. In other cases, the different heritage is 'simply' a direct reflection of a different history compared to the mainland. In all cases, we were impressed by the breadth and richness of the island experience and with the insights that they can bring to those who work on the mainland.

LONGEVITY OF MEMORY

To explore further the observation of the longevity of memory in small communities which are intimately related to their island status, we have identified three factors. First, the long-term role of 'guardians of memory', identified by one of the editors (G. C.) as those who seek to nurture and protect the war memory of their group, can have a disproportionate effect within a small population, where individuals can make a real difference and become well known. Their hegemonic control of war narratives, and the influence of and respect accorded to these same people, can also act to reject and repel outside interference, while ensuring the passing down of their mantle safely to the next generation.

Second, we might also observe a greater continuity of population in small islands, which helps to cement this phenomenon. This is not to deny that population movement takes place in small islands but rather to observe that such movement can often be within the same island rather than away from it and that those who leave will often return, allowing for a continuity of memory.

The third factor is the observation that the island nature of these places can also have a strong impact on the continued presence of war debris, and we cite this as being a key feature of small islands and of prime importance in their analysis, as will be discussed. Because of the difficulty and expense both of removing military hardware and of importing goods, recycling and adaptive reuse have historically been endemic in small islands, not to mention the opportunities for collecting that the continued presence of militaria provides. The continued presence of war debris has also often been assured by the rurality of many islands, where the spread of urbanism has not been as strong as on the mainland because of the difficulties and expense of importing building materials and sometimes the lack of population pressures. Countering this, the climate in small islands can hinder preservation. Being coastal environments, islands face wind and salt erosion, are more at risk from rising sea levels, and can also be prone to the predations of foliage because of their rurality and enhanced fertility.

There are, of course, war-affected small islands which do not conform to these observations. Such places can be silent about or reject their war memory, paying little heed to the war debris which litters their shores. This in itself is a phenomenon to be investigated and explored. Why do such places seem not to care about this aspect of their past, and is this really the case? Such lack of concern often has its roots in a belief that the debris does not form part of their war experience or their history. Sometimes this is because the debris belonged to a hated enemy who caused devastation to their island, too painful to bring to mind, or because it was dropped, constructed or left behind by mainlanders, foreigners, and/or 'colonial masters' from a long way away who seemingly cared little for islanders, their wishes and their sufferings. As some chapters in this volume seem to suggest, the war debris, heritage and memory of the war in the Pacific are neglected and of little interest to local people in Pacific islands. WWII was seen by them as a war fought by foreigners, and it is the foreigners who have instigated memorials in these islands. The only value of the relics of that war is in their potential to be translated into tourist dollars, to be either sold back to or displayed for tourists. Local people seem to have little interest in using war debris as part of their own cultural memory work, although there are exceptions in just a few areas, such as in the John Frum cargo cult which incorporates WWII ephemera within its cultural practices.

In the gap between the willing use and reuse of war debris for heritage and identity construction on the one hand, and its use only for exploitation for the tourist dollar on the other, lies a third pathway for attitudes towards war debris. In islands which have emerged from a colonial rule which held sway at the time of conflict, attitudes towards the colonial power's war machinery and militaria are more equivocal—or at least can be subject to competing factions and narratives. Should such islands exhibit and showcase the war debris of their former colonial masters in order to earn money from tourism, especially when many of those tourists are from the former colonial power, or should such debris be discarded in favour of that which prioritizes the local experience of war? Such dilemmas are only further compounded by the overlap and interweaving of narratives when colonizers and colonials worked together and experienced the same trauma of war together.

WAR DEBRIS EXPLORED

We identify *war debris* in all of its forms as a vital evidentiary lens through which to explore and understand the island response to war memory and heritage. This debris forms the daily reminders scattered about islands, in museums, private houses and still sometimes lying where it fell or was abandoned—or has been moved by visitors. We refrain from referring to war debris as 'heritage', as this is not how it is always perceived. Rather, it is a *legacy* of conflict that can become heritage through deliberate action or that can remain abandoned as simple litter-like debris.

War debris comprises the memory objects which send the wartime generation off into a reverie of recall and reminiscences, the objects used to teach the next generation about the war. And it is also present in the invisible, intangible and unspoken—the traumatic war memories which are passed down to the next generation in the form of post-memory, as defined by Marianne Hirsch in relation to Holocaust memories (1997, 2008), as a process of retrospective familial remembrance whose second-generation stories have come to assume a life of their own.

War debris is a good tool with which to think. But what do we mean by *war debris*? For the purposes of this chapter, we see it as the prime tool to be analysed that can be used to unlock and decode the long-term island response to war. We offer here a definition and subdivision of the term, and all are of equal importance. For us, the *primary* (military) debris of war refers to discarded, fragmentary and scattered large items of war matériel and machinery of war, such as aircraft, bombs and unexploded ordinance, ships, machinery and other war-related maritime heritage. There is also civilian debris, and this refers to destroyed civilian buildings and possessions, typically caused by bombing raids. Such debris is not always cleared away but is sometimes kept as a political statement, a statement of identity, or simply a physical manifestation of an island's war narrative, such as the ruins of the opera house in Valletta, capital of Malta, bombed on 7 April 1942 by the Luftwaffe.

The *secondary* debris of war refers to the infrastructure of war (destroyed or intact), including airstrips, piers, jetties, bunkers and fortifications—but also prisoner of war camps and air raid shelters. Many such sites are often rapidly reused for civilian domestic or tourist purposes but might also be used for war heritage purposes. Examples of this include the many restored German bunkers in the Channel Islands or the aircraft hangers at Ta'Qali in Malta that now host a craft village.

We might also speak of *tertiary* debris, and within this we could cite mainly portable conflict material culture, such as uniforms, medals, flags, helmets, pistols and military paperwork. Trench art is also included in this tertiary debris, a form of recycled war matériel that has been studied in detail by Nicholas Saunders (e.g. 2003, 2004 and, with Paul Cornish, 2009, 2014). Island examples of trench art emanate from a number of sources, such as the POW camp in Orkney discussed in this volume. We might also include within tertiary debris the testimonial objects discussed by Hirsch and Spitzer (2006), which were left behind by those exiled in the Greek island of Ikaria and are curated by local people.

Our fourth, or *quaternary*, and final form of war debris is that which is bodily and which can be subdivided into the (overlapping) tangible and intangible. While the former includes the graveyards, massacre sites and watery graves of military and civilian casualties of war, it also speaks of the injuries upon the living, both of the body (tangible) and the mind (intangible). Such bodily debris includes everything from (tangible) bullet wounds to (intangible) post-traumatic stress disorder (PTSD), which, as delegates at the conference in Cambridge learned, still haunts the war generation of Okinawa.[1] A related form of bodily debris might be said to include the children of soldiers, legitimately or illegitimately born to civilian women, a phenomenon explored later in this volume with respect to Fiji. For these bodily or biological legacies of war, the children of soldiers, their intangible heritage is carried around in their bodies and DNA and also in the memories of the mothers, who themselves become the (sometimes unwilling) guardians of memory of a very particular group, one that rarely speaks about their experiences, such is the shame or taboo that such heritage carries with it in small communities.

Other examples of intangible bodily war debris are the trauma, ghosts or memories of war which continue down the generations in the form of post-memory.

Intangible debris also includes the commemorative ceremonies of war, and here tangible and intangible heritage interacts, as both objects, people and places of war are used as part of the ceremonies, as can be seen in the island of Vanuatu. Oral tradition, too, is a vital part of intangible war debris. This is the case in every island examined in this collection. Because of the centrality of oral history to cultural memory, it is important for researchers to listen to islander voices, narratives and memories and not to force mainlander or outsider interpretations upon history, tourism and heritage. To a large extent, the keeping place of war memory resides with the islanders. Collectively, they can function as guardians of memory. Further, in many places there has been a lack of acknowledgement of the contribution of the indigenous populations to the war effort, whether that contribution is their military service or their degree of suffering, service and sacrifice. Island societies are often permanently changed by such suffering and service of their citizens, and this often becomes a post-conflict legacy which has repercussions up to the present day.

While *debris* is a word filled with connotations of discard, of unwanted rubbish, of leftover pieces, we reclaim and refashion the word here. While the term, as we use it, very much embraces the sense of that which is left over and left behind, we perceive it more as a *legacy*—items which are often cherished, reused or of absolute primary importance to communities and populations in telling the story of what happened to them and who they are today—or are there to be exploited for the tourist economy. They have the potential to be, for some, manifestations and symbols of identity. For others, they represent symbols of identity which are now unwanted or which have outlived their value or have become the focus of contestation between those who look to the past and those who look forward. Again, we might draw upon the example of Malta, where, as the Maltese have emerged from their colonial past, so they have rejected symbols associated with the erstwhile motherland such as the George Cross.

Yet with war debris defined and subdivided, the core of its value for researchers is to examine its local meaning and its biography—its changing role through time. How has it been treated since the war in question? How has it been used, reused, recycled, redefined, received, perceived or rejected by each post-war generation? The groups of debris just listed are presented in their 'unadulterated' original state; yet war debris is often recycled, reused, domesticated, tamed and refashioned. Trench art is perhaps the best known example of such a process.

The site of certain types of debris can become of prime importance in islands; they can become sites of memory and places where memorials are erected and commemorations take place. They can also become sites of deliberate forgetting and can be shunned by the community. Debris can also become hallowed and curated in museum exhibitions and even used in memorials. But what is the degree of overlap and the interrelationship between the types of war debris, tangible and intangible, when used, discussed and curated in heritage responses to war? How are they used, as a group, to tell the story of war to the next generation and to tourists? These are two very different populations, and there is often a difference between what can be learned from parents and grandparents around the metaphorical fireplace and what is displayed for public consumption on tourist routes. In fact, it seems that in some islands the heritage would not exist or would not be curated and displayed at all if it were not for the tourist interest—for example in the Solomon Islands. What, then, does this say about war debris and the use to which it would be put if it were not for heritage and tourism?

DEFINING SMALL ISLANDS

In this volume and in the original conference from which this volume emerged, we repeatedly refer to 'small islands'. We have been unwilling to define this phrase too closely; we did not want to be too prescriptive in terms of population size or island dimensions and thereby inadvertently exclude certain revealing or important case studies. We are probably guilty of stretching our own definitions of 'small' for the same reasons.

We have, however, been strict in excluding places, such as Gibraltar, which are on the edges of bigger countries but are still connected to them via a land bridge, despite being thought of as an 'island' of Britishness. Also rejected was Cold War Berlin, which was an 'island' of Western Europe within East Germany. But just as these places are not 'real' islands, we might also consider factors which arguably threaten an island's status (e.g. Baldacchino *et al.* 2007) or increase feelings of insularity (e.g. Vannini 2011), such as bridges, tunnels and causeways, and non-fixed links, such as ferries, flights and the Internet, respectively.

We are used to borders enclosing nations, ethnicities and cultures, with defined centres and peripheries, and seats of power, authority, and governing elites in the nucleus, with marginalized communities at the edge (Sofield 2006, 102), but small islands present something rather different. Not only is the concept of centre and periphery hard to identify within an area of only a few miles in diameter, but the idea of being bounded by nationality, ethnicity or culture can be a complex one when discussing small autonomous or semiautonomous states or archipelagos. Geographical and political boundedness in an island context is further complicated by issues of immigration, diaspora and, during the past twenty years, the arrival of the Internet to remote communities. There is thus fluidity in island communities in terms of how they are constituted geopolitically. This means that islands are separate when they want to be but joined up (both to the rest of their island groups or the mainland) at other times.

Our main concern in putting together this volume was to fight against the mainstream, mainland perception of these 'small' places, as well as the implications of something that was insignificant, easily dismissed and legitimately excluded from academic debate. Attention was drawn to such perceptions by Hau'ofa (1993), who argued that the prevailing attitude towards island societies was that they were too small, at a remove from key economic activities, and often ill-equipped with limited natural resources, which meant that they were unable to function autonomously. This was a Western-determined narrative that ultimately led to island communities being stereotyped as economically dependent and fatalistically consigned to the role of client states of powerful, often former colonial masters. Hau'ofa concluded that it was this attitude or metaphor that constituted a belittling language that was both potentially disruptive and misrepresentative of small island life. While Hau'ofa wrote explicitly about Pacific societies in a postcolonial context at the end of the twentieth century, his overarching point—that islands are often depicted as too small, culturally and economically insignificant, and often at a remove from the key drivers of power and influence—is an enduring one amongst small island communities throughout the world.

For the purposes of this volume, then, we contend that small island communities are ones that are typified by a relatively small scale and intimate cultural practices and accordingly are more vulnerable to external events (such as invasion or colonization) to a greater extent than mainland communities. Lest it be thought otherwise, small islands are not immune to civil war, as a number of case studies in this volume show. Yet if a small island is typified by smallness of scale and closely bound communities, then what do we mean by the term *island*? It is often bound up in the baggage of assumptions that comes with certain islands, namely issues of exoticism, untouchedness, 'otherness', wilderness, paradise playgrounds for mainlanders, remoteness, isolation and stereotypes of unproblematic lives typified by tranquillity. Yet in many cases, islands have been key theatres of war during the twentieth century and have experienced unprecedented levels of conflict and violence, where the residue of war continues into the present day both physically and psychologically.

Those of us who conduct fieldwork in small islands see and know the value in the very 'smallness' of these places; we know the richness of the case studies and their wider implications for places that are much larger. For us, small is certainly beautiful and has only positive connotations of close-knit communities, which has led to intact and strong war memories, narratives and heritage, as well as to material richness in terms of war debris, militaria and conflict material culture. Thus, we are interested in islands that are small enough to share these attributes and that are culturally rich in war heritage; in fact, our observation seems to suggest that these are attributes that many—if not most—post-conflict small islands seem to share. As argued earlier, it is their very *smallness* and all of the associated characteristics of such islands that have contributed to the rich survival.

IDENTITIES IN SMALL ISLANDS

Islands often choose or claim alternative identity markers to their mainland cousins, and different war narratives can be an important part of this. However, we must question this apparent 'choice'. Is identity a deliberate choice, or does it just happen naturally and organically, based on different war experiences? Is this, in fact, a false dichotomy? Who, in the first scenario, are the decision makers, and how do they agree upon and create narratives and consensus? In a free democracy, identity construction is often a series of decisions made by many along the way, rather than a preconceived grand plan manipulated by a single individual, although such examples are not unknown. In almost every case study in this volume, identity has developed as a series of 'organic decisions' made at different times by individuals and influenced by different events, people and cultural trends. When these events and influences are past wars and their impact (an impact which we must remember can be disproportionate in small islands), identities of conceptually or geographically bounded peoples can be built around post-war narratives such as victimhood, suffering, endurance and victory. The memories and legacies which inform and are informed by such narratives will shape the resulting heritage. This heritage, which takes its form in museums, memorials and commemorative ceremonies, makes identity claims for and on behalf of the local population. Just as they construct the heritage to proclaim who they are, so that same heritage will teach and influence the next generation. While subsequent generations will often modify the way that heritage is presented through time, drawing new narratives into that heritage as they are exposed to outside influences or embroiled in later conflicts, some war narratives can be extremely slow to change because of both the inward-looking nature of many island communities and the influence of the guardians of memory as discussed earlier in the chapter.

While we can hypothesize about how islanders see themselves and construct their identities in different ways after conflict, we must also consider how outsiders or mainlanders perceive them, as well as how this can change during and after conflict. Certainly we can observe how otherwise almost-ignored islands, which hover on the very outer edges of the awareness of the inhabitants of larger nations, can suddenly acquire new, more central positions, as happened after the Falklands conflict for people living in the United Kingdom. Islanders can find that they are treated with a new respect after conflict or move to the forefront of national consciousness if their ownership is contested. Inevitably this will impact upon tourism strategies and the particular face that they wish to present to newly inquisitive outsiders, who can be drawn to these islands in search of 'dark tourism' experiences (as defined by Stone 2006). But what aspects of war, war heritage and war identities are showcased, if at all, and who decides this? Decisions are not made only at the local governmental level: tourism officials and local entrepreneurs can both independently decide what to promote. There are also stories of war that are kept firmly within the community, the family and the home, and not shown to outsiders. How different are these two faces, or are they simply two sides of the same coin?

Perhaps it is more important to ask why anyone would choose to promote war heritage: is this a decision based on shrewd money making and the exploitation of outsiders, or is it actually simply unavoidable, given that heritage and identity are interlinked? Does such war heritage proclaim an unveiled this-is-who-we-are or a selective and manipulated this-is-what-we-wish-to-show-outsiders? How much clear blue water separates these statements?

These questions also impact upon written history (such as school textbooks) and heritage planning documents and legislation for use within island communities. These texts and documents are influenced by whether they are written by islanders for islanders or by outsiders from the mainland with little reference to the role of, or input from, islanders.

It is clear that multiple potential interest groups and narratives are at play here, and this is not limited just to islanders and mainlanders. There are also the different groups within the island communities to consider, such as the island officials/elites and non-officials/non-elites; there are also the different generational, religious and, potentially, ethnic groups; and there are different memory groups and their associated guardians of memory. If there is more than one island within an archipelago, then there are also inter- and intra-island identities to grapple with. To complicate matters further and to exacerbate (or to conceal) the clashes among these groups, when it comes to heritage and the representation of post-conflict identities, there are both heritage professionals and amateurs from each of these different groups to contend with. The problem only gets worse if we add into the mix the legacy of more than one historical conflict fought on island soil or the problems of colonialism. Clearly, we can never talk of a singular island voice which we might use to contrast with or to counter strong memories and narratives from the mainland and which themselves are similarly contested and multiple. When it comes to small islands, diversity is our watchword.

STRUCTURE OF THE VOLUME

The sixteen papers chosen for this volume, selected from twenty-six presented at the conference with an additional three commissioned, share many commonalities in terms of the war island themes already highlighted by the editors in this introductory chapter. These include the key role of guardians of memory; the importance of population continuity; and the continued presence, use and reuse of war debris in its many forms. All of these have contributed, we argue, to the longevity of memory and to the prominence of the heritage of war in small islands, which is not to say that there are not notable exceptions that prove the rule. While this volume was initially prompted by the observation of the editors of an unusual longevity of memory in the islands where we both carry out fieldwork, one of the central questions that we asked authors to address was what, if anything, is special about war memory and heritage in small islands that can contribute to debate in war heritage and memory studies as a whole? What, beyond the case study itself, can be carried forward to inform those who work in and with larger nations? We have been inspired by the answers we received, which were rich, diverse, valuable and fascinating.

While it will not escape the notice of readers that the majority of the chapters concern WWII, this was not a stipulation given to authors; rather, it is a reflection of the research specialism of those who carry out fieldwork in small islands. As WWII impacted almost every place on earth, unlike wars before, it is not surprising that small islands were also impacted. As this war is on the edge of living memory, the number of researchers now working in this field to collect data before it is too late is not surprising. We are also working in a time of anniversaries—the year of the publication of this volume marks seventy years since the end of WWII—and during such periods, public and academic attention is often highly concentrated. Similarly, readers will observe that a high number of chapters involve Greek islands or Pacific islands. We make no apology for this; not only are these places where fieldwork is concentrated, but they are also locations which provide us with fascinating insights, and the volume would have been the poorer for their omission.

Broadly speaking, the main theme of each chapter in this volume can be categorized into at least one of three overlapping sections presented here, and they cross-cut the trio of small island commonalities previously listed. These sections address the kinds of war memory and heritage that typify small communities; the importance of war heritage tourism for island economies; and the profound impact of a dark war heritage within the heritage and memory of small communities.

Our first section, 'Islands of Memory, Islands of Community', comprises chapters that focus primarily upon or compare selected communities of memory. Our first chapter, by Jacqueline Leckie, reveals the often overlooked experience (and intimate memories) of Fijian women who had encounters with American servicemen during the war, producing children. These women, historically invisible due to the masculinized military history of Fiji, were silent about their encounters, seldom revealing details to their offspring about their fathers. Leckie thus reveals that memory can also be about burying, eliminating and ultimately forgetting.

Hazal Papuccular's chapter addresses the different ethnic communities that make up the Dodecanese islands of Rhodes and Kos: the Greeks, the Turks and the Jews. While the islands were occupied by the Germans, leading to potential brutality, starvation and migration for all, the three groups perceive and remember this period in very different ways, which has led to different sites of memory within the islands. Even within small islands, Papuccular's work shows us, memory and heritage are far from monolithic.

The next chapter, by Elena Mamoulaki, takes us to another Greek island: Ikaria, a place of exile for political detainees during the Greek civil war. While this could have led to the island becoming a place of traumatic memory and stigma for generations to come, Mamoulaki shows us how the hospitable Ikarians have treated this memory as a source of pride and identity formation, leading to community commemorations and heritage construction.

Gilly Carr's chapter considers one of the central themes of this volume, namely the role of 'guardians of memory', a concept and epithet that she develops here with respect to the Channel Islands. These figures comprise mainly the senior members of the community who lived through the war and, through dint of their age, good memory and eloquence, are held up and respected as the ultimate authorities for the period. She considers their position with respect to the university-trained heritage professionals and the outside researchers, identifying how their mantle will be passed on and the impact that they have had and will continue to have on heritage and heritage structures in these small islands after their generation has passed on.

Finishing this section is the contribution of Rob van Ginkel, who examines the Dutch island of Texel. What makes this island unique is the formation of a community of memory and inextricable linking with Georgia. During WWII, the contingent of Georgians within the German army occupying the island rose up and killed their German comrades. Many Texelians were caught up in the fight and killed alongside a number of Georgians. Despite the potentially controversial presence of these hero/villain 'ambiguous dead' in the 'Russian' graveyard in the island, Texelians have long looked after it and hosted visiting dignitaries from Georgia. Although the population is split, the relationship with Georgia is cherished because it gives the Texelians a unique identity, different from the mainland Netherlands.

Van Ginkel's chapter provides a smooth segue to the first chapter of the second section of the volume, 'Islands of Tourism, Landscapes of War', which begins with Daniel Travers' assessment of Orkney. Like Texel and Georgia, Orkney has sought to differentiate itself from the mainland through its wartime connections with outside visitors to the island. In this case, the role of Italian prisoners of war, who built a beautiful and still-maintained chapel out of Nissen huts, has cemented ties with the Italian town from which the POWs came. Orkney has used its unique heritage to strengthen its own sovereignty and differentiate itself from its more powerful neighbours.

One of the core themes emphasized by many chapters in this section of the volume, in addition to the presence of war debris in the landscape of these tourist islands, is the importance of the indigenous perspective and local understanding of this material culture. Moving to the Western Pacific, especially Melanesia, Keir Reeves and Joseph Cheer's chapter emphasizes the local value of war heritage, which lies in its commercial potential rather than in its historical significance. A distinctive legacy of WWII is the monetization of communities that were previously custom-focused. Islanders have felt little need to conserve or curate the war debris unless there has been a commercial imperative to preserve it, regardless of its heritage significance to former combatant countries such as America, Japan, Australia and New Zealand.

Sandro Debono takes us away from the Pacific and back to the Mediterranean, where he considers the situation in his native Malta, the most bombed place on earth during WWII. Once a colony of the British but independent since 1964, the islanders today grapple with the importance of rival symbols to represent themselves as a postcolonial people. On the one hand, the Maltese feel a pride in having received the George Cross in 1942 for their solidarity, courage, fortitude and determination in times of difficulty—values which they still recognize and embrace today. On the other hand, this symbol of war carries the associated and complicated baggage of colonialism, unlike the preferred Maltese Cross, which has intertwined, associated and variously complementary and competing meanings. The non-British Maltese war heritage is a community one and revolves around their deceased kin, but as an island which receives many British tourists, the islanders instead promote an official pro-British war memory of a time when the island showed its finest attributes but which also brought about its destruction.

Returning to the Pacific, Lamont Lindstrom explores Efate island in Vanuatu, which hosted thousands of American soldiers during WWII. Now, two generations later, Americans are once again returning as tourists. This boom to the economy has spurred local entrepreneurs to collect war relics, which they use to share stories in the island's war museums. However, rather than this having a purely economic impact on the island, it has led to the salvaging of history alongside the salvaged war debris. War memory is being kept current through these artefacts and relics. As Lindstrom puts it, 'Vanuatu war memory today adheres in relics and the stories they revitalize'.

Tony Pollard, too, stays with the subject of war debris within the landscape of war of the British Falkland Islands. Invaded and occupied by the Argentineans and then liberated by the British in 1982, this chapter is one of the few which does not concern WWII. However, Pollard examines the way in which the material culture of conflict from different battlefields was gathered, recycled and used by soldiers and later differentially curated in the landscape by veterans and tour guides, giving us an insight into how similar strategies of curation and 'manicure', as he calls it, may have impacted WWII debris that we still find lying in the landscape today in other places.

The final chapter in this section by Geoffrey White emphasizes once again the indigenous perspective, although this time exploring attitudes towards memorials rather than the war debris that litters the Solomon Islands. Like the artefacts which were 'not their war', islanders feel little towards memorials, which have largely been the initiative of foreign powers due to a lack of local interest. Where memorials have included indigenous people, they, like the written histories, have tended to cast them as 'loyal natives' and subjects in other people's narratives. Thus, White's oral testimony project looks to indigenous logics to record indigenous perspectives, narratives and histories and to find a way to engage Solomon Islanders in their own history, heritage and war tourism.

Our third and final section of the volume, 'Islands of War, Islands of Dark and Difficult Heritage', looks at how islands overshadowed by particularly traumatic war memories and sites of violence, oppression or internment have tackled the difficult issues surrounding heritage presentation in these islands. Our first chapter, coauthored by Neil Price, Rick Knecht and Gavin Lindsay, examines Peleliu island in Palau, Micronesia. Like the chapters in the previous section, these authors also take the war debris in the landscape as the theme of their chapter. However, such was the high death rate on this tiny island, where American deaths were more numerous than those of the Allies on all five Normandy beaches on D-Day put together, that the darkness of this chapter is addressed in this third section of the volume. The authors examine the various reasons for souveniring from the dense scatters of militaria in Peleliu. While the local people take items of 'lethal litter' for pragmatic reasons (which include taking items for museums for tourists), the collecting behaviours of the visiting Americans and Japanese focus on memorializing the battle. While Americans tend to focus on Japanese souvenirs as war trophies, the Japanese perceive these items as sacred and leave them as offerings on shrines.

For Nota Pantzou, the darkness of the exile islands of the Aegean such as Makronisos, Giaros and Ai Stratis lies not in their material culture but resonates in the camp buildings that still remain in these remote places. For Pantzou, the dark narrative of exile jars with the beauty of the Aegean and the national narrative of resistance during WWII, leading to the neglect of tourism, especially dark tourism, in these islands of difficult history.

Pantzou, however, sees change afoot, with young Greek people showing an interest in preserving and presenting the heritage in these islands. The first green shoots of future dark tourism to these islands are, she believes, beginning to emerge after decades of neglect.

Staying in the Mediterranean, Irene Lagani's chapter addresses the Greek island of Kythera, where the silence and neglect of memory, along with the absence of a WWII heritage, has led to disastrous results. While this silence has been encouraged as a way to promote reconciliation and heal old wounds, Lagani shows that a different result has been achieved. A policy of forgetting, as well as the ignorance of the island's young people of their own history, has led to a vulnerability to neo-Nazi movements, namely Golden Dawn. With the current economic climate in Greece exacerbating problems, historical ignorance and the neglect of heritage have come at a high cost.

Maria Kagiadaki also deals with the German occupation of a Greek island: Crete. In a departure from other authors, the form of heritage that she addresses is art, and primarily the artistic forms employed in Cretan war memorials. Kagiadaki particularly focuses on how memorial art is used to construct and reflect Cretan memory and narratives of war, putting a positive spin on traumatic scenes of execution and fighting and turning them in to opportunities to depict, instead, brave episodes of Cretan resistance—a national characteristic that chimes with official war narratives. When designing the memorials, the Cretans drew on earlier modes of depiction of heroes of the Greek Revolution and of Byzantine motifs in art, thus drawing on the past to validate the politics of the present, regardless of historical accuracy.

The chapter which brings to a close our discussion of dark heritage is by Michael Leach, who examines Timor-Leste. Here the situation—and the heritage—is complicated by not just one war, but by 450 years of Portuguese colonialism which ended in 1974, the Japanese occupation in WWII and finally a traumatic twenty-four-year struggle against Indonesian occupation, which ended in 2002. These layers of trauma, each with its own sites of memory of massacre, political imprisonment and torture, and with their own guardians of memory, has led to fissures in the patchwork of memory claims and sites among different generations. This has not been a recipe for harmony in nation building in this post-conflict island.

Themes of heritage and memory of war are certainly not unique to small islands, although the editors believe that small islands have their own unique way of dealing with them. The impact of events of such magnitude as war cannot be hidden, neglected, forgotten or suppressed for long in small communities, especially while there are still many alive who remember them. We hope that this volume will result in the repositioning of fieldwork, case studies and concepts emanating from small islands into the mainstream of heritage and memory studies, and that the chapters here will be considered and assessed alongside better known case studies in centres of power across the world by professionals and practitioners alike.

NOTE

1. Paper by Ryoji Aritsuka, "Post-Traumatic Stress Syndrome (PTSS) in the battle in Okinawa islands in Japan 1945," Islands of War, Islands of Memory conference, University of Cambridge, 5–7 April 2013.

BIBLIOGRAPHY

Baldacchino, G., MacDonald, E. and Spears, A. (2007) *Bridging Islands: The Impact of Fixed Links*, Charlottetown, Prince Edward Island: Acorn Press.

Cornish, P. and Saunders, N. (2014) *Bodies in Conflict: Corporeality, Materiality and Transformation*, London and New York: Routledge.

Hau'ofa, E. (1993) *A New Oceania: Rediscovering Our Sea of Islands*, Suva, Fiji: University of the South Pacific.

Hirsch, M. (1997) *Family Frames: Photography, Narrative and Post Memory*, Cambridge: Harvard University Press.

Hirsch, M. (2008) 'The generation of postmemory', *Poetics Today*, 29 (1), 103–28.

Hirsch, M. and Spitzer, L. (2006) 'Testimonial objects: memory, gender, and transmission,' *Poetics Today*, 27 (2), 353–83.

Saunders, N. (2003) *Trench Art: Materialities and Memories of War*, Oxford and New York: Berg.

Saunders, N. (2004) *Matters of Conflict: Material Culture, Memory and the First World War*, London and New York: Routledge.

Saunders, N. and Cornish, P. (2009) *Contested Objects: Material Memories of the Great War*, London and New York: Routledge.

Sofield, T. (2006) 'Border tourism and border communities: an overview', *Tourism Geographies*, 8 (2), 102–21.

Stone, P. (2006) 'A dark tourism spectrum: towards a typology of death and macabre related tourist sites, attractions and exhibitions,' *Tourism*, 54 (2), 145–60.

Vannini, P. (2011) 'Constellations of ferry (im)mobility: islandness as the performance and politics of insulation and isolation,' *Cultural Geographies*, 18 (2), 249–71.

Section I

Islands of Memory, Islands of Community

1 Islands, Intimate and Public Memories of the Pacific War in Fiji

Jacqueline Leckie

INTRODUCTION

War is among the most celebrated and painful of collective and individual memories. This ranges from the world wars that are remembered globally, often through a national lens, to less known localized wars. The Pacific War evokes vivid memories of tenacious combat on small islands in Western Melanesia and Micronesia—in 2010 brought to the television screen in *The Pacific*.[1] This war also had a profound impact on the small Pacific Islands[2] that were outside the battle zone. New Hebrides (Vanuatu), New Caledonia, Fiji, Cook Islands, American Samoa, Western Samoa, Tonga, Society Islands, Wallis, Bora Bora and New Zealand were all support and rear bases. These were crucial to the operations of the Pacific War and the recuperation of the Allied personnel. The contribution of Pacific peoples, however, continues to be overlooked when remembering WWII, despite the significance of Islanders as combatants and the stories of civilians in these islands coming to light through academic research (Camacho 2011; Falgout *et al.* 2008; Laracy and White 1988; Poyer *et al.* 2001; Ravuvu 1988; Toyoda and Nelson 2006; White *et al.* 1988; White and Lindstrom 1989). Here only one group of the Pacific Islands is discussed—the Fiji Islands (Figure 1.1)—as part of a project that has traced the children born to Pacific Islands women and American servicemen during WWII.[3] This focused on the uncelebrated war memories—intimate encounters that are not usually part of the narratives of islands in the Pacific.

This chapter explores how the war is now being publicly remembered and represented in Fiji and then focuses on intimate memories that are often secret but rarely forgotten.[4] These are not the combatants' memories but those of Fijian women who had close relationships with American servicemen and became mothers of their children. These fragmented memories have been passed on to create war memories for these children, their descendants and others.

The images in this chapter evoke different memories of the Pacific War in Fiji. Figure 1.2 is from the seventieth anniversary in 2013 of the departure of the Ai Matai (the 1st Battalion and 1st Commando) from Suva on 13 April 1943 to fight in the Solomon Islands. In many ways this image corresponds to conventional war commemorations, with the ceremony held at

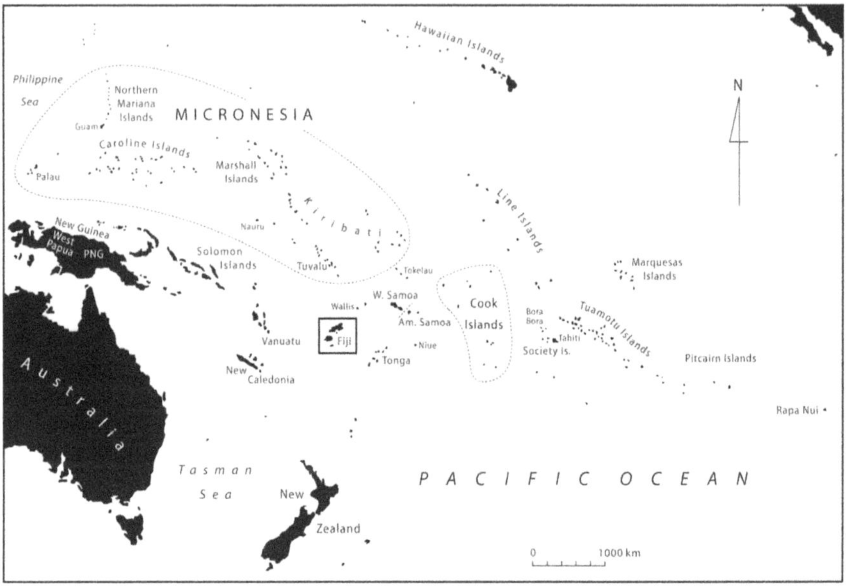

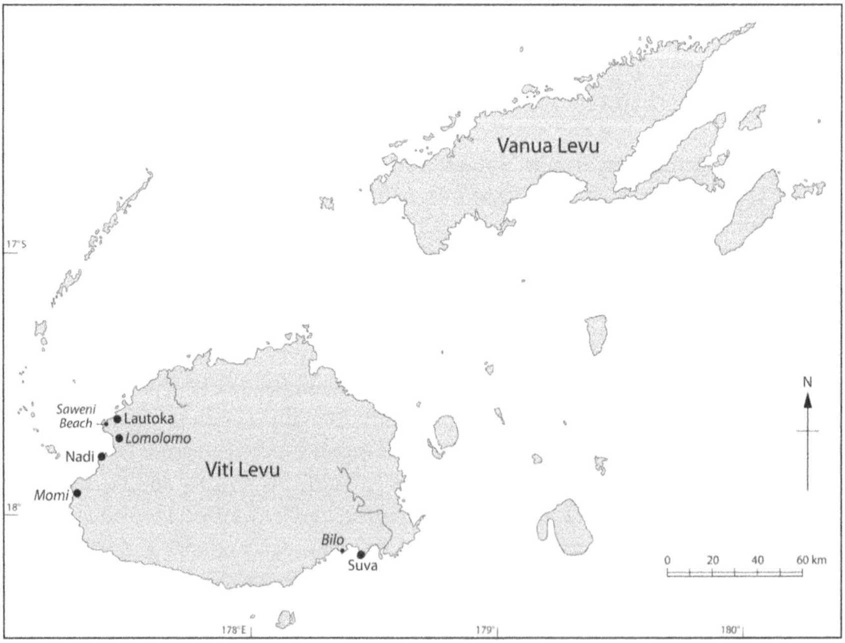

Figure 1.1 (a) Map of the Pacific Islands; (b) map of Fiji (Image copyright and courtesy of Les O'Neill).

the austere National War Memorial on Battery Hill in Veiuto, Suva. While the 'Last Post' was played, the 'Men of war—current and ex-servicemen—stood erect under a tent from the mid-morning sun. They were like statues' (Seru 2013).[5] Fiji Returned Soldiers and Ex-Servicemen Association president, Major (retired) Ratu Peni Volavola declared this as '[a] day that will be forever etched in the hearts of all patriotic Fijians' (Seru 2013). This shows how national war memories become woven into narratives of the nation and resonate with the present. Although the material remains of war may be forgotten, the hidden human connections of war have been kept alive. The guns of Bilo were considered by local villagers to be the biggest 'in the southern hemisphere',[6] but the locale, once 'a hive of activity during the war' (Tagivetaua 2011), has since became overgrown with vegetation. The ruins of the gun turrets, lookout post, tunnels, cellars and ammunition stores, have been used as 'a drinking spot, a lovers' lane and scenic lookout'; as a hurricane shelter; and as a film location (Tagivetaua 2011). The restoration of Bilo and some war sites in Fiji is now caught up in heritage projects that carry the allure of the tourist dollar. The story behind the photo of a mother and her son, Nate[7] (Figure 1.3), remains private, not from forgetting but due to the sensitivity of the intimate memories from WWII. Nate's memories, like those of many children born to American GIs, are among those that are both the most secret but also the most enduring of the war's impact in the Pacific Islands. There are several terms for these children, including *Kai Merika* in Fiji.

Figure 1.2 Ai Matai 70th Anniversary, 2013. National War Memorial, Suva, Fiji (Copyright and courtesy, Fiji Ministry of Information).

Figure 1.3 Nate and his mother, c. 1946 (Private source. Supplied courtesy of Marsa Dodson).

In considering examples of intimate and public memories, this chapter also touches on issues of how war might be remembered differently on small islands. Yet in Pacific cultures and Pacific studies, 'small' is a hotly contested concept—notably in Epeli Hau'ofa's provocative essay, 'Our Sea of Islands' (1993). He rejected negative and belittling perceptions of smallness of the Pacific Islands in order to assert pride in small island identity. Hau'ofa disturbed Eurocentric perceptions of separate small islands to stress their ecological, genealogical, cultural, linguistic and spiritual connections. The Pacific War demonstrated these linkages as well as the ties of small communities to global events. Some of the war's biggest battles were fought on tiny islands in Micronesia. Pacific material realities depend as much upon the sea as the land, and Pacific identity draws as much from 'The Ocean in Us' (Hau'ofa 1998). This contrasts with the preconceptions foreign soldiers had of places like Fiji as small, faraway, isolated and exotic islands. In fact, Fiji comprises over 300 islands, although most people live on the island of Viti Levu—hardly tiny (10,388 kilometres long and 146 kilometres wide) with a densely forested mountainous interior.

Hau'ofa's insight into Pacific concepts of time also has bearing upon how war might be remembered differently within the Pacific Islands. He contrasted ecological, circular and sequential notions of time within Pacific cultures with more Western 'linear progression in which the past is behind us' (Hau'ofa 2000, 459). Suzanne Falgout, Lin Poyer and Laurence Carucci (2008, 20) regard the latter as a 'file drawer' model of memory, where lived experience is stored to be opened only at specific places (such as at monuments) or times (such as during commemorations). Within many Pacific cultures, memory is a lived experience and legacy, and this is especially pertinent to intimate war memories. 'The past is alive in us, so in more than a metaphorical sense the dead are alive—we are our history' (Hau'ofa 2000, 460). Yet, as this chapter indicates, individuals and groups also bury aspects of the past—forgetting can be a strong response to intimate and traumatic events of the war—including the birth of children.

In addition, Falgout, Poyer and Carucci (2008, 26; also Camacho 2011) discuss how cultural politics may influence how war is remembered in the Pacific Islands. In Fiji, cultural politics are entwined with national politics, which bear upon interpretations of the nation's war history and its legacy.

THE WAR AS NARRATIVE OF THE NATION

In the Pacific Islands, WWII, as elsewhere, was a catalyst to social, economic, political and environmental change (see Bennett 2009, Firth 1997, Laracy 1994). At the national level in Fiji, it can be argued that rather than the direct impact of battles or events, the most significant legacy was how the war fed into the construction of national narratives of Fiji. This took on a markedly racialized nature, and such wartime narratives, although subtly changing, still feature within contemporary nation building.

The enduring war memories and legacies that became part of a racialized narrative of the nation resonates within the political and ethnic tensions in colonial and contemporary Fiji. As historian Brij Lal (1991, 18) emphasized, 'the war became a powerful symbolic event in Fijian history, bequeathing a legacy of suspicion, division, distrust, and hostility among the different communities in Fiji'. These different communities are indigenous

Fijians (*i-Taukei*), Indians (Indo-Fijians or Fiji Indians) who originally settled as indentured labourers (*Girmitiyas*) and Europeans who dominated Fiji when it was a British colony between 1874 and 1970. Fiji also has significant communities of mixed ethnicity, as well as Pacific Islanders from outside Fiji. By 1946, within a total population of 259,638, Indo-Fijians comprised approximately 46 per cent of the population, Fijians 45 per cent, but Europeans less than 2 per cent.[8]

This divisive legacy had its origins in Fiji's racialized colonial history that predated WWII, but the war years reinforced powerful ethnic stereotypes. Disloyal and cowardly Indians were contrasted with heroic and loyal Europeans and Fijians—the latter two being comrades within the Fijian Military Forces (Howlett 1948; Ravuvu 1988).[9] This narrative implies that the sacrifice of indigenous Fijians in saving the nation from alien invasion was also evidence of Fijian loyalty to the Empire. Ratu Sir Lala Sukuna called for Fijians to sacrifice their blood, which *John Kelly (1995*, 491) suggests brought 'blood credentials'. Lal argues that Fijian loyalty was reciprocated after the war with a strengthened and reformed Fijian administration and the continued British protection of the paramountcy of Fijian interests. Thus the war cemented the narrative of a special bond between Fijians and Europeans, adding fuel to Fiji remaining a colony until 1970. This narrative of a racially divided nation, which excluded treacherous Indians from power, was reignited within the racial discourse of the 1987 and 2000 coups in Fiji.

Remembering the Pacific War and the sacrifice of Fijian soldiers has been incorporated into those narratives of contemporary nationalism that emphasize Fijian nationalism and militarization. Fijian sacrifice during the war has also been expressed as not just for nation and empire but also more proactively to protect Fijian culture from an external threat. Looking back on village life during the war, 84-year-old Epeli Raganivatu recalled that '[t]he Japanese would have changed the Fijian system and way of life but we thank the Fijian soldiers who answered the call to battle and some of them gave their lives' (Tagivetaua 2011). The Fijian military contribution in WWII is also synonymous with the Fiji Military Forces.[10] This war legacy has become foundational to its identity and role as the most powerful body in Fiji today (see Firth and Fraenkel 2009).

Figure 1.2 shows the recently established National War Memorial in Fiji on the site of the WWII Suva Battery. The choice of this space is significant as it is also where a new Fiji parliament was built by the military regime after the 1987 coups led by Lieutenant Colonel Sitiveni Rabuka. Local businessman George Speight and his followers occupied the same parliament and grounds when he led a coup and took hostage the government for nearly two months in 2000.

It is tempting to see the funding and national hype given to the National War Memorial as a direct consequence of the military regime and interim government led by Commodore Frank Bainimarama after he instigated Fiji's fourth coup in 2006. He strongly articulated national unity and symbols that would signify this (see Firth *et al.* 2009). Plans for a national war memorial and military museum predate 2006 and began in 1999 when Rabuka was prime minister but were put on hold by the 2000 coup. In 2002, the government and Fiji's Great Council Chiefs re-endorsed the construction of a war memorial for 'Fiji's warriors' (Radio Australia 2002). The goals included repatriating the remains of Fijian war hero Corporal Sefanaia Naivalu from Papua New Guinea, along with other Fijian soldiers who died during WWII. Naivalu was awarded the Victoria Cross in recognition for sacrificing his life to save his comrades when fighting the Japanese in Bougainville in 1942. By 2010, these plans widened to include a clubhouse, a statue of Fijian war hero Corporal Sefanaia Sukanaivalu (Tuimaleali'ifano 2001), a war museum, a water garden, a 400-seat pavilion and marketing the memorial as a tourist attraction (fijivillage.com 2010).

The construction of the National War Memorial fits with perceptions of war and the military as a dominant narrative of the nation, although by 2010 this narrative had shifted from one of racial divisiveness or exclusivity to that of uniting the nation. The Ministry of Defence envisaged that the memorial would facilitate both the memory of the dead and national unification: 'a place where the people of Fiji can honour their Servicemen assisting in the healing process which will contribute to a more stable nation' (fijivillage.com 2010). Some considered this a means of distracting attention from pressing political and economic problems. As one blogger posted, 'Military Government spends $8.2 Million dollars on a war memorial while the economy goes down the gurgler' (Fiji Today 2010). War commemorations in Fiji have served to highlight not only heritage but also the agenda of national unity. At the 2013 Ai Matai anniversary, 87-year-old Filipe Matavesi, one of the two surviving members of Fiji's 1st Battalion, urged people, saying, 'Today there are lots of problems in Fiji. I call on Fijians to unite and help this Government and whatever government will come to move us forward' (Seru 2013). A documentary of the ceremony was produced by the Ministry of Information and screened on Fiji TV under 'The Nations [sic] Business'.[11] It commented on both the past and the present, urging more recognition of those who sacrificed themselves for Fiji and reminding the viewer that the current regime had taken on the plight of all servicemen by increasing the budget allocation to them and their families.

BURIED MEMORIES

Despite recent efforts to preserve and reconstitute wartime sites, there is limited interest in these among many Pacific communities.[12] More war relics are estimated to be in Micronesia than in Europe, but Falgout, Poyer and Carucci (2008, 32–3) found that Micronesians attribute little historical significance to these when constructing war memories. Many Islanders value their wartime experiences, but they perceive WWII as a war fought by foreigners. War relics are part of this cultural divide.

> [W]ar relics have a relatively weak connection to Micronesian memories of World War II. Although American and Japanese veterans and their descendants have erected monuments and held memorials at key sites, most Micronesians rarely use wartime remains or locales for cultural memory work about the war.
>
> (Falgout *et al.* 2008, 33)

Because Fiji was not ravaged by combat, far fewer war relics are there than in Micronesia. However, parts of Viti Levu (Fiji's main island) were intensively occupied by military from New Zealand and the United States when there was a fear of a Japanese invasion and when Fiji was the fortress protecting the Eastern and Southern Pacific (Lowry 2006). Between 1940 and 1942, the Kiwis laid extensive defensive and protective infrastructure, including:

> Fiji's forgotten guns of the second World War . . . Sadly though, historical places like the gun batteries at Bilo outside Lami, and Momi and Lomolomo in the western division have moved on to the endangered list. They have become places we must save.
>
> (Wesley 2008)

Journalist Fred Wesley's emotive plea to preserve Fiji's war sites linked memory to material remains: 'Structures and memory sometimes combine to remind us that certain places are special' (2008). This was realized at Bilo in 2010 in a restoration project involving the Fiji Museum, villagers and a grant from the United States government (Vula 2010). Earlier, the Momi guns had been restored by the National Trust, aided by a donation from American entrepreneur and former WWII navy officer, Robert Trammel.[13]

The large guns and commanding sea views at Momi impress tourists, but they can only imagine how the surrounding hills were transformed by troops during the war. Throughout Fiji, most material evidence of any military impact is buried in the past. This includes the institutions established by the Americans after they assumed command in 1942, when Fiji had become pivotal in provisioning the offensive against the Japanese and was a sanctuary for rest and recovery. Institutions such as camps, stores and hospitals left little permanent evidence. In the Nadi area after the war, many camps were quickly restored to civilian use, temporary buildings were sold or removed and defences were filled. Material was dumped into the sea; sites were covered by rapidly regenerating vegetation or reclaimed for agriculture (Lowry 2006, 109). Colonial administrator Philip Snow (1997, 201) recalled that 'by about five years after the war, one could hardly see any signs of there having been such a preponderance of Americans and New Zealanders in the country.' I often heard Saweni camp[14] mentioned during my own research, as the adjacent beach was the site of intimate encounters between Americans and Fijian women, but today there is no visible evidence of the American presence.

MOTHERS' DARLINGS

While recently public commemorations and some of the forgotten material remains of WWII have attracted interest in Fiji, our project, 'Mothers' Darlings', focused on uncovering the intimate legacies of war.[15] These possibly had a more enduring impact, including the birth of thousands of children to foreign soldiers and Island women. Yet much of the military history of the Pacific Islands is entrenched in combat—of battles, heroes and soldiers (e.g. Dunnigan and Nofi 1998). These are mostly masculinised memories with little space for women except when they came under Hollywood's gaze (Brawley and Dixon 2012; also Bennett 2009, 11–48), transposed to the Islands as 'the thoughts of beautiful dusky Hula Hula girls dancing in grass skirts or wearing sexy sarongs' (Henne n.d.). Although researchers have noted that 'indigenous peoples, are "missing in action" from the written accounts of WWII' (Falgout *et al.* 2008, 1), our project addressed the historical invisibility of women and their children born to foreign soldiers.

This lacuna is incredible because over two million American servicemen occupied the southern Pacific Theatre with a huge social impact. Mothers' Darlings sought to trace the children of American servicemen and the broader structural and cultural context of indigenous societies, as well as the military, civilian and colonial authorities that impinged on these intimacies. We focused on personal and family war memories that may have been passed on to descendants or that may have become buried in the Islands' past. We sought to understand the meaning for these children whose identity had been partly formed by these fragmentary and distorted intimate war memories that are riddled with silences and secrets.

In Fiji, the *Kai Merika* we traced, who had formed 'mind pictures' of their fathers, had been given positive recollections from their mother and kin. Frankie's mother, Neve, passed on loving but embellished memories of his American father, Frank:

> [S]he used him for me as an inspiration, like if I did well in school he did better you know, if I ran fast and I won my heat in school he was faster than me, and if I climbed this tree he was there, if I fish he was better and I thought my God, this guy I can't catch him! He seemed a superman.[16]

Her stories were enriched with the couple's letters, exchanged after Frank was forced to leave Fiji. Neve passed these letters and Frank's photograph to Frankie. Other *Kai Merika* had far more sketchy information about their fathers. Adi Romera, born in 1944, knew only that her father was Dick F. Henry, Commander of the American Navy in Fiji, and that he wanted to smuggle her mother Lusiana in a trunk and take her to America.[17] This image comforted Adi Romera as she grew up in a Fijian village. David, also born in 1944, had no such glimpses into his paternity when, aged 12, he asked his mother who his father was. Her response was violent and explosive; she smashed three plates over his head and remained silent to her death about this.[18] David grew up in families burdened with secrets that stem from his birth during the Pacific War. Likewise Meli, who gave birth to Api during WWII, had no contact with his father after he left Fiji: 'no more—finish the memory'.[19]

LOST MEMORIES: SILENCES AND SECRETS

Both the silence over David's father and the acceptance of ex-nuptial births point to the difficulties in tracing *Kai Merika* and whether such war memories will ever be known or acknowledged.[20] Culture and war conditions mediate memory and time. Silences and secrets are not specific to small islands, but culture influences how affective relationships are suppressed or celebrated.

In contrast to the certainty of statistical data concerning war casualties or troop numbers that are cited at public commemorations, the number of children fathered by American servicemen in the Pacific Islands is unknown. In Fiji, most *Kai Merika* were born and raised within their mother's kinship group and ethnic community. Those with indigenous Fijian mothers were usually socially integrated and registered into their mother or stepfather's *mataqali* ('clan') with rights to land and other indigenous resources.

Fiji has a long history of interethnic liaisons and marriages dating back to early colonial contact. Part-Fijian, part-European, half-caste, Euronesian, mixed race, *dra rua* ('two blood') and *Kailoma* are among the various terms for these communities (Riles 1997). Nate's American heritage can be traced to the early nineteenth century when his American great-grandfather married a Fijian woman.[21] Nate recounted vivid tales of this ancestor that were passed on to him by his grandmother. She had a relationship with an Australian, and from this union Nate's mother was born. The American presence recurred with Nate's birth in 1945, although a Chinese stepfather raised him (Figure 1.3). His family, like many in Fiji, is ethnically complex, making it impossible to pinpoint him as an American child. Adoption adds to the difficulty in tracing children with American fathers from WWII (see e.g. Lee 2011, 170). Moreover informal, customary and open adoption is widely practised in the Pacific Islands (Pulea 1986, 140–6).

Our research into intimate war memories was also problematic because Fijian culture and protocol are layered with many forms of silence (Nabobo-Baba 2006, 94–8, 100–25). Rumours about paternity and sexual relationships may be public secrets, but these are not always directly discussed, especially with a stranger. This does not imply that any births were shameful but instead reveals that biological paternity was often irrelevant within the communities the children were part of. It can be culturally offensive and stigmatizing to pry into a Fijian's American parentage as it may mark or re-mark a person who has been culturally accepted and integrated as an 'outsider'. Frankie reminded me: 'We don't go round asking, "you got some half sister in the war?"' but also emphasized that in his experience 'No it wasn't a big shame or anything.'[22]

Although local communities accepted ex-nuptial births or those known not to be the biological offspring of their father, there could still be considerable shame about paternity. This varied between families and cultures and reflected status, class and religion. As Nate was well aware, shame was borne by the mother: 'they have to keep it under their breath . . . keep it hidden.' Sex is joked about in the Islands but is surrounded by ambivalence, embarrassment and secrecy when directly spoken of:

> Yeah, same like in Fiji now, in the village they'll talk all about it like someone's pregnant, all hush-hush like the village gossipers you know, and it stays as a gossip, and . . . and then they'll play and no-one talks.[23]

Speculation over the circumstances surrounding war babies' births could be associated with illicit sexual activities, notably prostitution. There is only limited evidence of this in wartime Fiji, but rumours of immorality soon silenced questions about a child's father.[24]

Memories of intimate encounters were also hidden because other wartime activities were secret. The media were censored, and the presence of the Americans in Fiji (referred to by locals as 'Our Friends') was a secret until *Time* (1942) published the headline 'Yanks in the Cannibal Isles.'[25] GIs were ordered to write their address as 'Somewhere in the Pacific'. Captain Henne (n.d., 56) recalled, 'The men joked that the rules reduced what they could write to, 'Hey Mom, I'm healthy and doing fine.' Island women heard only fragmentary information about their lover's war activities or destinations or about the reality of their civilian lives, including any wives, girlfriends or children. The military culture of secrecy pressured GIs to silence their island lives to their American families. The GIs were prohibited from many places and activities in Fiji, so transgressions, including consorting with native women, might be quickly forgotten or buried among the unspoken memories of war.[26]

The Americans' island sojourn could abruptly end with secret and unpredictable troop movements. Frankie believes his father was shipped to Guam because he intended to secretly marry Neve. A sudden break would have been traumatic for young pregnant women, especially if the family condemned the pregnancy. Nate's mother was already married when she became pregnant by his father. Thus both the broader context of war and personal circumstances added to a forgetting or reinvention of intimate activities during the war.

GLOBALIZING ISLAND MEMORIES

War memories both have been confined to and have reached beyond islands such as Fiji. As shown, there are complex cultural reasons why war memories, bound up in silences and secrets, have remained on islands—within communities and families. Military, colonial and immigration regulations, as well as racial prejudice, meant that many intimate relationships in the Islands remained there. Marriage was almost always prohibited between Americans and indigenous women. The American consul informed Flora Bryson that 'full blooded persons of Fijian and/or Polynesian descent are "ineligible to citizenship"', while those of mixed blood 'must possess more than one half of white blood'.[27] She was denied a visa to join Paul Stultz, her baby's father. In desperation, he appealed to President Truman:

> I want [sic] to San Francisco Calif, last week to try and work my way back to her and the baby Sir but I couldn't make any head way out there so I thought it might be best if I wrote to you about this Sir and please Sir help me that I may win them both I will do any thing you have to say Sir.[28]

Thirty tears after the war, another American GI, Lewis Benesh, did return to Fiji to marry his sweetheart. She, their son and his family emigrated to America. On Memorial Day, his granddaughter, Maria, posted on Facebook that 'he was the most kind person I know, his name will be carried on . . . thank you GOD for one of the greatest gift[s] . . . will see you in heaven'. Maria's use of Internet-based social networking to commemorate her grandfather is a recent example of how Island memories have extended beyond the Pacific and are sometimes globalized.

The quest to trace their fathers took some *Kai Merika* to the United States, but this was financially prohibitive for most. Greater international mobility and new communication channels have accompanied the recent spread of the Fijian diaspora. When Martha began to search for her father, Samuel, during the 1980s, she was limited to the advice of the American embassy in Suva.[29] They could only suggest that Martha should travel to the United States, but this was not feasible. It was not until her daughter Anne settled in the United States and through the Internet traced Samuel, that Martha was united with him in Massachusetts in 2002. Adi Romera travelled to the States but was unable to locate her father. In 2012, at her request, we traced her father, grandmother and surviving American relatives through Ancestry.com, the globe's largest genealogical resource. Her father died in 1977, but Adi Romera plans to erect a memorial to him in Fiji.

Popular culture has also brought the war back to the Islands to be remembered in new ways. This can probe past memories, such as when Frankie was deeply moved by the series, *The Pacific*. He recalled a night on a tiny offshore island when his father shared bittersweet war memories:

> [I]t was just him and I, and that's the only night when he opened up and talked . . . what war was about, how the Japanese, 'you probably see one of your mates hanging up on a spear stuck up with his mouth cut hanging [open], you know dead, blood . . . oh it's so gory, we had to cut . . . you know bury him' . . . he said he got really scared . . . he said when he went home you prayed a lot, he said if he didn't believe in God he believed in God then.[30]

Frankie's vivid, emotional recall of his father's horror and fear in Guadalcanal was contrasted by Frank's memories of the beauty in Fiji:

> He did talk about his time in Suva as the best time in his life you know . . . capture the moments all the times he had, beautiful nights with cars and women, couldn't believe the hula skirts, he said 'this is the life'.[31]

CONCLUSION

This chapter has offered glimpses into war heritage and memories on small islands within the vast Pacific Ocean. WWII is remembered in Fiji in differing ways at the national, community and personal levels. National memories are entangled within racialized narratives and embedded within Fiji's colonial past. While war memorials may be serving to bolster Fiji's contemporary culture of militarization, they also appeal to public interest in commemorating those who served the nation, albeit in a selective way as to what and who are highlighted.

Despite the recent prominence given to the National War Memorial in Fiji, there have been few national war commemorations there or throughout the Pacific Islands.[32] The legacy of WWII has mainly fed into national narratives of sacrifice and the preservation of Fijian culture. These memories say as much about internal racial tensions before and during the war as about the battles Fijians bravely fought in. Little evidence remains of the material impact the Allied forces had on Fiji. A few guns and tunnels have been restored, but there is limited public interest. The possibility of developing such sites for tourism has been mooted, but Fiji has far more enticing attractions than war heritage. When the past is stirred up through epic histories, monuments and celebrations, it may evoke personal memories[33] which often remain secret and silent. Memory and history is also about burying, eliminating and ultimately forgetting.

Intimate memories of the Pacific War are still quietly celebrated and passed on by *Kai Merika* to their descendants. Martha was in tears as she recalled how she always knew that there must be others like her with American fathers but, until she participated in our project, was unable to connect with other wartime children with American fathers. She grew up in a close community on a small island yet felt isolated in the knowledge of her war heritage. Over time, she and some others discovered that their isolated memories extended far beyond the Island into new connections that go back to the war and create memories for the future.

NOTES

1. A transnational production (HBO, Seven Network Australia, Sky Movies, Playtone and DreamWorks) with leading producers (Steven Spielberg, Tom Hanks and Gary Goetzman). Hugh Ambrose (2010) wrote the accompanying book.
2. Capitals are used here for Islands and Islanders when these refer to the Pacific, as these are terms of pride and identity.
3. Sincere thanks to Alumita Durutalo for research collaboration, Marsa Dodson for research assistance and Tui Clery for comments on the draft. *Vinaka vakalevu* to research participants.
4. See Bennett, Leckie and Wanhalla (2015), regarding silences, secrets and Pacific War babies.
5. Also posted on Fiji's Ministry of Defence, Security and Immigration website: www.defence.gov.fj/index.php/news/28-a-day-to-hounor-the-brave (accessed 26 October 2013).
6. According to Lowry (2006, 89) the Bilo guns were not even the largest in Fiji.
7. Nate is a pseudonym. Interviewed by Marsa Dodson, March 2010.
8. CSO (Colonial Secretary's Office), Council Paper 35/47 (Fiji National Archives).
9. See Lal (1991) for a deconstruction of the stereotype that Indians were disloyal and explanations of why they did not serve in the Fiji Defence Force. See also Kelly (1995, 481–4; 489–91).
10. See the Republic of Fiji Military Forces website: www.rfmf.mil.fj/History/History.html (accessed 26 October 2013).
11. "Ai Matai, We Shall Remember Them," www.youtube.com/watch?v=zu1exFFUPwo (accessed 25 September 2013).
12. See Lowry (2006, 109–14) for list of WWII sites in Fiji.
13. See www.nationaltrust.org.fj/momi_guns.html (accessed 25 October 2013).
14. Operated by the US Navy 3rd Construction Battalion (Seabees) 1942–44.
15. This was funded by the Marsden Fund, Royal Society of New Zealand. See Bennett and Wanhalla, (forthcoming); www.otago.ac.nz/usfathers (accessed 25 October 2013).
16. Frankie and Frank are pseudonyms. Interview between Frankie and Leckie, 28 September 2010.
17. Interview between Adi Romera Drodrovakawai (Adi Romera) and Durutalo 2010. Interview between Adi Romera and Durutalo, Leckie, August 2011.
18. Confidential information and names have been changed.
19. Interview between Meli and Durutalo and Leckie, 12 August 2012.
20. See Bennett, Leckie and Wanhalla (2015).
21. Interview between Nate and Dodson, March 2010.

22. Interview between Frankie and Leckie, 28 September 2010.
23. Interview between Frankie and Leckie, September 2010.
24. Archival evidence, both confirming and denying prostitution, is discussed in Leckie and Durutalo (forthcoming).
25. 'Yanks in the Cannibal Isles' (1942).
26. Fiji Islands, National Archives and Records Administration (NARA), San Bruno, RG 313-58-3000, Box 1.
27. Winfield Scott, American Consul to Bryson, 21 November 1946, NARA, College Park, Maryland, NND 775115, Record Group (RG) 84, Box 15.
28. Stultz to Truman, 1 June 1947, NND 775115, RG 84, Box 15.
29. Interview between Martha Naua and Leckie, October 2010.
30. Interview between Frankie and Leckie, 28 September 2010.
31. Interview between Frankie and Leckie, 28 September 2010.
32. Camacho (2011) examines war commemorations in the Mariana Islands.
33. White (2000) analyses the spaces between public remembrance and private memories at the Pearl Harbor memorial in Hawai'i.

BIBLIOGRAPHY

Ambrose, H. (2010) *The Pacific*, New York: New American Library.

Bennett, J. A. (2009) *Natives and Exotics: World War II and Environment in the Southern Pacific*, Honolulu: University of Hawai'i Press.

Bennett, J. A., Leckie, J. and Wanhalla, A. (2015) 'Mothers' Darlings: secrets and silences in the wake of the Pacific War', in C. Twomey and E. Koh (eds), *The Pacific War: Aftermaths, Remembrance and Culture*, New York: Routledge, 214–32.

Bennett, J. A. and Wanhalla, A. (eds). (forthcoming) *Mothers' Darlings of World War Two: Stories of the South Pacific*, Honolulu: University of Hawai'i Press.

Brawley, S. and Dixon, C. (2012) *Hollywood's South Seas and the Pacific War: Searching for Dorothy Lamour*, New York: Macmillan Palgrave.

Camacho, K. L. (2011) *Cultures of Commemoration: The Politics of War, Memory and History in the Marianas Islands*, Honolulu: University of Hawai'i Press.

Dunnigan, J. F. and Nofi, A. A. (1998) *The Pacific War Encyclopedia*, New York: Checkmark.

Falgout, S., Poyer, L. and Carucci, L. M. (2008) *Memories of War: Micronesians in the Pacific War*, Honolulu: University of Hawai'i Press.

Fiji Today. (2010) 'Fiji Today'. Online open forum blog, 11 June. Available HTTP: http://fijitoday.wordpress.com/2010/06/11/military-government-spends-8-2-million-dollars-on-a-war-memorial-while-the-economy-goes-down-the-gurgler/ (accessed 22 October 2013).

fijivillage.com (2010) 'War memorial construction to finish next year', 11 June. Online. Available HTTP: www.fijivillage.com/?mod=archivedstory&id=11061053531cf0b1fcd01d43cc649c#, (accessed 22 October 2013).

Firth, S. (1997) 'The war in the Pacific', in E. Denoon, S. Firth, J. Linneken, M. Meleisea and K. Nero (eds), *The Cambridge History of the Pacific Islanders*, Cambridge: Cambridge University Press, 311–23.

Firth, S. and Fraenkel, J. (2009) 'The Fiji military and ethno-nationalism: analyzing the paradox', in S. Firth, J. Fraenkel and B. V. Lal (eds), *The 2006 Military Takeover in Fiji: A Coup to End All Coups?* Canberra: ANU E-Press, 117–38.

Firth, S., Fraenkel, J. and Lal B. V. (eds). (2009) *The 2006 Military Takeover in Fiji: A Coup to End All Coups?* Canberra: ANU E-Press.

Hau'ofa, E. (1993) 'Our sea of islands', in E. Waddell, V. Naidu and E. Hau'ofa (eds), *A New Oceania. Rediscovering Our Sea of Islands*, Suva: University of the South Pacific/Beake House, 2–16.
Hau'ofa, E. (1998) 'The ocean in us', *The Contemporary Pacific*, 10, 395–410.
Hau'ofa, E. (2000) 'Epilogue. Past to remember', in R. Borofsky (ed), *Remembrance of Pacific Pasts: An Invitation to Remake History*, Honolulu: University of Hawai'i Press, 453–71.
Henne, C. (n.d.) *The Fiji Experiences of the 145th and 148th Infantry Regiments*, unpublished manuscript. Copy held by Robert Lowry and kindly supplied to Jacqueline Leckie.
Howlett, R. A. (1948) *The History of the Fiji Military Forces, 1939–1945*, Fiji: Government Printer.
Kelly, J. (1995) 'Diaspora and world war, blood and nation in Fiji and Hawai'i', *Public Culture*, 7, 475–97.
Lal, B. V. (1991) 'For king and country: a talk on the Pacific War in Fiji', in G. M. White (ed), *Remembering in the Pacific War*, Honolulu: Center for Pacific Islands Studies, University of Hawaii at Manoa, Occasional Paper series 36, 17–25.
Laracy, H. (1994) 'World War II', in K. R. Howe, R. C. Kiste and B. V. Lal (eds), *Tides of History: The Pacific Islands in the Twentieth Century*, Sydney: Allen & Unwin, 149–69.
Laracy, H. and White, G. (eds). (1998) *Taem Blong Faet: World War II in Melanesia*, Special issue of *O'o: A Journal of Solomon Islands Studies*, 4.
Leckie, J. and Durutalo, A. (forthcoming) '*Kai Merika*! Fijian children of American servicemen', in J. Bennett and A. Wanhalla (eds), *Mothers' Darlings of World War Two: Stories of the South Pacific*, Honolulu: University of Hawai'i Press.
Lee, S. (2011) 'A forgotten legacy of the Second World War: GI children in post-war Britain and Germany', *Contemporary European History*, 2 (2), 157–81.
Lowry, R. (2006) *Fortress Fiji: Holding the Line in the Pacific War, 1939–1945*, New South Wales: Sutton.
Nabobo-Baba, U. (2006) *Knowing and Learning: An Indigenous Fijian Approach*, Suva: Institute of Pacific Studies, University of the South Pacific.
Poyer, L., Falgout, S. and Carucci, L. M. (2001) *The Typhoon of War: Micronesian Experiences in the Pacific War*, Honolulu: University of Hawai'i Press.
Pulea, M. (1986) *The Family, Law and Population in the Pacific Islands*, Suva: Institute of Pacific Studies, University of the South Pacific.
Radio Australia. (2002) Online. Available HTTP: www.radioaustralia.net.au/international/2002-11-14/chiefs-consider-war-memorial-for-fijis-warriors/599946 (accessed 26 October 2013).
Ravuvu, A. (1988) *Fijians at War, 1939–1945*, Suva: Institute of Pacific Studies, University of the South Pacific.
Riles, A. (1997) 'Part-Europeans and Fijians. Some problems in the conceptualisation of a relationship', in B. V. Lal and T. R. Vakatora (eds), *Fiji in Transition*, 1, Suva: School of Social and Economic Development, University of the South Pacific, 105–29.
Seru, M. (2013) 'A day to honour the brave,' *Fiji Sun*, 14 April. Online. Available HTTP: www.fijisun.com.fj/?p=148492 (accessed 26 October 2013).
Snow, P. (1997) *The Years of Hope: Cambridge, Colonial Administration in the South Seas and Cricket*, London: Radcliffe Press.
Tagivetaua, P. (2011) 'Memories', *The Fiji Times Online*, 29 January. Online. Available HTTP: www.fijitimes.com/story.aspx?id=164858 (accessed 26 October 2013).
Tagivetaua, P. (2011) 'The guns of Bilo'. *The Fiji Times Online*, 20 November. Online. Available HTTP: www.fijitimes.com/story.aspx?id=186399 (accessed 26 October 2013).

Toyoda, Y. and Nelson, H. (eds). (2006) *The Pacific War in Papua New Guinea: Memories and Realities*, Tokyo: Centre for Asian Area Studies, Rikkyo University.

Tuimaleali'ifano, M. (2001) 'Sefanaia Sukanaivalu. Soldier', in S. Firth and D. Tarte (eds), *20th Century Fiji—The People Who Shaped This Nation,* Suva: USP Solutions, University of the South Pacific, 79–80.

Vula, T. (2010) 'War site upgrade', *The Fiji Times Online*, 29 January. Online. Available HTTP: www.fijitimes.com/story.aspx?id=138621 (accessed 27 October 2013).

Wesley, F. (2008) 'Fijis forgotten guns of the second World War', *The Fiji Times Online*, 17 February. Online. Available HTTP: www.fijitimes.com/story.aspx?id=81548 (accessed 26 October 2013).

White, G. M. (2000) 'Emotional remembering: the pragmatics of national memory', *Ethos,* 7 (4), 505–29.

White, G. M., Gegeo, D. W., Atkin, D. and Watson-Gegeo, K. (eds). (1988) *Bikfala Faet: Olketa Solomon Aelanda Rimembarem Wol Wo Tu* [*The Big Death: Solomon Islanders Remember World War II*], Honiara/Suva: Solomon Islands College of Higher Education and Institute of Pacific Studies, University of the South Pacific.

White, G. and Lindstrom, L. (eds). (1989) *The Pacific Theater: Island Representations of World War II*, Honolulu: University of Hawai'i Press.

'Yanks in the Cannibal Isles'. (1942) *Time,* 26 October, 34.

2 Fragmented Memories

The Dodecanese Islands During WWII

Hazal Papuccular

INTRODUCTION

Memories have the power to reveal the past and present at the same time; they can emphasize, multiply, complete or challenge stories already unveiled. When researchers deal with the memories of people during a particular period of time, they usually take into consideration geographical location, historical facts, class, gender, identity and other variables that could make a difference to research outcomes. This study focuses on two of those variables: namely, geographical location and communal identities. In terms of geographical location, it is asserted here that the status of being an island (or group of islands) is very influential not only in determining the course of events in history but also in how this history is remembered by the inhabitants. In parallel with general assumptions about island geographies which indicate the vulnerability of those regions vis-à-vis occupation and colonization (Stratford *et al.* 2011, 116), the wartime histories of islands, specifically strategic ones, often focus on military attack and occupation. However, geographical conditions such as isolation can also be an issue and bring with them associated acute problems such as famine.

The history of WWII in the Dodecanese Islands[1] and the memories of the people of the period display a similar pattern. The occupation, blockade, famine and migration left its mark both on the history of the islands and on the minds of the Dodecanesians, the majority of whom emphasize almost the same themes in their narratives of war. However, the second focal point of this study, the impact of the collective identities on the memories of the islanders, shows that even in a small geographical area such as that occupied by these islands, different memories can be produced that are associated with the aforementioned themes when the geographical area under consideration comprises different ethnic groups. Two of the islands in the Dodecanese group, Rhodes and Kos, which are multiethnic in character, demonstrate well how the experiences of these diverse groups can vary and how these memories are constructed vis-à-vis one another

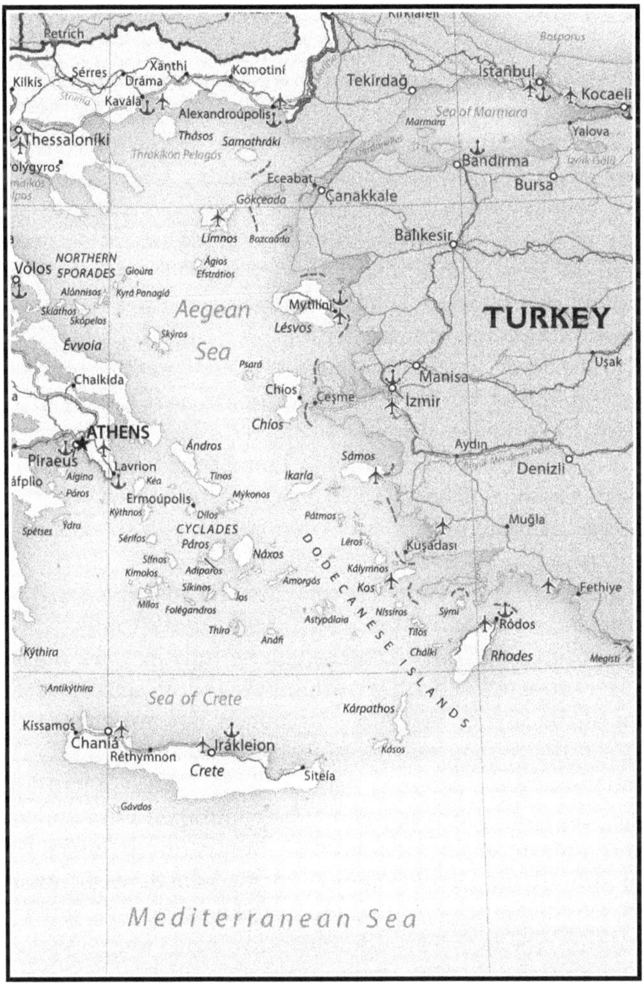

Figure 2.1 Map of the Aegean Archipelago (Courtesy University of Texas Libraries).

in a similar way. As Assman (2010, 114) observes, 'memory is knowledge with an identity-index'. Based on the two variables of geographical location and identity, this chapter reveals the contested WWII memories of the distinct ethnic groups which left fragmented and competing heritage in the islands. In order to consider these competing memories, this study draws upon sixteen semi-structured interviews with the first and second generations, conducted by the author, along with other oral testimony present in secondary sources.

HISTORICAL BACKGROUND

Placing the islanders' war memories into an appropriate context necessitates an emphasis on several facts of Dodecanesian history. The seizure of the Dodecanese by the Ottomans during the sixteenth century was an important turning point for the future of the islands.[2] Under Turkish rule, the relationship between the islands and Asia Minor altered, the demographic composition of the islands began to change based on the settlement of the Muslims and the Jews from Spain in Rhodes and, accordingly, a new administrative system was adopted. One of the major features of this administrative organization was the *Millet* ('Nation') system, which foresaw a certain degree of juridical and fiscal autonomy for different ethno-religious groups, which were organized by the administrator of each community that was responsible for his nation (Braude 1982, 69). The reflections of the *Millet* system implemented in the Ottoman Empire (including the Dodecanese) were twofold. On the one hand, since it gave certain autonomies to different ethno-religious communities, it was supposed to constitute a relative state tolerance of these groups, even if it fell short of equality.[3] On the other hand, it could be argued that the system perpetuated the absolute separation of the various societies from one another, both legally and socially.

After the occupation of the Dodecanese by Italy in 1912 during the war between the Ottomans and Italy over Libya and its seizure of the *de jure* sovereignty of the islands in 1923, the Italians did not change the general framework of this communal administrative system (Orakçı 2012, 75). The continuation of the *Millet* system in the Dodecanese had implications for both the interwar and war years. The communal lines and separations became an important tool for the Italian administration, which tried to balance the power of the Greek majority in the islands with other minority communities.[4] This position of the administration vis-à-vis different groups continued during the war. This administrative behaviour, together with the divided nature of society, became one of the significant elements regarding the different experiences and memories of the islanders about specific subjects related to war and occupation.

The strategic importance of the Dodecanese had already been acknowledged during the 1930s. Therefore, when WWII broke out, the plans of the Allies to capture the islands were complete.[5] Although these plans were not carried out, even after the entry of Italy into the war in 1940, the British bombed some of the islands such as Kastellorizo, Leros, and occasionally Kos.[6] However, despite the Allied raids on these islands, the Aegean Sea continued to be an Axis base, owing to the Italian control of the Dodecanese and the Nazi seizure of northern Greek islands such as Chios and Lesvos, which were part of the Axis occupation of Greece. The 'closure', or blockade, of the Aegean archipelago by the Axis powers posed strategic problems for the Allies, especially with respect to the situation in the Balkans and the Mediterranean. British reports during 1940 emphasized the unfavourable conditions in the islands due to the severe shortages related to the blockade and the consequent reluctance of the Italian soldiers to fight, leading to serious rumours that Italy would surrender the islands.[7]

The expected surrender came not in 1940, but in 1943 with the defeat of Italy—the turning point for the history of the islands. Accordingly, the Allies undertook an unsuccessful campaign against the Nazis in the Dodecanese between September and November 1943.[8] In the initial phase of the campaign, Britain had captured some of the islands, including the militarily important Kos and Leros. However, once the Germans secured Rhodes, which should have been strengthened as the hub of the Dodecanese at the expense of other minor islands (Smith and Walker 2008, 69), they expelled the British forces from the whole of the Dodecanese, island by island. Therefore, by the end of 1943, the days of uninterrupted German occupation for the Dodecanese had begun. It is important to note that both the campaign and the German occupation constituted one of the most vivid memories of the Dodecanesians, regardless of their ethnic identities, since this epoch is symbolized as the years of brutality, starvation and migration to the countryside or to Asia Minor for the islanders.

After the German surrender in 1945, a British administration was formed in the region. The fate of the Dodecanese was complicated by both Greek and Turkish claims, together with the Soviet desire for bases after the war. This was resolved by the Allies in peace talks which ended in favour of Greece, who took the control of the islands in 1947.[9] It should be stated that this transition period, from British to Greek control, was another important era in the minds of the islanders, who were asked to analyse the period with respect to their identity. In this sense, the evaluations of the people about the transition period reflect the idea that memories are the thoughts not of the time but of a group (Halbwachs 1992, 52), which also has an impact on how memories are later reconstructed.

SIMILAR THEMES AND FRAGMENTED MEMORIES

Since this study mainly focuses on the variations in the memories of the different communities of Rhodes and Kos, the population distributions are important to consider. Italian official documents suggest that in 1936, 78 per cent of the total population of 120,000 was Greek (i.e. about 93,600 people), around 12,200 were Muslims of Turkish origin and 4,666 were Jews.[10] The Turks and Jews lived mainly in Rhodes and Kos. Thus, while the story of those islands reflects different identities and memories, the narratives of other Dodecanesian islands may indicate a more monolithic story, at least in terms of a communal approach to the issues.

When Italy entered the war, the islands found themselves in a position of dealing with a scarcity of food. Both the historical studies about the Dodecanese and the reminiscences of the islanders emphasized this problem.

In *Myth and Memory in the Mediterranean*, Nicholas Doumanis (1997, 56) suggests that, after the Italian involvement in the war, locals mainly complained about the introduction of the strict food rationing. The interviewees reported that they were not allowed to consume even their own crops. The island status of the Dodecanese was important because the Dodecanese economy was not self-sufficient. It had long depended on trade with Anatolia, with whom the contacts during the war were either limited or completely frozen.[11] Therefore, the shortage in basic needs which were, instead, stockpiled for the military still survives in the memories of the islanders, both those who experienced the period themselves and the second generation who internalized the experiences of their relatives and groups through the narratives or photos of them in an example of Hirsch's concept of postmemory (2008, 106–7). Thus, the shortage of food is a common memory in the history of the Dodecanese. All of the members of the Turkish community with whom I conducted interviews similarly emphasized the scarcity of food and the strict control that the Italians had of the supplies. Sadi Nasuhoğlu, who lives in Rhodes and lived there through the war years, remembers the occupation of his family's farms by the Italians in order to control the agricultural yield.[12] However, in spite of the shortage of food and strict regulations, hunger was mostly identified in the minds of interviewees with the German occupation, during which deaths from starvation occurred, as will be seen.

Apart from the shortage of food in the first years of the war, another important issue to mention is the relationship between the administration and the people, where the memories of people appear to go hand in hand with their group identities. Doumanis (1997, 56) suggests that ethnic friction took place between the Greek islanders and the Italians after the attack of the latter on Greece. As might be expected, the Turkish community did not seem to attach any negative connotations to the Italians concerning the war years. On the contrary, some excerpts from memoirs suggest that the Italian soldiers were associated with positive emotions.[13] Likewise, while evaluating the wartime narratives of his mother, Ismet Buyukyasar, who lives in Izmir in Turkey, emphasized to me that the Italian attitude during the war years was more understanding and compromising towards the Turks.[14] This relatively positive image of the Italian soldiers in the minds of the Turkish islanders also coincides with the traditional policy of colonial administration during the interwar years, when the minority groups of the islands had been regarded by the Italians as a way of providing a balance with the majority of the population.

The German occupation, on the other hand, symbolizes a turning point in the memories of islanders. Doumanis (1997, 58–9) rightfully argues that the Nazi period left a deeper mark on the collective memory of the islanders compared to any other period, including the Turkish or Italian reign, since this period was distinguished by violence and famine. When the war years are talked about, almost all islanders primarily discuss the starvation.

Suna Hamit, who was a child in Rhodes during the war years, witnessed the sudden death of a person in the street while she was out walking.[15] In fact, seeing people dying in the streets and victims of starvation oedema carried importance in the memoirs of the islanders, who particularly emphasize such events.

However, variations among different groups come to surface in the discussion of famine in the oral testimonies. Although it is acknowledged that firmer controls of the supplies, together with the prohibitions on consumption, were introduced by the Germans,[16] it is stressed that the Greeks were much more vulnerable to famine than other communities:

> During the war, there was famine. However, we did not suffer so much because we had agricultural fields in Rhodes. Turks were generally farmers and had historically large agricultural estates in the countryside. Therefore, the Greeks starved more than us.[17]

Comparison among the different ethnic groups exists not just because they were separated but because of the need to perpetuate these separations based on narratives which traced 'the linkages between past and present, locating self and society in time' (Bell 2006, 5). Rafet Faralyali, who was a baby in Rhodes during the famine, reiterates this by stating that 'in the era of famine the most disadvantaged group were the Greeks who came to Rhodes in a later period and who did not have agrarian property'.[18] During the famine, holding vineyards or agricultural fields was an enormous advantage in the battle to survive. Indeed, being in the countryside was vital not only in finding a basic minimum amount of food but also for being protected from the bombings which generally targeted the city centre.[19]

However, while evaluating the recollections of the first or second generation, it is necessary to emphasize that group divisions, although reflecting important data for the period, may not be the only criteria in reaching clear conclusions. Socioeconomic divisions, as well as the size of the community regarding its diversity, are also important measures, although this was stressed by few members of the Turkish community. The words of Faralyali display the problematic nature of making generalizations based on one criterion without taking the others into consideration:

> One of our Turkish neighbours, who we called Uncle Ali, told us later that he had removed the leather parts of his shoes, cut them into the small pieces, made a soup with them to eat . . . It is stated that the period of famine had the least impact on the Turks. I think the reason behind this idea is that they were rich because they had agricultural fields. But the poor people like Uncle Ali should not be forgotten either.[20]

The memories of famine give clues about the conditions of the soldiers during the war years. Sadi Nasuhoğlu states that the Italian and German soldiers were as desperate as the ordinary population to the extent that when all stocks were exhausted, they tried to find animals in the streets to eat, including dog and rats.[21] In a newspaper interview, the responder Ahmet Hüsamoğlu told a story in which a Nazi soldier in Kos tried to steal the remaining provisions of his family by acting as if he had been ordered by his commander to do so (Tokmakoğlu 2013). Memories of the islanders about Nazi rule state that there was not a huge gap between the occupying and the occupied after 1943—at least in terms of hunger.

Another important theme in the narratives of islanders is the mobility of the people due to the conditions of war and occupation, as well as of famine. During the war, Dodecanesians moved to different regions of their own islands, migrated to other countries and were, in the case of the Jews, subjected to deportation. The communal identities and their marks on wartime memories reflect a more poignant division in this theme, since the experiences of these groups in terms of their movements were varied. The least complicated was the movement of people from the city centre to the countryside, as we saw with the narratives of starvation. However, obtaining food from the fields was not the only reason behind this exodus. After the surrender of Italy, the war between the Nazis and the British over the islands in the last months of 1943 was crucial in this decision:

> The bombings are still in the minds of the people . . . In the city centre there were stone houses in which the islanders were living. The basements of those buildings were used as a shelter . . . One day, a bomb was dropped in front of our building and ten houses were destroyed. After the bombings, the flight of the islanders into the countryside began.[22]

The countryside was not totally protected from the bombardments either. People continued to take shelter in the basements of their houses for days on end, even in the rural areas.

Escaping from famine and the bombardments by moving from the city centre to the countryside was not a reliable option for everyone. Although the recent literature about the islands criticizes with justification the concept of 'islands as paradise, islands as prison' (Hay 2006, 21), the island status of the Dodecanese seems to symbolize the latter part of the contradiction in that large numbers of people, both the islanders (regardless of their ethnicity) and the soldiers, tried to flee from the islands to the nearby coast of Anatolia during this period. Maşazade explained to me that his family decided to go to Anatolia after spending five months in the countryside.[23] They travelled to Marmaris, opposite Rhodes, in a small boat in the middle of darkness. Narratives of escape from the islands were often based on similar themes; both Abdullah Kazimlar and Orhan Morali emphasized the impact of war in the decision to cross the sea in order to reach the neighbouring coast.[24]

The neighbouring coast of Turkey had to deal with serious economic difficulties throughout the war despite its non-belligerent status.[25] Although this period was one of the worst economic epochs for Turkey, islanders contrasted the better conditions in Asia Minor with the degree of famine in the Dodecanese.

While the Turkish Dodecanesians could be regarded as lucky since they were allowed to settle in south-western Anatolia, where they had opted to stay even after the end of the war, the trajectories for the other Dodecanesians were different. Many Greeks, too, as well as the Italian and German soldiers, left the Dodecanese. Both Greek refugees and Axis POWs were sent to Allied Middle Eastern refugee and POW camps, from which the majority returned to their home countries at the end of the war. However, a particular group of islanders could not come back to the Dodecanese: the Jewish community of Rhodes and Kos. The history and the memory of Dodecanesian Jewry transcend the local history of the Dodecanese and intersect here with the Holocaust.

The anti-Jewish laws had first been introduced by the Italian government of the Dodecanese in 1938, leading to a 50 per cent decrease in the Jewish population of about 4,000.[26] However, the turning point for Dodecanesian Jewry coincides with the occupation of the Nazis, who ordered the deportation of this community to Auschwitz in July 1944. Only 150 of them survived the death camps.[27] Thus, it is not surprising to note that the war memory and heritage of the Dodecanese Jews focus on both deportation and the traumatic experience of the Holocaust, rather than placing any emphasis on starvation and hunger. Nathan Shachar's work (2013, 217–19) shows how the survivors remembered the days following the Nazi deportation orders. According to oral testimony in his study, the Germans ordered the people to gather in the city centre, together with their valuables. They then locked up these Jewish families in a building, took all their money and gold under threat of death, and sent them to the boat which marked the first step on their journey to Auschwitz, which they would reach within a couple of days.

However, deportation was not the only theme in the narratives and memories of Dodecanesian Jewry. A common feature of almost every document or testimony is the mention of the Turkish Consul General in Rhodes, Selahattin Ülkümen, whose story remains one of the most vivid images of the war period. Ülkümen is considered to be a Turkish Schindler because he saved forty-two Jews from the death camps by preparing Turkish passports for them and by helping them to escape to Anatolia. The testimony of Giamila Tarica, cited in Shachar's work, describes the narrative of the escape with the help of the consul, stressing the risks that he took:

> Only those who have trembled in the presence of the SS are able to fully appreciate the courage of Signor Selahattin Ülkümen. To put himself, humanely, in the way of the ferocious beasts—*las belvas ferozes*—and

> to pull their victims away from their claws was an act of supreme heroism. Il Signor S. Ülkümen was one of the few who ever experienced that satisfaction.
>
> (Shachar 2013, 227)

While the deportation of the Dodecanesian Jews is mentioned, Consul Selahattin Ülkümen is usually referred to as the second focal point of the narratives. This is not just because of his heroic acts but also because the testimony of those who were helped by the Turkish Consul and thus survived comprises an important part of the anthropological and historical studies of the Dodecanesian Jews. In his memoirs, Ülkümen (1993, 115) refers to the July 1944 deportations by stating that 'it was not only my diplomatic mission, but also my debt to humanity and my responsibility to my conscience above all'. The name of Ülkümen lives on both in the recollections of the island's Jews and those in Israel, where he was declared as one of 'Righteous Among the Nations'.[28] Ülkümen has also been honoured in Turkey where, in 2012, a school in the eastern part of the country was constructed by the local Jewish community after an earthquake and given the name of the consul. The head of community said that 'by naming this school after Ülkümen, we wanted to express our thanks to him once again.'[29]

While the central subject of Jewish memory has been deportation, the Holocaust and Nazi violence during the war, a difference in the memories exists among certain Dodecanesian groups regarding certain aspects of Nazi rule after 1943.

For example, Hüsamoğlu, the son of a Turkish farmer in Kos, while emphasizing German harshness, also spoke about an episode in which a German soldier asked him whether he was a Turk, warned him that the German army would bomb the city that night, and suggested that all the community should go to the shelters (Tokmakoğlu 2013). The account of Nasuhoğlu (2013) reflects a similar understanding:

> Because Turkey had not yet entered to the war, the German forces did not behave with hostility to us. For example, the German General and his cortege, who visited our Fethi Pasha Library in Rhodes, ordered chemicals from other countries in order to restore the 600 year-old Holy Koran. Because the Germans stayed in Rhodes only one and a half years, they did not usually behave with hostility to the indigenous people . . . Because of the starvation which reigned during the last years of the war, the Nazis fled to Turkey with the Italian soldiers and others in order to escape from the hunger. The Germans were not strong enough to commit outrages because they were starving; they ate the cats, dogs and rats in the streets as well as all animals in the zoo.[30]

Obviously, memories towards the Germans were as conflicting as the personal and group experiences. In other words, the nature of the German occupation had different meanings for the Jews, Greeks and Turks and generated different memories.

When the WWII ended in the Dodecanese, the Greeks and Turks had both experienced famine and migration, as had the Italian and German soldiers. The majority of the Jewish community had been executed. However, the end of the war did not bring instant relief to the Dodecanese. After the German surrender, the uncertainty about the sovereignty of the islands became one of the most influential issues of the region. Despite a high probability of union with Greece, the British rule that started in 1945 and endured until 1947 led to concern among the Greeks about its potential permanence in the islands (Doumanis 1997, 59). This transitional period of sovereignty after the war caused confusion about the future for the population, whose expectations and hopes were shaped by their ethnic identities. While the Greeks grew impatient for a union with Greece as the diplomatic negotiation process took its course, the Turks were hopeful of Turkish sovereignty. An interviewee, Hüsamoğlu, stated that 'the British decided to give the Dodecanese to Greece but we were not told this. When the ships that were carrying the new administrators of the islands came, the Greeks said they were Greek and the Turks said they were Turkish' (Tokmakoğlu 2013). Indeed, this confusion over expectations constitutes one of the most remarkable narratives of the oral testimonies. Ahmet Kirevliyasi suggested that, for a long time after Greek possession of the islands, people did not abandon their hopes regarding sovereignty.[31]

The memories of this transition period were divided along the lines of group ethnic identities. While this period symbolizes one of thorough contentment for the Greeks, it meant absolute disappointment for the Turks. The Turks soon began to experience some problems over minority identity rights after the Greek seizure of the islands, and these difficulties continue to the present day (Dilek 2008, 96–131). Thus, Turkish memory of the end of the war and the subsequent transition period is coloured by the contemporary situation.

WAR HERITAGE IN THE DODECANESE: FRAGMENTED MEMORIALS

Today, the traces of WWII in the Dodecanese can be seen through the narratives of the islanders who have constructed their past with similarities, fragmentations and sometimes contemporary concerns. However, their connection to the wartime past does not manifest itself solely through these narratives; the projection of these narratives into a material culture, in the form of museums, statues, cemeteries and commemoration activities, also occupies an important place in the mark that the war left on the islands. As Millar (1989, 13) argues, heritage creates a sense of belonging and historical continuity; the monuments, museums and commemoration events related to

WWII in the Dodecanese provide a link that binds the past with the present day in the region. Since the essence of a heritage is a political construction of history and memory (Richter 2010, 258) which may converge or diverge, as has been shown in this chapter, it is not surprising to note that war heritage and commemorative activities are fragmented in the region. However, the fragmentation of heritage in the Dodecanese is a much more diverse phenomenon than merely the fragmented narratives of different groups, both in terms of identity and of geography, since the remnants of the war transcend Rhodes and Kos, spreading out to other islands, specifically Leros as the naval base of the group. Therefore, war heritage and its utilization in the Dodecanese take the fragmented nature of the legacy of war in the islands still further.

The Holocaust constitutes a significant component of war heritage in the Dodecanese. Obviously, the fate of the Dodecanesian Jewry was the most tragic one among the different ethnic groups of the islands, in parallel with their mass extermination throughout Europe. Therefore, the struggle of the remaining few Jewish people to provide continuity in their local history through memorials and commemorative activities has produced an elaborate war heritage in the region. The Monument of the Victims of the Holocaust, on which is written 'Never Forget', is one of the most important memorials in Rhodes situated in Seahorse Square. Mcdowell (2008, 38) stresses that 'place' has an important meaning for the identities in different levels—familial, communal, and national—and for the heritage related to them. In this sense, the existence of the majority of the Jewish heritage in Seahorse Square shows not only the importance of space in memory formation but also the spatial fragmentation of heritage in Rhodes because this square is an important place for Dodecanese Jewish history. On the one hand, it had long been within the Jewish quarter of the island, and on the other it was the place that Jews gathered just before their deportation to concentration camps. Therefore, although the Jewish heritage is not totally confined to the area, Seahorse Square serves both the spatial fragmentation of heritage in Rhodes in general by the domination of the Jewish memory in the region and the understanding that emphasizes the importance of place in heritage management.

In fact, the symbol of Seahorse is so important for the Jewish community of Rhodes that the website of the Rhodes Jewish Museum,[32] opened in 1997 next to the Khalal Shalom Synagogue, again in the Jewish quarter of the town, uses it as an emblem. Museums are places to remember, to conserve, and to preserve history as well as to mourn for a haunted past (Kirshenblatt-Gimblett 1998, 138–9). The Jewish museum is no exception, with its exhibitions ranging from photos of families who once lived in Rhodes to the stamp that was used on identification papers by the administration to identify the Jews in 1943. The genocide of the Dodecanesian Jewry is also remembered through commemorative events, which perform the duty of referring to a shared history with which the whole group may identify (Ben-Amos and

Figure 2.2 Monument of the Victims of the Holocaust in the Jewish Martyrs Square, Rhodes (Copyright Hazal Papuccular).

Bet-el 2005, 169). Therefore, these commemorative activities in Rhodes, which begin on 23 July every year, the day of deportation, are actually a call for all the descendants of the Jewish community who are now scattered around the world to reunite in the island. The activities vary from memorial services taking place in the synagogue to film screenings and walking tours in Rhodes and Kos.

War heritage in Rhodes is not the monopoly of the Jews. The reflection of the fragmented group histories in the Dodecanese can be seen in different monuments and the meanings that they carry. For instance, the Greek war memorial, the Statue of Victory, has been dedicated to the British, French and Greek soldiers who played a role in the liberation of the islands in the northern and southern Aegean Sea between 1943 and 1945 (Figure 2.3). This dedication shows the importance of the unification of the Dodecanese with Greece after the end of the war since it links the history of the region, which followed a different path of development until 1947, to its motherland, making the archipelago a monolithic entity. The geographic location of the statue is as meaningful as the Jewish heritage because it is situated in the harbour, the entrance point to the island and one of the most visible and crowded places of Rhodes. It is the symbol of the dominant position and discourse of the Greeks in the Dodecanese; the Jewish heritage, by contrast, is mostly limited to the old Jewish quarter. In addition, the position of Dodecanese identity as an important part of this dominant discourse today is celebrated with parades and cultural activities by the islanders in the city centre every year on 7 March, the day of integration and a day of national holiday.

While Greek heritage focuses on liberation and the Jewish emphasis is on the Holocaust, it may be observed that any war memorial specifically concerning the Turkish community is almost non-existent in Rhodes or Kos. Obviously, when the preceding narratives of the Turkish community are kept in mind, this fact seems understandable. The Turks were neither deported like the Jews, nor did they fight side by side with the Allies against the Axis occupation like the Greeks, who had a national aspiration to unite with their motherland. Therefore, it is not surprising to see that the heritage that the Turkish community deals mostly with is associated with the Ottoman era as a reflection of a vivid past, rather than WWII.

The diversity of war heritage in the Dodecanese is restricted neither to the local groups nor to Rhodes. On the contrary, the British, Greek, French, Canadian and South African soldiers lie in two major war cemeteries located in Rhodes and Leros, which are important commemorative places that transcend the narratives of the local ethnic groups.[33] Armistice Day on 11 November, which remembers the Commonwealth citizens who died during two world wars, is commemorated in both cemeteries. The importance of Leros, especially for the British citizens, is much broader since the Battle of Leros in 1943 symbolizes a defeat for the British in the

Figure 2.3 Statue of Victory, Rhodes (Copyright Hazal Papuccular).

face of Nazi air superiority, which had played a significant role in the sinking of the Allied destroyers *Queen Olga* and *Intrepid* on 26 September (Gooderson 2002, 11). Therefore, in addition to the Leros War Museum,[34] which displays military equipment in a tunnel from WWII, and the cemeteries, the war memorial of Leros commemorates this tragic event. Parades and memorial services still take place in Leros in memory of the destroyers every year from 26 to 29 September.

It should be emphasized that all these memorials and commemoration activities in the Dodecanese cannot be separated from tourism. Heritage has become a big tourism industry since the 1970s with museums, artefacts and historical sites adding to the expansion of cultural production and consumption (Aitchison *et al.* 2002, 97). The Dodecanese are no exception in this sense. The creation of major museums in the islands, such as the Leros War Museum and Rhodes Jewish Museum, as well as the erection of the monuments and the statues such as the Monument of the Victims of the Holocaust coincides with the transformation of heritage as a concept. Therefore, although heritage is important in the perpetuation of local histories and identities, the dimension of tourism cannot be ignored. It is this that leads the tour companies of Leros to sell three- or four-day tour packages that include mostly tourist activities such as diving around the shipwrecks of WWII; the formal commemoration ceremony occupies a very small place in the programme. Likewise, although Holocaust-related tourism seems at first glance to be inappropriate (Prentice 2010, 246), the activities in Rhodes to commemorate deportation are not so divergent from heritage tourism elsewhere.

CONCLUSION

The bombings, the Nazi occupation, the starvation, the migration and the transition period constitute the key common themes in the Dodecanesian memory of WWII. Being on an island, with all of the associated implications of limited resources and susceptibility to occupation, has had an influence both on the historical course of events in the region and on collective memory. Despite the commonality of these fundamental themes, this chapter has attempted to argue that, even in small geographic areas, memories can be fragmented by certain variables. In the case of the multiethnic islands of the Dodecanese, such as Rhodes and Kos, the most important element would appear to be ethnic differences. I have shown how group identities and boundaries in these islands, where the islanders lived separately with their own legal system for centuries, has resulted in the formation of distinct memories based on distinct experiences. This is why the image of the Nazi soldier holds different meanings for the Greeks, the Turks, and the Jews. The material reflection of these diverse experiences within the islands has been a fragmented heritage of war, both in terms of local identities and in terms of geography. This fragmentation in museums, commemoration activities and monuments seems to provide variety for the islanders who have succeeded in combining the perpetuation of war memories with the tourism industry.

Contemporary issues also have a role to play in the formation of memory, and this causes an overemphasis on certain themes. While the constant Turkish reference to fraternity among the groups during the interviews has obviously been shaped by their contemporary identity concerns in the islands, the memories of famine referred to by the Greeks can also be interpreted as a reflection upon the recent economic crisis in Greece. The Greeks in Rhodes, especially the older generation, frequently compare the 'German-imposed' austerity measures with memories of the German actions in the war by stating that the Nazis killed a lot of people by causing famine.[35] In other words, the memories of the groups have been shaped and reshaped over time, bringing some themes into prominence at different times. The Dodecanesian narrative of WWII has never been a monolithic story.

ACKNOWLEDGEMENTS

I would particularly like to thank the Rhodes, Kos and Dodecanese Turkish Culture and Solidarity Association for their help during my period of interviewing.

NOTES

1. The Dodecanese are one of the island groups of the Aegean Archipelago, near south-western Anatolia. The group includes more than twelve islands. The major ones are Rhodes, Kos, Leros, Astypalaia, Halki, Kalymnos, Karpathos, Kasos, Nisyros, Patmos, Symi and Tilos. Kastellorizo (Megisti), though historically and geographically not in the group, has been included within the Dodecanese since the twentieth century.
2. The Ottomans began to conquer the islands in the Aegean Archipelago near Asia Minor from the beginning of the fifteenth century. The process was completed with the fall of Rhodes, the most important island of the Dodecanese group, in 1522. For more information, see Vatin (2000).
3. Even if the system ensured a certain extent of autonomy, the Ottoman state did not treat all the groups with a sense of equality. During the period of the Ottoman Empire, Muslims had prominence over other ethnicities. This situation continued until the nineteenth century, during which the equality of all citizens was ensured for the first time by the modern state.
4. The Italian Governor of the Dodecanese, Mario Lago (1922–36), occasionally emphasized the importance of the Dodecanesian Turks as a way of finding a balance with the Greeks, since the latter challenged Italian sovereignty and desired unification with Greece. In some of the official correspondence, Lago stated that Italy did not want a decrease in the number of Turkish citizens

and, for this reason, implemented ‘soft’ policies on this community. For more information about this rhetoric, see Archivio Storico del Ministero degli Affari Esteri, Rome, Italy, Pacco Dodecanneso 991, ‘Cancelliere del Consolato di Turchia in Rodi,’ 13 November, 1928.

5. The National Archives, London, CAB 84-8-14, ‘Plan for the Capture of the Dodecanese’, 10 October 1939.
6. *The New York Times*, 25 April 1941, 4; *The New York Times*, 29 October 1940, 10.
7. For these reports, see The National Archives, FO 195/2468.
8. In fact, the capture of the Dodecanese had been specifically desired by Churchill. According to him, it would not only constrain the Germans in Crete but would also compel the Turks, who were non-belligerent, to enter the war.
9. To gain an understanding of the discussions regarding the transition of the Dodecanese, see The National Archives, FO 371/48342.
10. Archivio Storico del Ministero degli Affari Esteri, Busta 16 Dodecanneso, ‘Il Possedimento delle Isole Italiane dell’Egeo’ (n.d.).
11. The National Archives, FO 371/29932, 20 January 1941, 31 December 1940, 10 February 1941, 13 February 1941.
12. Interview between author and Sadi Nasuhoğlu, 24 August 2013.
13. In a novel based on the memoirs of a family who emigrated from Rhodes to Smyrna during the war, the Italian soldiers were portrayed as a sympathetic people (Izmirli and Izmirli 2007).
14. Interview between author and Ismet Buyukyaşar, 5 August 2013.
15. Interview between author and Suna Hamit, 30 September 2013.
16. Interview between author and Orhan Moralı, 1 October 2013.
17. Interview between author and Suna Hamit.
18. Interview between author and Rafet Faralyalı, 4 November 2013.
19. Interview between author and Rıfat Hamit Maşazade, 2 October 2013.
20. Interview between author and Rafet Faralyalı.
21. Interview between author and Sadi Nasuhoğlu.
22. Interview between author and Rıfat Hamit Maşazade.
23. Interview between author and Rıfat Hamit Maşazade.
24. Interview between author and Orhan Moralı; interview between author and Abdullah Kazimlar, 9 June 2012.
25. Although Turkey was a neutral country with some responsibility towards the Allies, one million people out of a total population of seventeen million were mobilized, and the economy was channelled to the needs of the army. The WWII years symbolize one of the worst epochs in terms of the Turkish economy.
26. *Holocaust*. Online. Available HTTP: www.rhodesjewishmuseum.org/history/holocaust (accessed 13 January 2013).
27. Today, just more than thirty Jews live in the Dodecanese. See *Holocaust*. Online. Available HTTP: www.rhodesjewishmuseum.org/history/holocaust (accessed 13 January 2013).
28. An honour given by Israel to those who risked their lives in order to save Jews.
29. *Primary School for Van from the Jewish Community of Turkey*. Online. Available HTTP: www.turkyahudileri.com/content/view/2018/227/lang,en/ (accessed 28 October 2013).
30. Interview between author and Sadi Nasuhoğlu.
31. Interview between author and Ahmet Kırevliyası, 1 October 2013.
32. For more information about the museum, see its website: www.rhodesjewishmuseum.org/museum (accessed 13 January 2013).
33. These cemeteries are under the management of Commonwealth War Graves Commission, and information about both of them can be found at their website: www.cwgc.org/ (accessed 10 January 2014).

34. To see some photos from the museum, see www.leros.org/lerostouristhttp/leros-war-museum.html (accessed 10 January 2014).
35. 'Greeks Try to Keep the Peace with Their Dwindling German Tourists', *The Guardian,* 2 March 2012. Online. Available HTTP: www.theguardian.com/world/2012/mar/02/greeks-keep-peace-germany-tourists], (accessed 20 February 2013).

BIBLIOGRAPHY

Aitchison, C., Macleod, N. E. and Shaw, S. J. (2002) *Leisure and Tourism Landscapes, Social and Cultural Geographies*, London, New York: Routledge.

Assman, J. (2010) 'Communicative and cultural memory', in A. Erll and A. Nünning (eds), *A Companion to Cultural Memory Studies*, Berlin and New York: De Gruyter.

Bell, D. (ed). (2006) *Memory, Trauma and World Politics: Reflections on the Relationship Between Past and Present,* Basingstoke and New York: Palgrave Macmillan.

Ben-Amos, A. and Bet-el Ilana. (2005) 'Commemoration and national identity: memorial ceremonies in Israeli schools,' in A. Levy and A. Weingrod (eds), *Homelands and Diasporas: Holy Lands and Other Places*, Stanford, CA: Stanford University Press.

Braude, B. (1982) 'Foundation myths of the Millet system', in B. Lewis and B. Braude (eds), *Christians and Jews in the Ottoman Empire: The Functioning of a Plural Society*, New York: Holmes and Meier.

Dilek, B. S. (2008) *Ege'nin Unutulan Türkleri* [*Forgotten Turks of the Aegean*], Istanbul: Cumhuriyet Kitapları.

Doumanis, N. (1997) *Myth and Memory in the Mediterranean: Remembering Fascism's Empire,* Houndmills, Hampshire: Macmillan; New York: St. Martin's Press.

Gooderson, I. (2002) 'Shoestring strategy: the British campaign in the Aegean, 1943', *Journal of Strategic Studies*, 25 (3), 1–36.

Halbwachs, M. (1952 [1941]) *On Collective Memory*, Chicago: University of Chicago Press [L. A. Coser (trans), 1992].

Hay, P. (2006) 'A phenomenology of islands', *Island Studies Journal*, 1 (1), 19–42.

Hirsch, M. (2008) 'The generation of postmemory,' *Poetics Today*, 29 (1), 103–28.

Izmirli, Z. and Izmirli Y. (2007) *Rodos'tan Karşıyaka'ya, 1685 Sokak* [*From Rhodes to Karsiyaka, Street 1685*], Istanbul: Neden Kitap.

Kirshenblatt-Gimblett, B. (1998) *Destination Culture: Tourism, Museums, and Heritage*, Berkeley: University of California Press.

McDowell, S. (2008) 'Heritage, memory, and identity', in B. Graham and P. Howard (eds), *The Ashgate Research Companion to Heritage and Identity*, Hampshire, Burlington: Ashgate.

Millar, S. (1989) 'Heritage management for heritage tourism', *Tourism Management,* 10 (1), 9–14.

Orakçı, M. (2012) *Rodos Müslümanları: Selam Gazetesi 1926–1936* [*The Muslims of Rhodes: Selam Newspaper 1926–1936*], Istanbul: Kitap Yayınevi.

Prentice, R. (2010) 'Heritage: a key sector in the "New Tourism"', in G. Corsane (ed), *Heritage, Museums and Galleries: An Introductory Reader*, London; New York: Routledge.

Richter, L. K. (2010) 'The politics of heritage tourism development: emerging issues for the new millennium', in G. Corsane (ed), *Heritage, Museums and Galleries: An Introductory Reader*, London and New York: Routledge.

Shachar, N. (2013) *The Lost World of Rhodes: Greeks, Italians, Jews, and Turks: Between Tradition and Modernity*, Brighton and Portland, OR: Sussex Academic Press.
Smith, P. C. and Walker E. R. (2008) *War in the Aegean: The Campaign for the Eastern Mediterranean in World War II*, Mechanicsburg, PA: Stackpole Books.
Stratford, E., Baldacchino, G., McMahon E., Farbotko C. and Harwood A. (2011) 'Envisioning the archipelago', *Island Studies Journal*, 6 (2), 113–30.
Tokmakoğlu, B. (2013) 'Alman İşgali ve İstanköy'den Kaçış [German Occupation and Escape from Kos]', 20 January. Online. Available HTTP: blog.milliyet.com.tr, http://blog.milliyet.com.tr/alman-isgali-ve-istankoy-den-kacis/Blog/?BlogNo=398347 (accessed 25 February 2012).
Ülkümen, S. (1993) *Bilinmeyen Yönleriyle Bir Dönemin Dışişleri: Emekli Diplomat Selahattin Ülkümen'in Anıları* [*Foreign Affairs of a Period with Unknown Aspects: The Memories of Retired Diplomat Selahattin Ülkümen*], Istanbul: Gözlem Gazetecilik Basın ve Yayın A.Ş.
Vatin, N. (2000) *Rhodes et l'Ordre de Saint-Jean-de-Jerusalem*, Paris: CNRS Editions.

3 From Poetic Anamnesis to Political Commemoration

Grassroots and Institutional Memories of the Greek Civil War on an Aegean Island

Elena Mamoulaki

Ikaria is a middle-sized, mountainous Greek island surrounded by some of the wildest waters of the Aegean Sea. In contrast to the neighbouring islands, which all have natural harbours, the sharp coastline of Ikaria does not include gulfs protected from the strong winds. Thus, ringed as it is by a frequently turbulent sea, it has been often quite isolated (Figure 3.1). Historical and archaeological findings indicate that Ikaria—like many other islands of the Aegean Sea—was used as a place of exile[1] for the political opponents of different regimes during its long history (Melas 2001; Papalas 2002, 2005).

During the Roman and the Byzantine empires, it seems that Ikaria was also used as a place of exile for noble political opponents. Little evidence is found for the practice of exile on the island under the Ottoman Rule until the beginning of the twentieth century. After its unification with the modern Greek state in 1912,[2] Ikaria was used as an exile[3] location during the Ioannis Metaxas[4] dictatorship of the 1930s for some 120 left-wing high-ranking military personnel and government officials. During the Greek Civil War period of 1946–9, Ikaria—which at that time had fewer than 10,000 inhabitants—was used as an exile camp for more than 10,000 political detainees. The latter were Greek citizens from other parts of the country exiled to the island by the authoritarian government due to their expressed or suspected leftist political leanings. During this period, exiles and islanders lived together in the same community, often under the same roof guarded by dozens of gendarmes and police officers.

This paper—based on extensive fieldwork on Ikaria—is about the passage from the personal/intersubjective memories to the public/institutional commemorations of that cohabitation between locals and political exiles. Compared with other places of exile, what is especially interesting in Ikaria is that people have transformed what would in most contexts be considered a traumatic memory into a collective heritage with positive connotations which, according to them, constitutes an important part of their Ikarian identity. The representation of this local identity is based on the values of

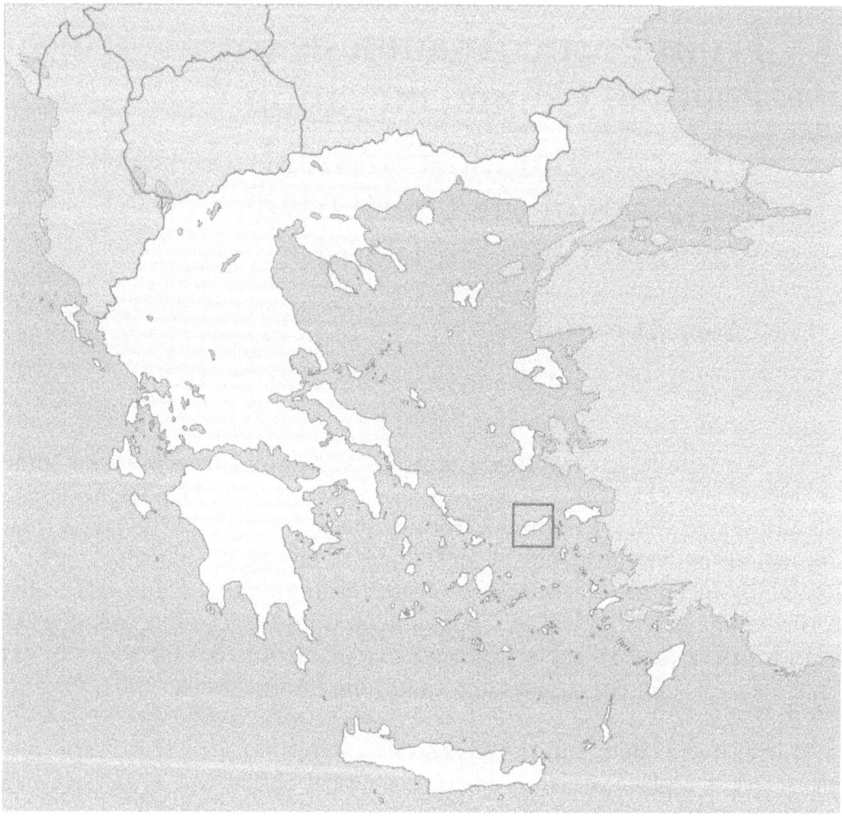

Figure 3.1 Map of Greece with Ikaria highlighted (Creative Commons, Pitichinaccio, amendments by the author).

solidarity and resistance in a broad political sense. All my Ikarian informants stated their pride in the island's history as a place of exile and the hospitality practised by their ancestors or by themselves. They also expressed their respect for the exiles who stayed in their island throughout its long history. Many Ikarians claim that they, as descendants of the famous Ionians,[5] are a doubly democratic and civilized people. Thus, they argued that hospitality is an innate property of the Ikarian 'race' (they often said 'it is in our blood') since it is their ancestors' tradition which they are proud to continue. As 64-year-old Katerina from Kampos said, 'We have always been *filoxeni* ['hospitable']; we are Ionians.'

So today almost all islanders evaluate positively their experience of exile and cohabitation with the political detainees.

> Ikaria has always been a place of exile. This is an honour for us, because we had the chance to live with the cleverest; the exiles were the most intelligent and avant-garde people because those were the most

> dangerous for the regimes and the tyrants. Why should they send an idiot into exile? We lived together with the exiles and that is why we are more progressive and open-minded than other Greeks.
>
> (76-year-old Nikos from Evdilos)

When talking about the fact that the island was used as an exile camp, many persons mix pride and complaint: pride because of the sympathy and admiration towards the exiles, but also complaint because of the indifference or hostility of the state.

> We were treated as exiles even when we did not have exiles to live with us. We learnt how to organize our lives, how to survive and above all how to resist each time greedy invaders, call them pirates, Turks or Athenian politicians, who only wanted to take our few belongings away and our children as soldiers. We got to know how to hide and help each other and that's what we also did when living with the exiles.
>
> (86-year-old Kostas from Arethousa)

ISLANDS OF EXILE: MEMORY AND OBLIVION

The memory of a civil war is often a dividing factor within societies. The handling of such legacy usually triggers conflicts that can derive not only from the past civil war but also from the present interests and conflicts in the context of the remembrance. In Greece, as in other places with similar historical experiences, the memory of civil war is usually either the victim of an attempted suppression or channelled by official authorities.[6]

The long history of exile in Greece shows that islands played a crucial role in the development of this particular phenomenon. Not only the political conditions but also the geographical distribution of the country, with hundreds of islands, was essential for conceiving and realizing the practice of political punishment and exclusion.

We know that only a few remote islands of the mainland were exile destinations only for a very limited number of persons. Islands, on the other hand, have often received swarms of exiles, sometimes even exceeding in numbers the local population (as in the case of Ikaria, Ai Stratis etc.). So, despite the fact that in other historical periods—when transportation was very difficult even in the mainland—the phenomenon of exile was not endemic in the islands, in the specific period of the Greek Civil War, the small, remote Aegean islands, like Ikaria, were the only 'secure' (in terms of surveillance and difficulty to escape) destination for the deportation of thousands of political detainees. This leads us to the fact that the islands used as places of exile were marked by the practice of exile and the legacy left behind by successive batches of detainees.

What is particularly interesting and requires further research is that although various neighbouring Aegean islands were used as exile camps in several periods from the 1910s to 1970s, the way former exiles, locals and their respective descendants remember (or forget) and evaluate this 'coexistence' ranges from pride to shame and from reification and political instrumentalization to total suppression and silence. An in-depth comparative analysis would shed light on these important discrepancies, allowing us to understand the specific historical and social conditions that shape the collective memory of the Greek Civil War on the Aegean islands.

The memory of that period in Ikaria stands in vivid contrast to the memory's silencing and to its monopolization by the political elites. Moreover, while in other parts of Greece the memory of this period is perceived as a potential triggering event for conflicts inside the community, it seems that in Ikaria, the memory of the Civil War era serves as a unifying factor among the population of different political ideologies and interests. Furthermore, while on other Greek islands with a similar history, the memory of the exile is considered a stigma that could hinder its touristic development and is therefore suppressed, on Ikaria it is mostly a source of pride and an object of collective commemorations initiated by the local community.

So the main hypothesis of this paper is that, in the case of Ikaria, we face a type of social memory of the exile period that contrasts with the conventional and institutional memory of civil war more generally and more specifically of the Greek Civil War of 1946–9. I will argue that this difference is strongly related to the past experience of the cohabitation between locals and exiles which was based on to the *longue durée* historical process of the shaping of what Bateson (1958) would call *ethos* and Wittgenstein (2009) *form of life*. If we also follow E. P. Thompson (1995) and Scott's (1977) concept of *moral economy*, it seems that in Ikaria, it involves a great deal of informal exchanges of products and services within the community (among family members, neighbours and friends), as well as a sense and practice of social justice and solidarity dependent on the members of the community and not on an official institution outside the community. Characterized by a *form of life* related to the resistance against what is contrary to the Ikarian values of humanism and commitment to the common good and a *form of life* of frugality and autonomy, the Ikarian historically determined *ethos* is related to the island's political history and its periods of independence and self-reliance. This ethos is also reflected in the political discourse of Ikarians as well as in their ways of remembering—and forgetting—the Civil War era on the island. On the other hand, in the *ethos* and the ideology of the political exiles, the values of class solidarity and struggle against oppression were crucial.

Over the past few years, as members of the older generation pass away, we have begun to see a transition from personal, intersubjectively conveyed memories of the Civil War period to public, institutionally communicated histories. The content of this collective memory of the Civil War era on the island is not related to an event (such as a battle) separated from people's ordinary lives, but to the everydayness of the community in the context of the exceptional state of exile and cohabitation with those thousands of their 'unexpected guests'. Thus, I will try to describe how this memory is currently at a transitional point: from being personally related by people who had had a firsthand experience of it to their family and friends, to a public discourse mediated by commemorations that are not yet formally and totally 'crystallized' into an officially recognized institution. Thus, in this study, based on ethnographic methodologies and an architectural perspective, I attempt to describe what happens when the diffuse memory of a past that had been embedded in social memory and practices is being crystallized institutionally into a specific object or site, recognized by the majority of the community involved with that past. We are dealing here, in other words, with a 'homeless' memory, in the sense that there is no conventional monument or museum imposed by an official authority to stand for the memory as a pointer, a marker of the past, for which some wish to find a home.

This opens up a number of questions. Where does such memory 'dwell' before it becomes a concrete material monument? How is it related to people's everyday lives? Which are the changes in the perception and representation of the places of memory in such moments of transition from personally lived to collectively institutionalized memory?

In the broader academic discourse about the complex relationship between space, materiality, memory and identity,[7] this chapter provides some insight into these questions by examining the various aspects of memory in its transition from the experienced (personal/intersubjective) to its institutionalized (public/collective) constitution. In particular, it explores the ways in which specific sites correlate experientially—through interpersonal narratives and activities—with past events, while new sites are marked or created through collective actions in order to *spatialize* (*materialize*) that past. I argue that the preservation or elimination of community memory and the ways in which this is effected are related to what happened in the past as well as to the historical process of the memory within this community.

The analysis of the experiential constitution of space through everyday practices of memory (involving personal narratives, *ethos*, representations, rituals) and of the institutional construction of memory (through the management of space, monuments, museums, reconstructions) shows that the transition from the first to the second marks, in a certain way, the transition from interpersonally performed and narrated poetical anamnesis to publically substantiated political commemorations.

In short: this work is the result of an encounter between a complex and particular phenomenon of internal exile, related hospitality and the resulting links of sociability and alliance, on the one hand, and, on the other, contemporary Ikarian society which offers valuable insight into the transmission from experienced to institutional memory.

EXILE ON IKARIA[8]

As mentioned, during the Greek Civil War of 1946–9 more than 12,000 left-wing Greek citizens were sentenced to internal deportation to Ikaria, as punishment for having opposed the authoritarian post-WWII Greek government, which was made up, among others, of right-wing, anti-Communists and former Nazi collaborators, who were supported by the British and American governments. Most of the political detainees had participated or were suspected of having participated in the leftist movement of the national resistance against the Axis occupation of 1940–4, which was mainly organized by the Communist Party of Greece and supported by a large part of the population. Among those chosen by the government for internal exile were often also some of the relatives or other possible 'supporters' of those who had taken part in this resistance movement or who were suspected of being pro-Communists.

These prisoners, accompanied by dozens of gendarmes and police officers, were sent to more than thirty-five Aegean islands, both inhabited and uninhabited. Thus, while the experience of internal exile was common in Greece (see Birtles 2002; Georgiadis 2004; Gritzonas 2001; Kenna 2001; Mastroleon-Zerva 1986; Oikonomopoulos 2004; Panourgia 2009; Sarantopoulos 2000; Staveri 2006; Tsakiris 1996; Voglis 2002), the case of Ikaria was distinct in that it involved a particularly close and intensive form of cohabitation of local and exile communities. Although political exiles also lived on other inhabited islands, on Ikaria their large number and the fact that many of them shared houses with the locals, created many special conditions. In 1947, in a period of just a few months, more than 10,000 political opponents of the right-wing Greek government were exiled to Ikaria—without any provisions for housing, medical care or even much in the way of food supplies (Figure 3.2).

At that point in time, Ikaria had only about 10,000 inhabitants, who themselves were just managing to scrape by when the government decided to use it as an exile camp. According to the authorities, the main reason behind this decision—and all internal exile—was to protect the country from the 'communist threat' (see Voglis 2002, 52–73). While the conflict between the Right and Left was widespread in many parts of Greece (see Margaritis 2002; Mazower 2000; Voglis 2002) not only on battlefields but also within families, on Ikaria the conflict between the local royalists and the communists was dealt with in such a way that involvement in the Civil War was essentially avoided on the island.

Some Ikarians claim that due to their democratic and open-minded 'nature' grounded in Ionian culture, they were socially mature and responsible enough to avoid the bloodshed on their island. Others explain that the Civil War was avoided due to the relatively small size of the island and the prevalence of the Left who did not want to put in danger the exiles who were kept as hostages in the hands of the government's gendarmes.

Figure 3.2 The arrival of hundreds of political exiles by a warship at the shore of Aghios Kirikos, Ikaria, in 1947 (Reproduced courtesy of H. Malachias).

By the end of the Italian and German occupation of WWII, the majority of the locals supported the communist-led EAM (National Liberation Front) (Papalas 2005, 227), and a significant proportion of the population was actively involved in the anti-fascist organizations of EPON (United Greek Youth Organization) and the local Communist Party committees. The rest of the islanders supported the centre and right-wing parties and the king.

Despite belonging to political parties from widely different locations on the political spectrum—with a then remaining dominance of the Left on the island—Ikarians for the most part responded to the influx of exiles by overlooking their internal differences in order to 'absorb' the exiles through an overwhelming wave of hospitality (Mamoulaki 2008, 2010). It seems that the negative reactions against the exiles were very much limited. Actually, none of my twenty exile informants could recall any. Some of the locals referred to specific persons who were appointed by the government to create problems but who were soon isolated by the community. The great majority of the local population tried to protect and support the exiles. 'Ikarians embraced us'['*οι Ικαριώτες μας αγκάλιασαν*'] is what most of the former exiles said when asked about the locals' reaction upon their arrival (numerous interviews from 2006 to 2010). Given that there were no prisons or concentration camps on the island, the locals opened their houses to accommodate exiles—either in a family house not in use or in a room in the family home. They provided food and free land for cultivation to the recently arrived exiles and tried to incorporate the newcomers into the social life of the community in various ways, such as through creating kinship bonds of god-parenthood.

> We couldn't believe that apart from the many leftists on the island, many of our hosts were ardent rightists and royalists and they were offering us their houses to stay in and their few products to eat. It is thrilling even after so many years to think about that.
>
> (Former exile, 87-year-old Kostas)

The exiles, for their part, took initiatives to reciprocate the hospitality of the locals in a number of different ways. Working groups were organized—for example of engineers, lawyers, artisans, craftspersons, workers, farmers, agronomists—and free services were offered by exiles in all professions to locals as well as to one another. Actors and theatre professionals presented plays; musicians taught and performed. Doctors treated both locals and exiles; professors taught local children as well. The exiles also took on technical projects such as building cisterns, roads and water infrastructure for the villages.

As these actions demonstrate, the forced cohabitation among locals and exiles was dealt with within a context of widely shared 'rules' of hospitality and reciprocity, as well as the adoption of the respective roles of the host and the guest. Beyond these acts of reciprocity, the exiles also drafted and imposed upon themselves a list of ten articles indicating proper behaviour within their group and in relation to the locals (as was common practice among exile groups during Metaxas's dictatorship of the 1930s, see Kenna 2001, 45). Probably the most discussed article was the one that forbade exiled men to have any contact with local women. The exiles had to obey the Political Exiles Coexistence Groups (OSPE; for more on the function of these, see Kenna 2001 and Gritzonas 2001) and suppress their feelings and desires in order to respect the balance and *ethos* of the local population. Offenders were severely punished by exclusion from the OSPE, which was created to cover their basic daily needs.

While the exiles had their articles as a guide for conduct, the locals treated the exiles as their 'unexpected' and harassed 'guests' who needed their assistance in order to survive. This was not always simple, not only because of the extreme poverty of the locals due to the devastated local post-war economy and the lack of supplies but also because of the repression by the gendarmes and the prohibition of contact between locals and exiles even when living in the same house. Thus, many Ikarians were persecuted and punished—at times themselves being sent into exile on other islands—for supporting the exiles.

THE MEMORY OF EXILE AND COHABITATION

In trying to understand how that cohabitation was made possible and fruitful, I realized the importance of the *longue durée* Ikarian *ethos*, which has been little investigated in relation to the experience of exile on the island. Highlighting it is important for at least two reasons. First, it is crucial for what made the case of political exile on Ikaria unique for both the exiles and the locals who hosted them. Second, it continues to play an important role in the ways in which former exiles and Ikarians deal with the memory of the Civil War.

Gregory Bateson's definition of *ethos* as 'the expression of a culturally standardized system of organization of the instincts and emotions of the individuals', as presented in his classical work *Naven* (Bateson 1958, 118), is useful in this connection. Based on empirical data provided by both locals and former exiles, I argue that the fact that locals assisted the exiles was a result of the compatibility of their *ethoses* as performed in the reciprocal relationship of hospitality. Exiles and Ikarians, as actors in this 'forced cohabitation', through their repeated actions which asserted a particular kind of identity and a particular way of being (in other words, through their performativity (Butler 1990)), constructed an almost unanimous and uncontradicted picture of that peaceful life, a picture which has become fixed in memory with a remarkable degree of consistency. I also support that Ikarians and exiles formed one, united community that emerged as a result of the particular social and political osmosis during their forced cohabitation.

Following Bateson's definition of *ethos*, I want to stress its three crucial dimensions: social, moral and affective. I consider *ethos* less as a purely cultural—and even less an 'innate'—characteristic. Rather, my ethnographic material documents that *ethos* is a historical process, formed and reformed under specific social conditions. Today's relations of many locals with tourists document that Ikarian notions and practices of hospitality—their *ethos*—can shift over time.

According to historians' accounts (Melas 2001; Papalas 2005) and to oral tradition, Ikaria was a society without a state for long periods of time, and thus its members developed relations of solidarity and reciprocity. The relations with the exiles demonstrate that Ikaria has a different history with sufficient autonomy—including an ephemeral period of independence—to draw upon to empower its own society, self-help and determination in facing the contingencies of history.

Ikaria's long periods of relative independence and lack of central authority and institutions may be part of why Ikarians did not automatically obey authorities' rules or rely upon the public institutions outside the community itself for their social stability and safety. Ikarians questioned the alleged 'patriotic' measures taken by the government independently of their political affiliation. Their reactions show vividly how a latent ethical conception of justice, as it was spelled out in the *ethos* and the everyday practices of the Ikarians, was a way of standing up against a legalistic conception of rule following as promulgated by most governmental representatives and their followers.

Ikarian hospitality is related to that historically shaped *ethos* rooted in a rather egalitarian society of the epoch. Egalitarianism here refers to the fact (per the narrations of numerous informants) that in their everyday interactions people saw and treated each other—and to some extent still do (as I observed during my fieldwork)—as equals. And here I want to stress the importance of informality that is widespread in its performativity in Ikarian society then and now. Hospitality as performed by Ikarians has probably its origins in the social interaction, the mutual help and the informal exchanges of products and services among them in periods of autonomy and self-reliance.

Thus, egalitarian treatment could foster the creation of an economy largely based on informal exchanges instead of money. Today, Ikarians still practice several kinds of local customs based on reciprocal relations. It is not uncommon to exchange food, to assist each other, to organize feasts for the redistribution of local products and to dedicate much time to work for the common good. In short: still today, as in the past, collective mechanisms of social solidarity are practiced.

In relation to the handling of the memory of the Civil War, most Ikarians react by discarding the institutional memory of the various political agents (who seek to appropriate and monopolize any positive aspect of the past in order to increase their present political capital) because it is not related to their *ethos*. In other words, if during that epoch people did not follow orders of the politicians in order to cope with the situation, they do not accept that their memory of that past is monopolized and channelled by institutions outside the community. The majority of the locals, independently of their political affiliation, believe that this heritage should not be appropriated by any political party because Ikarians did not act the way they acted due to their belonging to one or another party or because the exiles were of a specific political ideology. If that were the case, the arrival of the political detainees would have divided the population and created conflicts due to political rivalries.

The conflict between those who believe that the memory of the exiles' period is inextricably connected to the legacy of the Greek Communist Party and those who think that the memory of the cohabitation with the exiles should not be used under a political party's agenda (usually voters of other political parties) for political propaganda was dramatically expressed in 2008. Occasioned by the inauguration of a monument for the political exiles erected by the Prefectural Committee of the Greek Communist Party, the clash was manifested with the vandalism of the monument two days after its unveiling ceremony. On the other hand, many Ikarians expressed their disappointment for the participation of well known right-wing politicians in the commemorations because, according to them, 'this was a painless way to be redeemed for their sinful past', said Manolis, a 65-year-old man from Raches.

EXILE IS NO STIGMA; IT IS OUR HERITAGE

What is especially interesting about Ikaria is that people consider exile not as a stigma that could potentially harm the island's image and representation but as an important—in moral and historical terms—heritage that has to be properly stored and celebrated. Over the years, many former exiles have continued to visit Ikaria with their families in order to share their past with their children and to maintain their relationships with the locals, seeking out their old island friends and often bringing presents as recognition of their past help. Nowadays, their children and grandchildren visit the island to retrace their descendants' steps. Thus, there are also hundreds of exiles (or members of their families) who do not consider this past as a stigma, for they often call it a *pilgrimage*, which shows the importance for them in moral and affective terms. Through these reciprocal gestures, the memory of these past social relationships is renewed in the same places where they were first formed. Ikarians living in Athens or elsewhere often organize feasts that are attended by former exiles. In this way, Ikarians and former exiles reconstruct a past community through dance and the sharing of food and wine, far from Ikaria.

There have also been innumerable incidents in which Ikarians meet former exiles or family members of former exiles, entirely by chance, in Athens or in other parts of Greece. In one case, an Ikarian had to go to a hospital in Athens many years after the end of the Civil War. When he told the doctor that he was from Ikaria, he was immediately taken to a better room and was attended very closely by the hospital staff for the rest of his stay. He was later told that the doctor had been an exile in Ikaria during the Civil War; the doctor's actions towards the sick man were a gesture of reciprocity in remembrance of the help that he himself had received from Ikarians during his exile.

There are many such examples of encounters between strangers characterized by spontaneous expressions of gratitude by the exiles and their family members towards the Ikarians (generalized reciprocity). In these instances, the memory of cohabitation is again re-enacted, and places of memory are being constructed through these actions, through people's will and intention to remember their common past experience under new circumstances.

Thus, former exiles and Ikarians often find themselves recalling their past experiences through chance encounters and in unexpected, unremarkable places. For example, it is not uncommon for Ikarians today to discover boxes with correspondence, photographs and small objects hidden by exiles in the walls of the houses they lived in. It is important to note that most of them carefully restore and keep these items together with the valuable objects (photographs, documents, etc.) of their ancestors, and they consider them as family heirlooms loaded with affective and moral significance. To the exiles and their families, as well as to locals and theirs, these objects are used as 'inalienable'[9]—a term often used by anthropologists about objects that are not to be sold but only exchanged or passed on as gifts to specific persons or institutions.[10]

In Ikaria, in contrast to other cases in which the building of a historical identity has been imposed from above, collective memory is actually also a grassroots memory: Ikarians and former exiles can recognize the lived past in the public initiatives that recreate it. This is a subjective, intentional and collectively shaped memory that attempts to formulate, interpret and actualize the past using an identity discourse. Memorials, public events, congresses, publications and museums are some of the forms of this becoming-institutionalized memory.

In 1997, the fiftieth anniversary celebration of the arrival of the exiles was organized by a group of locals and exiles. Speeches, feasts, concerts and the first official memorial service for exiles who died in Ikaria during 1947–9 were part of the activities. In 1998, the fiftieth anniversary of the UN's Declaration of Human Rights, a local village association decided to organize an event in which Ikaria would be declared the Island of Solidarity and Human Rights, attended by the president of Greece, various ministers and other officials, together with former exiles and intellectuals from throughout the country. In 2004, the cultural organization of an Ikarian town organized a conference titled 'Exiles' Cultural Voyage in the Aegean'. Currently, the citizens' association of another Ikarian village is preparing a publication of photographs and texts written by exiles and has also commissioned a sculptor (a former exile on the island) to create a big statue for the memory of that era.

FROM EXPERIENCED TO INSTITUTIONALIZED MEMORY

Comparing experienced/intersubjective and institutionalized/public memory, we observe that traces of the first are dispersed all over the island. They are often occasionally or accidentally recognized among family members or in social contexts in which memory narratives are conveyed and have a very restricted local range.

In contrast, institutional memory takes place at specific times and in specific spaces—preprogrammed and organized. They mark specific points in time and place that claim a broader range of acceptance and recognition.

Thus, intersubjective memory is characterized by everydayness[11] and informality, whereas institutional memory has a more monumental character. While intersubjective memories can be re-enacted by actors other than those originally involved (e.g. relatives in an instance of generalized reciprocity) and in different spatial contexts (such as in Athens or elsewhere), institutional memory is usually re-enacted by people who claim some degree of authority in relation to the past (e.g. former exiles, official representatives of the state or of the communities, professional historians, public intellectuals and so on).

In the context of intersubjective memory, there is no explicit obligation of continued recollection. In the context of institutionalized memory, the act of remembrance is converted into something akin to a duty, a fulfillment of a promise and a publicly shared commitment to remember.

MEMORY IN TRANSITION

Having pointed out the main characteristics of both experienced and institutional memory, I now turn to two examples of collective memory in practice: the annual memorial ceremony that takes place in Ikaria and a proposal for a museum dedicated to this period of exile on the island. Both the ceremony and the museum represent transitional moments in the practice of collective memory, lying somewhere between the experienced and the institutional, the poetic and the political.

Every August, a memorial ceremony takes place at the Mounte Monastery[12] in the town of Raches. During the Civil War, the monastery housed exiles suffering from tuberculosis; five exiles who died in the monastery are buried there. The ceremony is organized by the citizens' association of Raches, a complex of villages in the north-western mountainous part of the island. On the patio of the monastery are benches and tables. Coffee and refreshments are offered, and under the shadow of the pine trees people gather and talk. There are also panels of old photographs of the exiles and scenes from their daily life in Ikaria, together with photocopies of letters, documents such as warrants of arrest, orders and so on. Apart from the exiles and their relatives, the majority of the audience is locals of all generations. Ikarians who attend the ceremony listen to the speeches, look at the photographs, explain the events to their children and meet their friends as well as the exiles and their descendants.

Apart from the special memorial service in honour of the exiles who died at the Monastery, this ceremony touches on the entire history of all the exiles and their stay in Ikaria. Several speakers are invited: representatives of political parties and municipalities, members of organizations of former exiles or National Resistance fighters, young people from Ikaria who read poems or parts of letters from the exiles and so on.

Looking at the photographs, people recall and narrate various things of that past cohabitation: the places where the exiles lived, their daily life, their relationships, the role and attitudes of the police, different persons and events. An hour later, people head to the church where the requiem service takes place. After the priest performs the service, he gives a short speech about the courage and the values of the exiles, 'who gave their lives for their ideals and who are a model for the youth of today'. Afterwards, the crowd follows the priest to the tombs, where the priest closes the ceremony and people sing a hymn for the Communist fighters who died in the Civil War. After a minute of silence, the crowd applauds, and everyone gathers on the patio again and waits for the speeches to start (Figure 3.3).

This ceremony contains elements and expressions of both experienced and institutional memory. On the one hand, it belongs to the realm of experienced memory in the sense that it is performed in much the same way that a normal religious memorial would be performed by locals for their deceased relatives, their 'own people', a phrase locals often use to refer to the exiles. On the other hand, part of the ceremony takes the typical form of a public commemoration, with speeches from state officials, representatives of community organizations and so on.

During the first years of the ceremony, it was attended by many former exiles and locals who had first-hand experience of the period being

Figure 3.3 Snapshot from the memorial service at the site of the exiles' graves during the political commemoration ceremony in Mounte Monastery, Raches (Author's archive).

commemorated. In these early years, the place of memorialization—the Mounte Monastery—functioned less as an official site of remembrance and more as an opportunity for more socialized celebrations of shared memories: before and after the ceremony, attendees would go together to eat in the village square or directly to the houses where the former exiles and locals had lived together.

Over the past few years, the feast is becoming less and less lively; now, when the ceremony is over, there is not much continued interaction between the younger Ikarians and visitors, many of whom are relatives of former exiles or just visitors to the island. Because the social context and personal connections are disappearing, there is an increased need for a stable keeper and guarantor of the fading memory. Thus, we see that the ceremony becomes increasingly institutional—a commemorative event with a fixed, preprogrammed time and place in which the memory of exile can be vigilantly managed and maintained.

IN PLACE OF A MUSEUM

In 1947, Mikis Theodorakis[13] was sentenced to exile on Ikaria. While in exile, he lived in a small house (which he called the 'House of the Scorpions') in the village of Vrakades in the north-western part of island, where he wrote some of his first and most well known compositions. In 2009, the house was donated by its owners to the municipality, to be restored and converted into a museum. The local authorities (knowing that I was both an architect and anthropologist focused on this very issue) solicited a proposal for a museum that would commemorate the period of exile in Ikaria and that would house the existing archive (primarily photographs and documents) from this period.

I worked together with another anthropologist and two architects to come up with an idea for a museum that would take into account the importance of the vitality of experienced memory and that would try to transform it into an institution that would maintain this memory as a vivid part of local community life.[14] We proposed that, instead of a typical museum that would house the archive in a more or less traditional way, the house should keep its original function as a space of hospitality by housing a residency programme that would give people of various backgrounds the opportunity to live and work on the island for a certain period of time (from a few weeks to a few months, depending on the resident's proposed project). Applicants would have a proposal for a specific project; although the project should not have to specifically address exile, the Civil War or Ikaria, the residency programme itself is modelled after the affective relationships of reciprocity between islanders and foreigners that developed during the years of exile. In other words, the programme aims to bring residents to the island in order to foster creative and dynamic relationships of mutual respect and mutual benefit between Ikarians and their resident guests. For this reason, residents' projects must engage with the local community in some way and offer something back to the island. The residency programme is open to applicants of all professions; an artist might organize an exhibition on the island, or a healthcare professional might organize and run a health-awareness educational programme. Alternatively, the resident could be a person who is specialized in or has experienced exile in different periods and/or parts of the world. So, 'guests' would have the chance to share and communicate the political situation of other exile cases, the reasons for their exile, as well as their experience as politically persecuted citizens in different social and political contexts.

In general, residents will be encouraged to interact socially with the community as well. While residents will have plenty of time and space for their own work, the real emphasis in this residency programme is on the relationship between residents and island/community.

The suggested museological and architectural project ventures to balance the tension between living and institutionalized memory. More specifically, my empirical data shows that the way Ikarians remember is related to what was (and partly still is) the society that received the exiles during the Civil War. These ways of remembering (and forgetting) also provide hints as to the direction in which this society is turning—or could turn—in such moments of crisis.

Be that as it may, the restoration of the House of the Scorpions is stagnant. The reasons given for this inaction are the same ones given for the broader historical moment in which Greece finds itself and the dominant values associated with the victory of neo-liberal politics.

On Ikaria, significant traces of both the political and cultural past that the House of Scorpions evokes are still vivid. That can also partly explain why memorials of the period are popular, the personal relationships between locals and former exiles are strong, and left-wing parties have received more than 50 per cent in any election for the last sixty years. Now the political and social agents involved are required to face the past—the memory—as well as the future and the challenge to do something new, if that is possible. In the current historical context, doing something new might just mean returning to their Ionian roots.

For the moment, the restoration works have stopped, and it appears impossible to secure the financing for the functioning of the museum, which is still undecided. In present-day circumstances of 'economic crisis' and related 'structural changes', Greece is called on to leave behind any 'socialistic illusions' and face the dilemma to either amputate whatever reminds people of them or bow out of the so-called first-world economy. Thus *political culture* is transformed into *cultural politics*, politics now oriented towards tourist development as dictated by the interests of the current hegemonic economy and ideology.

Under these circumstances, the House of the Scorpions is not only a *ruin of the past* but also a *ruin of the present*—a present caught between a creed of patrimonial investment in tourism and an amputation of all public spending. What in the past could be marginalized by a lack of interest is now sidelined by a sort of counter-interest which defines itself as something necessarily superior.

This leads us to a paradox: on Ikaria, sixty-five years ago, amidst a dramatic shortage of financial resources and with a much larger population than the present one, it was possible for 20,000 people not only to survive but also to create a fruitful cohabitation. Nowadays, however, in a country supposedly at peace, it seems 'necessarily' impossible to provide a humble shelter for the memory of what increasingly, if nothing is done to alter this course, will look like a long-lost heroic past.

ACKNOWLEDGEMENTS

I would like to thank Professor Margaret Kenna and Professor Ignasi Terradas for their crucial comments. Special thanks to Hristos Malachias for giving me access to his photographic archive. Lois Woestman and Vasso Pappi kindly helped me in editing this paper.

NOTES

1. *Exile* in Greek, *εξορία* (ex-oria), literally means 'out of the border'. The first part, *εξ-ex*, comes from the prefix *εχσ*-(echs) which means 'out, off'. The second part, *όριο*, means 'border' or 'limit'.
2. From July to November 1912, Ikaria was declared an independent state with its own government, administration, police, currency, flag, anthem and stamp. In November 1912, Ikaria was annexed to the Greek state.
3. The Modern Greek state did not officially use the term 'exile' to speak of these politically-motivated internal displacements of Greek citizens. Government discourse, rather, has used the term 'deportation'—a measure and a term it has long applied to thieves and drug addicts. Political deportees have intentionally utilized the term 'exile' in all their references to the period. In so doing, they have attempted simultaneously to provide the political context for their displacement and avoid the connotations of the term deportation.
4. From 1936 to 1941, Greece was ruled by an authoritarian regime under the leadership of General Ioannis Metaxas, a regime akin to the fascist regimes of the interwar period in Europe.
5. Ionians, according to the ancient historian and geographer Stravonas, were the first Greeks to colonize Ikaria and came from Miletus of Ionia, the cradle of the first pre-Socratic philosophers of Greece.
6. In the post–Civil War era that lasted at least until the fall of the Dictatorship of the Colonels of 1967–74, the memory of the defeated—that is of the leftists—was severely suppressed together with other civilian rights in work, education and the like. On the other hand, one way or another, the memory

of the Civil War has often been used for the political propaganda of most of the leading parties.

7. For more on these topics, see Argenti (2013), Kenna (2014), Mamoulaki (2007), Munn (2013), Navaro-Yashin (2009, 2012), Yalouri (2001).
8. Most of the information comes from the extensive number of interviews with locals and former exiles while in fieldwork at different times from 2006 to 2010. Other sources include Dalianis (1999), Kalo (1998), Lountemis (2000), Papalas (2005), Papageorgakis (2003), Kamarinou (2005), Karimalis (1992), Kassimatis (2002), Mamoulaki (2010), Mavroyorgis (1996), Pantzou (2011), Theodorakis (2000), exiles' handwritten newspapers and the local newspaper *Nea Ikaria* and unpublished exiles' correspondence and diaries from Hristos Malahias's archive.
9. For more, see Mauss (1967), Weiner (1992), Godelier (1999), Bofill Poch (2002), Moreno (2002), Narotzky (2002), Terradas (2002).
10. For more on the role of the photographs in the shaping of the memory of the exile era on Ikaria, see Mamoulaki (2014).
11. For more information see Mamoulaki (2013, chapter 1).
12. The monastery was normally used for religious purposes after the exiles left. When they decided to organize a (political, as they call it) memorial for the exiles, Ikarians chose that specific place which included the most quintessential monument of the presence of the exiles: their graves.
13. Mikis Theodorakis (1925–present) is a famous Greek composer and politician.
14. The working group was formed by Toby Lee (anthropologist), Fotini Lazaridou-Hatzigoga (architect) and Thanasis Zagorisios (architect).

BIBLIOGRAPHY

Argenti, N. (2013) 'Emptiness of Anavatos, ruins of Kidianta: displacement, commemoration and melancholia in Greece', working paper presented at the workshop Re-assessing Trauma and Violence: Re-framing the Bodily, the Spoken and the Self, Bergen University 28–30 June.

Bateson, G. (1958 [1936]) *Naven: A Survey of the Problems Suggested by a Composite Picture of the Culture of a New Guinea Tribe Drawn from Three Points of View*, Stanford, CA: Stanford University Press.

Birtles, B. (2002) *Εξόριστοι στο Αιγαίο* [*Exiles in the Aegean*], Athens: Filistor.

Bofill Poch, S. (2002) 'Organización forestal communal y sustentabilidad en San Juan Nuevo (Michoacán): poder y conflict en torno a los usos locales, nacionales e internacionales de la reciprocidad [Forest community organisation and sustainability in San Juan Nuevo (Michoacan): power and conflict over the local, national and international uses of reciprocity]', *Éndoxa*, 15, 61–74.

Butler, J. (1990) *Gender Trouble. Feminism and the Subversion of Identity*, New York and London: Routledge.

Dalianis, D. (1999) *Το σανατόριο εξορίστων Ικαρίας 1948–1949* [*The Sanatorium for the Exiles in Ikaria 1948–1949*], Larissa: Ella.

Georgiadis, M. (2004) *'Στα χρόνια της συμφοράς'. Αναμνήσεις από την εξορία. Αϊ Στράτης-Μακρόνησος, 1947–1950* [*'In the Years of Affliction': Memories from the Exile: Ai Stratis, Makronissos, 1947–1950*], Kavala: Istoriko & Logotechniko Archeio Kavalas.

Godelier, M. (1999) *The Enigma of the Gift*, Chicago: University of Chicago Press.

Gritzonas, K. (2001) *Ομάδες Συμβίωσης: Η συντροφική απάντηση στη βία και τον εγκλεισμό* [*Symbiosis Groups: The Comradeship as a Response to Violence and Confinement*], Athens: Filistor.

Kalo, K. (1998) *Όσα δεν πήρε ο άνεμος. Η αυτοβιογραφία μιας θεατρίνας* [*What Was Not Gone with the Wind. Memoirs of a Theatre-Actress*], Athens: Agra.

Kamarinou, K. (2005) *Τα πέτρινα πανεπιστήμια. Ο αγώνας για τη μόρφωση στις φυλακές και στις εξορίες, 1924–1974* [*Stone Universities. The Struggle for Education in Prisons and Exiles, 1924–1974*], Athens: Sygchroni Epochi.

Karimalis, S. (1992) *Η Νικαριά στην Αντίσταση και στις φλόγες του εμφυλίου πολέμου* [*Ikaria in Resistance and in the Flames of the Civil War*], Athens: Efstathiou.

Kassimatis, M. (2002) *Ενθύμιον Εξορίας* [*Exile's Mementos*], Athens: Etherovamon.

Kenna, M. (2001) *The Social Organization of Exile: Greek Political Detainees in the 1930s*, Belfast: Harwood Academic.

Kenna, M. (2015) 'Rituals of forgiveness and structures of remembrance: memorial services and bone-depositories on the island of Anafi, Greece,' *History of Religions*, Vol. 54, No. 3 (February 2015), pp. 225–259.

Lountemis, M. (2000) *Το κρασί των δειλών* [*The Cowards' Wine*], Athens: Ellinika Grammata.

Mamoulaki, E. (2007) 'Η διαχείριση της μνήμης και η πόλη: το παράδειγμα της Βαρκελώνης [The handling of memory and the city: the case of Barcelona]', in S. Stavridis (ed), *Μνήμη και Εμπειρία του Χώρου* [*Memory and Experience of Space*], Athens: Alexandria Publications.

Mamoulaki, E. (2008) 'Memories of cohabitation and exile on the island of Ikaria', unpublished master's thesis, Barcelona, University of Barcelona.

Mamoulaki, E. (2010) 'Εθιμικές και ιδεολογικές μορφές αντίστασης: η συμβίωση ντόπιων και εξόριστων στην Ικαρία 1946–1949 [Customary and ideological forms of resistance: the symbiosis between locals and exiles in Ikaria 1946–1949]', in *Κριτική Διεπιστημονικότητα: Κοινωνικές Διακρίσεις* [*Critical Interdisciplinarity: Social Discriminations*], Athens: Nissos, 85–104.

Mamoulaki, E. (2013) 'An unexpected hospitality: narratives, practices and naterialities of the memory of exile and cohabitation on the island of Ikaria', unpublished doctoral thesis, Barcelona, University of Barcelona.

Mamoulaki, E. (2014) 'Pictures of exile, memories of cohabitation: photography, space and social interaction in Ikaria Island', in P. Carabott, Y. Hamilakis and E. Papargyriou (eds), *Camera Graeca: Photographs, Narratives, Materialities*, London: Ashgate.

Margaritis, G. (2002) *Ιστορία του ελληνικού εμφυλίου πολέμου 1946–1949* [*History of the Greek Civil War 1946–1949*], 2 vols, Athens: Vivliorama.

Mastroleon-Zerva, M. (1986) *Εξόριστες. Χίος, Τρίκερι, Μακρόνησος* [*The Exile Women's Martyric Triangle: Chios-Trikeri-Makronissi*], Athens: Sygchroni Epochi.

Mauss, M. (1967[1923]) *The Gift: Forms and Functions of Exchange in Archaic Societies*, Norton: Norton Library.

Mavroyorgis, C. (1996) *Οδοιπορικό στο κακοτράχαλο μονοπάτι* [*Travelogue Along the Rugged Path*], Athens: Self-published.

Mazower, M. (ed). (2000) *After the War Was Over: Reconstructing the Family, Nation, and State in Greece, 1943–1960*, Princeton, NJ: Princeton University Press.

Melas, I. (2001) *Ιστορία της Νήσου Ικαρία [History of Ikaria Island]*, Athens: Self-published.

Moreno Feliu, P. (2002) 'Organizar: suspensión de la moralidad y reciprocidad negative [Organise: suspension of the negative morality and reciprocity]', *Éndoxa*, 15, 97–128.

Munn, N. (2013) 'The "becoming-past" of places: spacetime and memory in nineteenth-century, pre-Civil War New York', The Edward Westermarck Memorial Lecture, 2003, *Journal of the Finnish Anthropological Society* (*Suomen Anthropologi*) 29 (1), 2–19. Online: *HAU: The Journal of Ethnographic Theory* 3 (2), 359–80. Available HTTP: http://www.haujournal.org/index.php/hau/article/view File/353/375.

Narotzky, S. (2002) 'Reivindicación de la ambivalencia teórica: la reciprocidad como concepto clave [Vindication of the theoretical ambivalence: reciprocity as key concept]', *Éndoxa*, 15, 15–29.
Navaro-Yashin, Y. (2009) 'Affective spaces, melancholic objects: ruination and the production of anthropological knowledge', *Journal of the Royal Anthropological Institute*, 15 (1), 1–18.
Navaro-Yashin, Y. (2012) *The Make-Believe Space: Affective Geography in a Post-war Polity*, London and Durham, NC: Duke University Press.
Oikonomopoulos, P. (2004) *Οι εξόριστοι* [*The Exiles*], Athens: Ardin.
Panourgia, N. (2009) *Dangerous Citizens. The Greek Left and the Terror of the State*, New York: Fordham University Press.
Pantzou, Nota (2011) 'Materialities and traumatic memories of a twentieth-century Greek exile island,' in A. Myers and G. Moshenska (eds), *Archaeologies of Internment*, New York: Springer, 191–205.
Papageorgakis, S. (2003) *Η Ικαρία στη θύελλα* [*Ikaria in the turmoil*], Athens: Sychroni Epochi.
Papalas, A. (2002) *Αρχαία Ικαρία* [*Ancient Ikaria*], Ikaria: Kalokairinos.
Papalas, A. (2005) *Rebels and Radicals. Icaria 1600–2000*, Mundelein, OH: Bolchazy-Carducci Publishers.
Sarantopoulos, A. (2000) *Οι εξόριστοι στον εμφύλιο* [*The Exiles During the Civil War*], Athens: Zacharopoulos.
Scott, J. C. (1977) *The Moral Economy of the Peasant: Rebellion and Subsistence in Southeast Asia*, New York: Yale University Press.
Staveri, O. (2006) *Το μαρτυρικό τρίγωνο των εξόριστων γυναικών: Χίος—Τρίκερι-Μακρονήσι* [*The Agonizing Triangle for the Exile Women: Chios-Trikeri-Makronissi*], Athens: Paraskinio.
Terradas, I. (2002) 'Acerca de un posible malentendido sobre la obligación de reciprocidad [Towards a possible misunderstanding on the obligation of reciprocity]', *Éndoxa*, 16, 113–38.
Theodorakis, M. (2000) *Οι δρόμοι του Αρχαγγέλου, Τομ. 2* [*The Archangel's Roads*], vol. 2, Athens: Kedros.
Thompson, E. P. (1995) *Costumbres en común* [*Customs in Common*], Barcelona: Crítica.
Tsakiris, K. (1996) *Σίκινος: Αναμνήσεις από την Εξορία 1936–1941* [*Sikinos: Memories from the Exile 1936–1941*], Athens: Odysseas.
Voglis, P. (2002) *Becoming a Subject: Political Prisoners During the Greek Civil War*, New York: Berghahn Books.
Weiner, A. (1992) *Inalienable Positions: The Paradox of Keeping While Giving*, Berkeley: University of California Press.
Wittgenstein, L. (2009) *Philosophical Investigations*, Oxford: Wiley-Blackwell.
Yalouri, E. (2001) *Acropolis: Global Fame, Local Claim*, Oxford and New York: Berg.

4 Islands of War, Guardians of Memory

The Afterlife of the German Occupation in the British Channel Islands

Gilly Carr

INTRODUCTION

The apparent longevity of war memory in small islands is the phenomenon explored in this volume and chapter alike. Those of us who work in or with small island communities will know that they can often be inward-looking and independent in character with little inclination to be receptive to outside influence and interference, especially from their larger neighbours or even from other islands within their own archipelago. Because of this, small islands are not easily influenced or affected by outside events unless they intrude directly upon them uninvited. Such momentous events, especially if traumatic and/or war-related, can have a big impact on island life, and the reverberations can often be felt for generations.

Only rarely do such events happen more than once to such unfortunate islands and, when they do, the original trauma can shape how it is remembered, commemorated and responded to with heritage. Sometimes the impact of the later trauma can supplant the earlier one in terms of how islanders shape their identity, and this ousting can also cause earlier heritage to be supplanted, changed or marginalized.

Those of us who carry out our fieldwork in small islands will have observed how the shape of war heritage, especially that which is related to events within living memory, is rarely produced by neutral, disinterested or non-stakeholder parties. A casual examination of the biography (in a Youngian sense, e.g. 1993, ix–x) of sites such as war memorials, museums and commemorations will reveal the hand of certain individuals who crop up repeatedly as instigators, agitators and what Jordan refers to as 'memorial entrepreneurs' (2006, 11). These same people will often be among the first to contact the new researcher, making sure that their narrative is heard. Other people will often defer to their version of events. We will return to the role of these important figures, as they are key to this chapter and vital, I argue, to understanding the longevity of war memory in small islands.

This chapter takes as its case study the British Channel Islands, comprising the five islands of Jersey (the largest), Guernsey, Alderney, Sark and Herm (the smallest), situated in the Bay of St Malo, within sight of the

French coast (Figure 4.1). The islands were part of the Duchy of Normandy and have been British crown possessions for over 800 years. The two largest islands of Jersey and Guernsey, which will be the focus of this chapter, have populations today of close to 100,000 and 65,000 and sizes of circa 45 and 30 square miles, respectively.

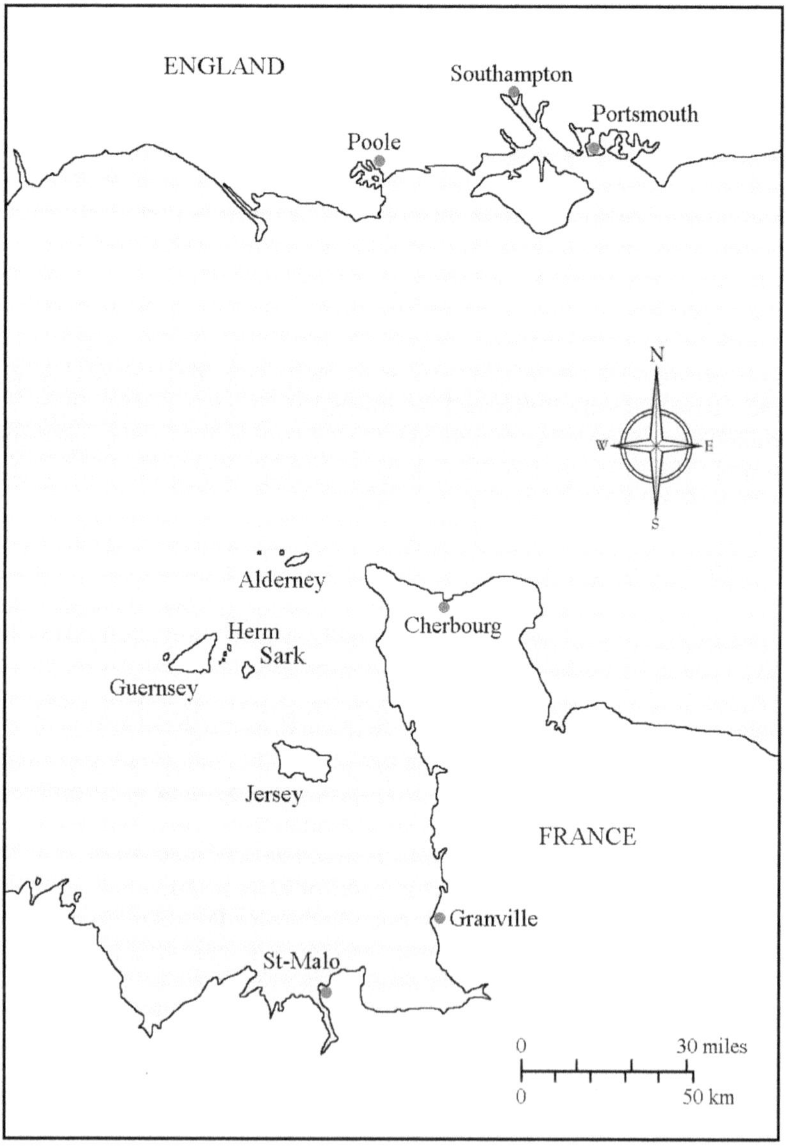

Figure 4.1 Map showing location of the Channel Islands (Copyright Gilly Carr, courtesy Ian Taylor).

Although there is a long-standing traditional rivalry between the two islands, they are united in their experiences of both WWI and WWII. Although conscription was not compulsory in the Channel Islands, they sent thousands of men to fight in the trenches (e.g. Ronayne 2009, 136). Just over twenty years later, they were to face their now-traditional enemy for a second time, when German forces invaded in late June and early July 1940, staying until 9 May 1945. The Channel Islands have the dubious honour of being the last place in Europe where German forces surrendered.

While it is important to state that, compared to other occupied countries in Western Europe, the Channel Islands were probably the safest place to be and had one of the least harsh occupations because Britain was still a fighting force and not a defeated nation, this must be tempered with certain historical facts. One in twenty of the wartime population (2,200 people in all) were deported to German civilian internment camps in 1942 and 1943. Those who committed acts of resistance were imprisoned locally (one in fifty of the population) and/or deported to Nazi prisons and concentration camps on the continent (about one in four of all those imprisoned). Of these, one in ten did not return (see conclusion of Carr *et al.* 2014). Orders were enacted against the Jews in the Channel Islands, and three Jewish women in Guernsey, who had arrived in the island in the 1930s (Cohen 2000, 22, 50), were sent to Auschwitz in 1942, where they subsequently died. British Jews were sent to civilian internment camps with other non-Jewish islanders. After the Allied invasion of Normandy in June 1944, the islands were cut off and the population slowly starved until the Red Cross ship, *Vega*, arrived once a month, packed with food parcels, from December 1944 onwards, saving the lives of thousands of islanders.

Despite the similar war experience of the Channel Islanders to their continental cousins, their war memory is firmly tied to that of the British (Carr 2012a). The experience (or myth) of those on the mainland was mostly that of being bombed but, against all odds, enduring until the final victory; or, as historian Paul Sanders put it, of 'sublime and unwavering steadfastness in the face of adversity' (Sanders 2012, 25), which led to what he called the 'Churchillian paradigm'. This states that 'the British were not a nation of victims, but of victors' (Sanders 2005, 256), which in turn has led to subsequent narratives that emphasized victory over victimhood. In the Channel Islands, this manifested itself from the earliest days and for the first decade after the war through a focus on the role of the military personnel (living and dead), victory, and patriotism (see Carr 2014 for a greater discussion of this). Although this led to Navy warships, RAF fly-pasts and battalions of soldiers being invited to the islands every 9 May to celebrate Liberation Day, the focus on victory also resulted in later decades in a prominence in heritage given to the victors' spoils of war–German militaria and bunkers, swastikas and mannequins in German uniform. The camera was rarely if ever pointed at Jews, slave workers, political prisoners or deportees. Memorials and museum exhibitions were not to feature their voices or experiences until as late as the fiftieth anniversary of liberation, a whole generation later than in mainland Europe (e.g. Rousso 2007). This silence was broken for the first time in 1995, at the fiftieth anniversary of liberation, by Jersey's then-Bailiff,[1] Philip Bailhache. He was the first Bailiff to actively support such victims of Nazi persecution by unveiling memorials to their memory and by referring to them in speeches. He was also prepared to discuss openly Jersey's occupation history, 'warts and all'. No island leader has yet been prepared to discuss this history in Guernsey; this island has lagged behind Jersey in acknowledging the experiences of victims of Nazism. While the reasons for this are manifold and complex, it is sufficient to observe that, in 1995, the then-Deputy Bailiff of Guernsey, de Vic Carey, was the grandson of the wartime Bailiff, Sir Victor Carey. De Vic later went on to become Bailiff himself in 1999. Since his retirement in 2005, he has been a Lieutenant Bailiff of the Royal Court. It is likely that his presence in the upper echelons of Guernsey governance has acted (through no fault of his own) to stifle debate or public discussion of the conduct of his grandfather during the occupation and his support (or lack of it) for Jews, forced labourers, deportees and political prisoners. While the public speeches of politicians—or the lack thereof—can have an important impact on the consciousness of a population and how they perceive their past, it is the role of ordinary members of the public that I wish to explore in the chapter.

GUARDIANS OF MEMORY, GUARDIANS OF COUNTER-MEMORY

The German occupation is a key aspect of Channel Islanders' self-identity today and is one of the main features that they will cite as a way of differentiating themselves from their English cousins on the mainland. This is hardly surprising, given the constant reminders of the Occupation in all aspects of Channel Island life and in their annual calendar. The coasts of the islands are covered in hundreds of concrete German bunkers, and the capital towns of Guernsey and Jersey, St Peter Port and St Helier respectively, have thirty-three Occupation-related memorials between them (Carr 2012b). Further, annual commemorative ceremonies which bring the occupation to mind punctuate the year (such as Liberation Day, Charybdis Day, Holocaust Memorial Day and Armistice Day). Occupation-related articles are also printed frequently in the local newspapers, the *Guernsey Press* and the *Jersey Evening Post*. These report on newly discovered unexploded bombs, new Occupation-related books or letters to the editor comparing aspects of daily life today unfavourably with the Occupation years, a time when everybody seemingly recycled their rubbish, didn't waste food, fuel or other resources, and shared what little they had. Obituaries and articles on notable local people and places will also invariably comment on what these people did and what happened in these locations during the Occupation years.

The local newspapers often print the memories of members of the Occupation generation, and there are certain people whose opinions and memories of the war years are always sought: people who have achieved notoriety because of their eloquence, good memory, activities during the war years and commitment to the legacy of the Occupation since 1945. Their longevity has also contributed to their local fame. These people have become—and are perceived as—what I term here 'guardians of memory'. These people are ordinary members of the public who are always consulted and feted, especially by local and national TV and radio. They are greatly respected. While all of these facets of the guardians of memory are important, it is their role in safeguarding the memory and heritage of the Occupation years that is particularly important. If any facts relating to the Occupation need to be checked, then the guardians of memory are consulted above and beyond (and either in addition to or instead of) any other source. If what they remember contradicts the archival record, then it is often the archival record which will be called into question. These are thus people who, over the years, have accrued great local power and influence. Their number is dwindling fast, and those still alive today are in their eighties and nineties. In fact, as time passes and the fewer guardians of memory there are still living, the more important the position and respect that is accorded to those few who remain. It might be observed that the position of guardian of memory is, arguably, a position which develops and whose power accrues only with the passing of time. Thus, in the post-war decades, when everybody could in theory function as a guardian of memory, it is likely that there was no such noticeable phenomenon to be observed. This is thus a position which has come into existence or solidified only after the passage of several decades—within a couple of generations—since the event that the incumbent remembers and safeguards.

Before we examine some examples of such men[2] in the Channel Islands, it is pertinent to examine a subset of the guardians of memory: the guardians of counter-memory. These men are often similarly well known locally, but the memories that they have safeguarded have been marginal or marginalized and have comprised a controversial narrative of events which has been deliberately kept out of the mainstream, often by those in positions of authority in local government. These men have been silenced and denied the same level of control over mainstream heritage. Instead, they have channelled their memories into other avenues or private heritage and have fought long-running battles for acceptance and recognition. Perhaps unexpectedly, guardians of memory and guardians of counter-memory are today on good terms with each other. While they each safeguard different memories and memory groups, they are, by definition, often from the same generation and, as time passes, recognize each other as among the few fellow survivors and witnesses of a period that is on the edge of living memory.

I argue here that guardians of memory are noticeable features of small islands; it is important to stress that in small communities it is easy to become well known and to exert authority and influence over memory and heritage projects in a way that such members of the public would find difficult (but not impossible) to achieve in larger communities. In larger nations, individuals are often anonymous, and influence can be difficult to wield without being in a high-ranking job or without certain qualifications and professional experience which legitimizes that position. Of course, there will always be exceptions to this rule: we might cite guardians of memory of the French Resistance, such as Lucie and Raymond Aubrac. Henry Rousso noted that 'by their very popularity and fame, [they] were considered untouchable, even by professional historians' (2002, 77), thus emphasizing the power that these figures can wield over the historical and archival record. I argue, however, that their hegemony of power over national war-related heritage projects in France does not compete with that of guardians of memory in small islands.

Guardians of memory often (but not always) work within groups of what Jay Winter (1999, 41), working with the memory of the Great War, calls 'fictive kin'—the social and experiential rather than biological 'families of remembrance' who come together to express and discuss memories of (in this case) occupation. For Winter, the fictive kin groups which formed after the Great War had the function of helping each other recover from the trauma of war (Winter 1999, 47). In the Channel Islands, these groups have formed instead to preserve and remember the legacy of war through its material remains and are often led by guardians of memory. Winter also suggests that the fictive kin occupy 'the space between individual memory and the national theatre of collective memory choreographed by social and political leaders' (ibid, 41). Here we might draw an important distinction which is appropriate to the small islands setting, where (in the Channel Islands at least) such groups have had the power to dictate collective memory through control, curation and creation of heritage. Like Winter's fictive kin, the Channel Islands' groups are 'agents of remembrance' (Winter 1999, 59). The group in question in this case study is the Channel Islands Occupation Society (CIOS), which has both a Guernsey and a Jersey branch and which was founded in 1961 in Guernsey and ten years later in Jersey. This society is a 'non-political, voluntary organisation, dedicated to the preservation and recording of all aspects of the German Occupation' during WWII.[3] Although it has not always been the case, today they are perceived as the authorities on the Occupation, and their opinion, support and voice can and have lent considerable weight to promote heritage projects which are carried out under their aegis and to block those which are not. While they have become the heritage 'amateurs' or 'enthusiasts' in the Channel Islands, their work contrasts with that of the heritage professionals in ideology, aims, methods and practice.

Guardians of memory have acquired their senior position over decades, but despite their longevity these men need successors who are groomed to one day take over their mantle. While the appointment of the heir apparent is not a public or official process, it is nonetheless a process that takes place over many years—even decades—of working closely together (Figure 4.2). The heir may well be a family member of the guardian who has spent years listening to wartime stories but may equally be a younger friend or colleague within their fictive kin who has worked with them publicly for many years on heritage projects such that their succession is undisputed. Not every guardian of memory has an heir—clearly, it is hard to find a substitute in someone who was not *there* during the Occupation—and at this point the work of the guardian finishes and their legacy dies with them, unless their mantle is picked up at a later date. However, unless those who pick up the mantle are accepted by the community and fictive kin group as being legitimate heirs who are likely to stay the distance and keep up the commitment to the community, they will potentially be spurned and ignored, which is why it is hard for an outsider—such as an academic—to take on this role.

Figure 4.2 Guardian of memory Michael Ginns standing beside his heir apparent, Paul Burnal (current president of the Jersey branch of the CIOS), outside a restored bunker at Noirmont Point in Jersey (Copyright Gilly Carr).

GUARDIANS, FICTIVE KIN AND THEIR MEMORY NARRATIVES

In the Channel Islands, the CIOS (traditionally led by guardians of memory) has been responsible for creating the heritage of the German Occupation. Elsewhere (Carr 2014), I argue that heritage is created through intervention in the legacy of an event - in this case, occupation. Those legacies can be intangible (in the form of memories) or tangible (e.g. bunkers or militaria), but rather than neglecting them and letting them become overgrown and metaphorically or literally hidden in the undergrowth, they can be converted to heritage through restoration, curation, collection or some other interruption in the process of forgetting.

A generation after the end of the Occupation, the CIOS (whose membership, for the most part, today and in the past comprises men born after the end of the Occupation but who were originally led by guardians of memory) was instrumental in restoring bunkers, collecting militaria, opening Occupation museums, organizing Liberation Day celebrations and, later, erecting memorials to the Occupation years. Every five years from the 1980s onwards, those memorials have remembered the important anniversaries of liberation. From the mid-1990s onwards, after memory began to encompass victims of Nazism, those memorials have commemorated other fictive kinship groups, such as deportees, evacuees, resistance, Jews and foreign forced and slave labourers. It is interesting to note that these later memorials have often been erected after a long battle by guardians of counter-memory. Their counter-narratives have been such that their fictive kin groups have, like their resulting memorials, existed on the margins and have not been as large in size because of the lesser popularity associated with their narrative.

Generally speaking, memory narratives as expressed through heritage have gone through four phases in the Channel Islands (Carr 2014): a primary phase during which the war dead, patriotism and military victory were expressed (1940s to early 1950s); a period of self-imposed 'amnesia' or silence about the Occupation (1950s to late 1970s); a period of Occupation nostalgia (1970s to present); and a fourth phase during which the victims of Nazism were 'remembered' and included in the memory of the Occupation (1995 onwards). In this section I would like to focus on the guardians of memory and counter-memory of the third and fourth phases of memory: Michael Ginns and Joe Mière in Jersey and Richard Heaume and Frank Falla in Guernsey, all of whom are or were locally well known and well respected men (Mière died in 2006, Falla in 1983).

Michael Ginns and Richard Heaume have been at the forefront of occupation heritage in the Channel Islands since the 1960s and 1970s. While Ginns was a teenager during the Occupation and was deported to a civilian internment camp in Germany with his parents in 1942–5, Heaume was born at the end of the Occupation. While Heaume might be disqualified for guardianship of memory status by not having experienced the Occupation, he has spent his life, since boyhood, collecting German militaria and Occupation memorabilia and has owned the German Occupation museum in Guernsey since the 1960s. For over fifty years, a steady stream of Occupation-generation islanders and, indeed, former soldiers have visited Heaume and told him their stories. He is thus recognized as the most knowledgeable man in the island when it comes to local Occupation history. Heaume lives and breathes the Occupation and is a good example of the way in which Marianne Hirsch's concept of postmemory manifests itself in the Channel Islands (e.g. 1996, 1999). For Hirsch, postmemory is a form of second-generation memory in which the memories and experiences of that generation are entirely taken over and dominated by the all-encompassing memories and experiences of the first generation. While Hirsch's concept was originally formulated to explore the experiences of the children of Holocaust survivors, a form of 'Occupation postmemory' is evident among many children of Occupation-generation Channel Islanders.

Heaume, like Ginns, was responsible for the restoration of selected German bunkers during the 1970s and 1980s onwards and for opening them to the public, the work of both conducted through the auspices of the CIOS. Both Heaume and Ginns are and have been leading lights in the CIOS. As guardians of memory, both have been presidents of the society at different times. The aim of bunker restoration in the Channel Islands is to refit them to look as they did during the Occupation. Mannequins of soldiers in uniform are placed in the bunkers, as are guns, swastika flags, bunk beds, plastic food, wartime radios and field telephones, and everything that an occupying soldier might need. The aim is 'authenticity'—to make it look as if the German soldiers have left the room for only a second. The mannequins are shown in various harmless poses, such as playing chess, eating soup, listening to the radio—or even manning a gun emplacement—but never in the act of harassing local people, deporting them or enforcing Nazi policies. Bunker restoration and the many Occupation museums set up by members of the CIOS, including Heaume, originally portrayed what might be described as a 'nostalgic' view of the Occupation, where islanders and occupiers behaved 'correctly', although in the last fifteen to twenty years these exhibits have been modified to become more inclusive of victims of Nazism.

Guardians of counter-memory, including Falla and Mière, were strongly affected by their own wartime experiences as political prisoners who suffered in continental prisons and Jersey jail, respectively, for acts of resistance. Both of these men subsequently made it their life's work, especially from the 1960s and 1970s respectively, to fight for recognition of the contribution of island men and women who committed acts of resistance and who were imprisoned or deported because of these acts. These decades were important ones during which Channel Islanders slowly emerged from decades of a strong reluctance to talk about the Occupation—a reluctance that has never entirely faded. These decades were not, however, a time for facing the full horrors of what some of their fellow islanders suffered. Rather, they were a time when the second generation broke the silence of their parents and explored for themselves what their parents had experienced. This was done through the material culture of Occupation, whether through bunker restoration or militaria collection, both of which were (and are) popular pastimes for boys and men of a certain age.

While the speeches and memorial unveiling of former Jersey Bailiff, Sir Philip Bailhache, were instrumental in ushering in a phase of memory in Jersey at least from 1995, Falla and Mière fought their battles before this date, with limited success. The reasons for this may seem strange given that the Resistance in continental Europe were among the first groups to be recognized with memorials, medals, recognition and honour. In the Channel Islands, however, people who had committed acts of resistance were not part of an underground network. The islands were so heavily occupied, with one soldier for every three members of the population on average (Sanders 2004, 128), with many of the soldiers billeted in islanders' homes, that organized, armed resistance was impossible (see Carr *et al.* 2014). Thus, resistance in the Channel Islands was carried out in absolute secrecy by individuals or extremely small groups comprising only a handful of people at most. This meant that after the war, there was no unified organization to campaign for recognition. Further, given the small size of the islands, resistance was seen to be absolutely suicidal and risked bringing fierce reprisals upon the population as a whole, as was seen at the massacre of 642 people at the village of Oradour-sur-Glane in France (e.g. Farmer 1999). While the threat of a massacre may not have been in people's minds, the German occupiers had other ways of punishing the population, all of which were carried out at different times, including fines, the reduction of already minimal rations and hostage-taking of senior members of the community. Thus, people who committed acts of resistance were seen to be 'stupid', 'criminal' or 'trouble-makers' and for this reason were not honoured after the war. Instead, the Bailiffs of Jersey and Guernsey received knighthoods; the British government couldn't be seen to be supportive of both the local government, who had been vocally strongly anti-resistance, and those who disobeyed that stance (Turner 2010, 273). In their lifetimes, both Mière and Falla found it hard to shake off the reputation for being troublemakers, and so they remained as guardians of counter-memory.

While guardians of memory and their fictive kin have intervened in the legacy of the Occupation to restore and create heritage that is now mainstream, prone to giving a nostalgic view of the past and representative of their sites of memory, so the heritage of the guardians of counter-memory have become (and still to a certain extent remain) sites of counter-memory. Combining the work of Foucault and Pierre Nora, Legg (2005, 181) has identified sites of counter-memory as those which challenge *lieux de mémoire*, or sites of memory. Such places 'mark times and places in which people have refused to forget' and which 'rebut the memory schema of a dominant class . . . providing an alternative form of remembering and identity'. For Nora, sites of memory were 'any significant entity, whether material or non-material in nature, which by dint of human will or the work of time has become a symbolic element of the memorial heritage of any community' (Nora 1996, xvii). These could be as diverse as places, songs, dates, books and ceremonies (e.g. Nora 1996, 1997). While the work of guardians of memory have seen the likes of restored bunkers and their occupation museums and the celebration of Liberation Day raised to the status of sites of memory in the Channel Islands, so the work of men like Falla and Mière have become sites of counter-memory and are not—or at least, were not until very recently—included in the pantheon of the mainstream memory of the Occupation.

Falla's main achievement was to help Channel Islanders who had committed acts of resistance and had been imprisoned to get compensation from the West German government as 'victims of Nazi persecution'. In his profession as a journalist and privately, he worked tirelessly to campaign on their behalf, writing testimonies for the surviving family on behalf of men who had died in the prisons and camps that he had experienced or composing statements for those who were too mentally, physically or emotionally affected, or too poorly educated, to do so. Falla's book, *The Silent War* published in 1967, detailed his own experiences in Nazi prisons and his work for victims of Nazi persecution. He also used the book as a vehicle for expressing his bitterness against the local government for their lack of help during either the Occupation or the period of the compensation claims (1964–6). In small communities such as the Channel Islands, people are rarely able to attack those in positions of authority or even stick their heads above the parapet to challenge the status quo without suffering the consequences of social ostracism or disapprobation in one form or another. Even though Falla's book was published, no person who committed acts of resistance ever received any public recognition or honour in the Channel Islands in his or her lifetime. Guernsey is still lacking any memorial to those who died in Nazi prisons or camps after being deported for resistance 'crimes'.[4] Falla's site of counter-memory was thus his book, which is now out of print although second-hand copies are not difficult to obtain.

In Jersey, Mière had the advantage of working as a curator at the popular German Underground Hospital, a private Occupation museum in a large underground concrete tunnel complex built by slave and forced labour during the Occupation. Mière set up a display using photographs and testimonies given to him by former political prisoners, first and foremost those who had been in Jersey jail with him and who became his fictive kin group. They would often visit him at the museum and give him material so that their stories could be added to those already on display.[5] Mière was denied access to closed files at the archives in Jersey, so he was never able to conduct proper archival research; his collection of testimonies was built up the hard way. It was not until after Bailhache's speech to mark the fiftieth anniversary of liberation that the political and memorial climate had changed enough in Jersey for Mière's campaign for a memorial for former political prisoners to be successful. In 1996, a small memorial plaque was erected on the external wall of the former prison, on the edges of the commercial district of St Helier. It was unveiled by Bailhache and is now neglected. The golden lettering is peeling, the ivy on the wall threatens to cover it, and it plays no part in annual ceremonies. If it wasn't for the fact that the Channel Islands are increasingly incorporating victims of Nazism into their heritage, then we might expect this site of counter-memory to become a *lieu d'oubli*, a site of oblivion that that 'public memory has expressly avoided because of the disturbing affect that their invocation is capable of arousing' (Wood 1999, 10).

GUARDIANS OF MEMORY AND THE IMPACT ON COLLECTIVE MEMORY

Since the 1960s and 1970s, the era when Occupation heritage began in earnest in the Channel Islands, heritage has been created and controlled by guardians of memory and their fictive kin. Only in more recent decades have heritage professionals been appointed in the island, and the work carried out by self-taught enthusiasts in the CIOS has been challenged by alternative and more inclusive narratives. Because the membership of the CIOS still controls the majority of the Occupation heritage in the Channel Islands (whether restored bunkers or Occupation museums), the heritage professionals are in charge of only a few Occupation sites. In Jersey this is restricted to three sites: a single bunker which has been turned into a memorial for forced labourers; the German Underground Hospital which has been rebranded as Jersey War Tunnels and refitted with a cutting-edge exhibition; and the Occupation Tapestry, finished in 1995. In Guernsey, the heritage professionals have worked with the amateur enthusiasts but have not challenged their heritage with their own professional narratives and ownership of sites. This should be seen as a reflection of the power of the guardians of memory in this island. Members of the CIOS have also often had positions in committees which organize Liberation Day celebrations and decide on the erection of Occupation-related memorials.

Because of the decades-long control over Occupation heritage of the guardians of memory and their fictive kin, their nostalgic vision of the Occupation has had a hegemonic influence and control over collective memory in the Channel Islands. Younger generations of islanders learn about the Occupation through heritage, Liberation Day narratives and the largest and most central monuments in St Helier and St Peter Port (which recall the act of liberation in 1945; the narratives which compete with these memorials are small, easily overlooked and often marginalized, as already noted). As a nostalgic vision of the past holds sway in heritage, this subtly influences the questions which children ask about the Occupation within their own families. While the younger generation are quite capable of reading Falla's book or visiting Mière's exhibition at Jersey War Tunnels (now reinterpreted by heritage professionals and placed within the museum's café), the narratives that are repeated and reinforced across the Channel Islands are not yet, by and large, the counter-narratives.

It is difficult for academics and other outsiders who wish to intrude on these memory conflicts to work within the Channel Islands without inevitably or inadvertently taking sides with either the guardians or the professionals. On the one hand, unless one's work is endorsed by a guardian of memory, then one risks it being rejected locally. Invariably, however, an academic is more likely to write in a style, paradigm and format of publication more favoured by the heritage professionals. The most serious sin is perceived to be those who write with an agenda which pays no heed to locally favoured narratives (of either guardians of memory, guardians of counter-memory or heritage professionals) and trespasses upon what are referred to under the blanket term of 'sensitivities'—the local shorthand for anything which touches even remotely upon the subject of collaboration. These 'sensitivities' are highly taboo. To write about resistance is also dangerous, as it highlights the existence of its polar opposite. Islanders are thus still reeling from guardian journalist Madeleine Bunting's 1995 work, *The Model Occupation*. This book dismissed much of the resistance that took place in the Channel Islands and focused instead upon collaboration, especially horizontal collaboration. Respected historian Paul Sanders is one of the very few people who have explored this subject objectively through archival sources, and he judged the Channel Islands to be a 'haven of sanity' compared to the 'delirious heights' to which collaboration rose elsewhere in Europe. There was no evidence of 'heart and soul' collaboration. The 'dominating influence', he wrote, 'was a definite 'submission on the grounds of superior force', interspersed with elements of 'shield philosophy' and 'tactical collaboration' (Sanders 2005, 97). Sanders, too, drew upon the wisdom of the 'wise men of the Expert Panel', all of whom were guardians of memory and guardians of counter-memory, and included Michael Ginns and Joe Mière (Sanders 2005, xiii). He recognized not just the importance of working with these men in order for his work to be accepted locally but also the very real value of the memory of these men who lived through and witnessed the occupation.

CONCLUSIONS: GUARDIANS OF HERITAGE, GENERATIONS OF MEMORY

This chapter has argued that, while guardians of memory are not a phenomenon restricted to small islands, it is only in these small, close-knit communities that they are able to wield a long-lived, hegemonic power and control over memory and heritage, spanning several generations, in a way that would be much more difficult in a large community or nation. People in small islands are also tied by intimate threads of friendship, kinship and locality in a way that is not so apparent elsewhere, which has the potential to perpetuate the influence of guardians after their death.

Today we are on the edge of living memory of WWII, meaning that we are observing a dwindling in the number of guardians and the end of their influence. But this is to underestimate the importance of occupation memory and identity in the Channel Islands. I have often been surprised at the strength of feeling of the second generation, and it is likely that here at least the mantle of the guardians of memory will be passed down successfully.

For Michael Ginns, now in his mid-eighties and no longer enjoying good health, there are men within his fictive kin in the CIOS who have worked alongside him for decades and who are carrying on his work of bunker restoration, even if Ginns is sometimes critical of their methods or complains about being sidelined now that he can no longer regularly attend meetings. As he owns his own museum, there is no pressure on Richard Heaume to retire; his health is good, he is an active member of the CIOS, and he is young compared to other guardians of memory.

The situation is rather different for our guardians of counter-memory. The fictive kin of Mière and Falla were their fellow political prisoners, and they did not form, either during the Occupation or after it, an organized society which recruited younger members to take on or aid them in their work. Joe Mière was among the last of Jersey's political prisoners to pass away, and his son has no interest in carrying on his work, resenting the time that his father spent on the cause rather than with the family. When Frank Falla died, his legacy died with him. His fictive kin comprised men who had been imprisoned with him and in other Nazi prisons and camps. These men would meet for dinner once a year on the anniversary of the liberation from their prisons, and they would raise a toast to their friends who did not survive (Figure 4.3) (Wood and Wood 1955, 233). Like Mière, Falla had nobody to continue his work of remembrance. While many of the families of these men live in the Channel Islands today, they witnessed the struggle of their fathers and, for a variety of reasons, decided that the burden of the fight had been their fathers' alone to carry. Those few who do wish to fight for their fathers' memory find themselves alone.

In 2010, Falla's daughter gave to the author her father's archive. It comprised the most important resistance archive ever to come out of the Channel Islands, containing many testimonies of experiences in prisons and concentration camps.[6] Such is the value of this archive that it will provide important research material for years to come. Some of the data from this archive is included in the first-ever academic volume on resistance in the Channel Islands during the Occupation (Carr *et al.* 2014). The authors hope that this work will trigger or at least ease the passage of a resistance memorial in Guernsey

Figure 4.3 Frank Falla (back row, first on the left) surrounded by his fictive kin during a reunion (Courtesy Sally Falla).

and wish to lend their support to any such endeavour. However, enquiries at the Department of Culture and Leisure, the body responsible for memorial projects in Guernsey, have revealed that in order for such a memorial to be erected, any campaign must have a 'weighty body' behind it, some sort of local 'group behind the initiative' who can 'achieve consensus within the community' and who are 'supported by significant persons.'[7] It seems that even today, the structure of a group of fictive kin headed by a guardian of memory has worked its way into the very structure of professional heritage creation in the Channel Islands. Without such a group to promote this cause, it is likely that this site of counter-memory will never come into existence. By requesting this structure in the creation of new heritage, any involvement by outsiders is effectively kept at bay. The status quo of the long-lived Churchillian paradigm and the nostalgic memory of occupation can stay in place, unchallenged at least in this culturally conservative island.

NOTES

1. The position of Bailiff in Jersey and Guernsey is appointed by the Crown after consultation with the island. He is president of the Royal Court and local parliament (or States) and acts as Speaker of the Assembly in the Westminster tradition.

2. At the time of writing, the author has observed mostly male guardians of memory during the period of fieldwork. This is not to say that female guardians have not existed in the past and do not exist today; rather it is to observe that men are more prominent in this role in the Channel Islands.
3. Text taken from http://ciosjersey.org.uk/about (accessed 18 December 2013).
4. See 'Heritage, Memory and Resistance in the Channel Islands' by the author in Carr *et al.* (2014) for more information on the work of Frank Falla and Joe Mière.
5. Jersey War Tunnels Research Files Box 5, document written by Joe Mière.
6. See the Youtube video 'Forgotten Heroes': www.youtube.com/watch?v=EHEHvUAI_B8.
7. Email from Guernsey Museums and Art Gallery employee to author 12 December 2013.

BIBLIOGRAPHY

Bunting, M. (1995) *The Model Occupation*, London: BCA.

Carr, G. (2012a).'Occupation heritage, commemoration and memory in Guernsey and Jersey', *History and Theory*, 24 (1), 87–117.

Carr, G. (2012b) 'Examining the memorialscape of occupation and liberation: a case study from the Channel Islands', *International Journal of Heritage Studies*, 18 (2), 174–93.

Carr, G. (2014) *Legacies of Occupation: Heritage, Memory and Archaeology in the Channel Islands*, Heidelberg: Springer.

Carr, G., Sanders, P. and Willmot, L. (2014) *Protest, Defiance and Resistance in the Channel Islands, 1940–1945*, London: Bloomsbury Academic.

Cohen, F. (2000) *The Jews is the Channel Islands During the German Occupation 1940–1945*, Jersey: Jersey Heritage Trust.

Falla, F. (1967) *The Silent War*, London: Leslie Frewin.

Farmer, S. (1999) *Martyred Village: Commemorating the 1944 Massacre at Oradour-sur-Glane*, Berkeley and Los Angeles: University of California Press.

Hirsch, M. (1996) 'Past lives: postmemories in exile', *Poetics Today*, 17, 659–86.

Hirsch, M. (1999) 'Projected memory: holocaust photographs in personal and public fantasy', in M. Bal, J. Crewe and L. Spitzer (eds), *Acts of Memory: Cultural Recall in the Present*, Hanover and London: University Press of New England, 2–14.

Jordan, J. (2006) *Structures of Memory: Understanding Urban Change in Berlin and Beyond*, Stanford, CA: Stanford University Press.

Legg, S. (2005) 'Sites of counter-memory: the refusal to forget and the national struggle in colonial Delhi', *Historical Geography*, 33, 180–201.

Nora, P. (1996) 'From *lieux de mémoire* to realms of memory', in D. Kritzman (ed) and Arthur Goldhammer (trans), *Realms of Memory: Rethinking the French Past. Vol. I: Conflicts and Divisions* (English language edition with a foreword by L. D. Kritzman), New York: Columbia University Press, xiv–xxiv.

Nora, P. (1997) *Realms of Memory: Rethinking the French Past. Vol. 2: Traditions* (English language edition with a foreword by L. D. Kritzman, Arthur Goldhammer [trans]), New York: Columbia University Press.

Ronayne, I. (2009) *'Ours': The Jersey Pals in the First World War*, Stroud: History Press.

Rousso, H. (2002) *The Haunting Past: History, Memory, and Justice in Contemporary France*, Philadelphia: University of Pennsylvania Press.

Rousso, H. (2007) 'History of memory, policies of the past: what for?', in K. H. Jarausch and T. Lindenberger (eds), *Conflicted Memories: Europeanizing Contemporary Histories*, New York and Oxford: Berghahn Books, 23–36.
Sanders, P. (2004) *The Ultimate Sacrifice*, Jersey: Jersey Heritage Trust.
Sanders, P. (2005) *The British Channel Islands Under German Occupation 1940–1945*, Jersey: Jersey Heritage Trust.
Sanders, P. (2012) 'Narratives of Britishness: UK memory and Channel Islands occupation memory', in J. Matthews and D. Travers (eds), *Islands and Britishness: A Global Perspective*, Cambridge: Cambridge Scholars Publishing, 24–39.
Turner, B. (2010) *Outpost of Occupation: How the Channel Islands Survived Nazi Rule 1940–1945*, London: Aurum.
Winter, J. (1999) 'Forms of kinship and remembrance in the aftermath of the Great War', in J. Winter and E. Sivan (eds), *War and Remembrance in the Twentieth Century*, Cambridge: Cambridge University Press, 40–60.
Wood, N. (1999) *Vectors of Memory: Legacies of Trauma in Postwar Europe*, Oxford and New York: Berg.
Wood, A. and Wood, M. (1955) *Islands in Danger*, New York: MacMillan.
Young, J. (1993) *The Texture of Memory: Holocaust Memorials and Meaning*, New Haven, CT, and London: Yale University Press.

5 Turncoat Heroes or Reckless Egotists?

The Ambivalent Memorialization of the 'Russian War' on the Dutch Island of Texel

Rob van Ginkel

INTRODUCTION

By the spring of 1945, WWII was nearing its closure in Europe. Many areas in the Netherlands had already been liberated from September 1944 onwards, and the inhabitants of the Dutch island of Texel were anxiously awaiting their turn. However, due to a series of contingencies, Texel unexpectedly turned into 'Europe's last battlefield'. A *Wehrmacht* infantry battalion consisting of Georgian 'volunteers' and German soldiers had been stationed there in February 1945. On 6 April, these 'Russians'—as the islanders dubbed them—rose up against the German military. Fierce fights ensued. The so-called Russian War (*Russenoorlog*) took the lives of almost 500 Georgians and at least 420—but probably many more—Germans (Bartels and Kalkman 1980, 307*ff*.). There were 89 civilian casualties. Material damage was enormous, particularly in Den Burg and in the northern part of the island. This episode and its heritage, remembrance and commemoration are replete with disagreements, ambivalences and revisions, particularly at a local level. This chapter focuses on the rebellion and on Texel's post-war memorial and commemorative trajectories in particular. It seeks to show how memorialization is embedded in and influenced by national discourses and global geopolitics.

In due course, the Russian War captured the imagination not only in the Netherlands but also in Germany, the Soviet Union and particularly in Georgia. It has been the subject of several historiographies, films, documentaries, children's books, novels and even an opera. Texel's Aeronautical and War Museum hosts a permanent exhibition on the Georgian uprising and a modest museum in the Georgian town of Manglisi, established by one of the rebels, is devoted to the uprising. Two websites present information, photographs and video footage. As concrete evidence and a permanent reminder, there is a Georgian war cemetery on the island, which is probably the only one outside the Republic of Georgia. Although the Georgian revolt has sometimes been designated a 'forgotten' war episode, this is in fact not the case, at least neither in the Netherlands nor in Georgia. In the larger scheme of global WWII events, the Georgian rebellion and its aftermath are a tiny footnote, but on Texel it is the single most important fact of twentieth-century history. How did the islanders come to terms with this peculiar war within the war? How is it remembered, memorialized and commemorated? And how did and do Texelians regard the Georgian rebels and their uprising? Before dealing with these questions, I shall first sketch the scene and what happened there during the war.

TEXEL, THE WAR AND THE GEORGIAN REBELLION

In May 1940, German troops invaded the Netherlands, including Texel, one of the Dutch Wadden Islands (Figure 5.1). Speckled with seven small villages and a score of hamlets, Texel was home to approximately 8,200 inhabitants who mostly earned a living in agriculture, fishing and a slowly expanding tourist sector. There was no combat on the island. The first German soldiers arrived days after the Dutch army had surrendered on the 15 May. The mayor warned his fellow citizens to behave in a 'calm and dignified' manner.

Whereas the nearby naval port of Den Helder suffered many German and allied air strikes, Texel seemed to be of little strategic importance. Initially, relationships between the islanders and the occupying forces were rather polite but aloof. Perhaps because there had been no violence or major damage, there seemed to be a tacit understanding to refrain from making each other's lives miserable. Initially, the occupation force numbered only 200 soldiers, a number that soon expanded to 1,000. Gradually a flood of regulations and other measures began tormenting the island population. The Germans impounded fishing vessels to convert them into *Vorpostenboote* ('advanced post boats'). As part of the *Atlantikwall*, Texel was heavily fortified. In November 1944, the Germans commanded 800 young Texel men to conduct forced labour on the mainland. They returned to the island only in March the following year. Nonetheless, the islanders seemed to be escaping atrocities, but this would soon change dramatically.

This had to do with previous developments in the European theatre of war, particularly the Nazi expansion into Russia, which demanded that German military were rushed to the Eastern Front. In compensation, the *Wehrmacht* needed ancillary troops elsewhere. They included *Ost-Legionen* consisting of Armenians, Turkmens, Volga Tatars, North Caucasians and Georgians who had either volunteered for ideological reasons or enlisted as an escape route from the inhumane conditions of prisoner of war camps. In September 1943, an infantry battalion of approximately 800 Georgians and 400 Germans was stationed in the coastal resort of Zandvoort, situated roughly 80 kilometres south of Texel. The Georgians, who had served in the Red Army, became the Germans' auxiliaries to avoid maltreatment and starvation in POW camps. After military training, they joined the 822 Georgian

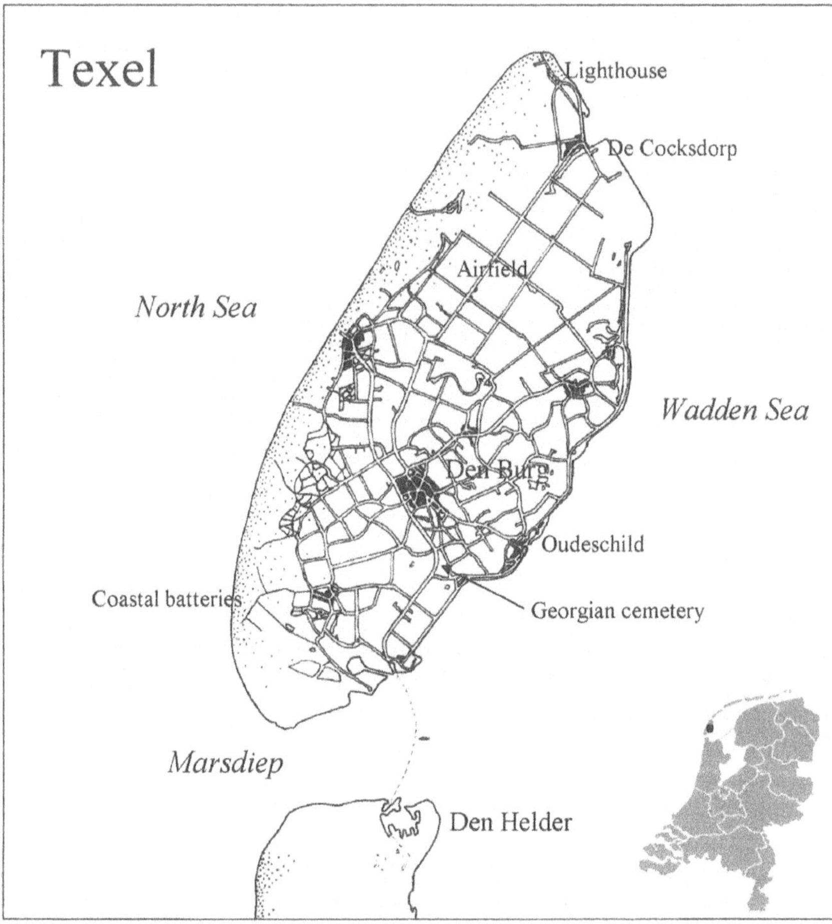

Figure 5.1 Map of Texel (Image copyright Rob van Ginkel).

Infantry Battalion Queen Tamara. One of them had served as political commissar in the Red Army: Evgeni Artemidze. Clinging to firmly held convictions, he and some other Georgians soon established contact with the communist resistance movement in the Zandvoort area and unfolded plans to rise up against the Germans. The resistance leaders knew that at this stage of the war, this would only result in bloodshed and failure. They convinced the Georgians, who thereupon gave up their plans. Communication was greatly facilitated because one of the resistance leaders, Annie Averink, was fluent in Russian, a language she had mastered during her training at Moscow's International Lenin School a decade earlier.

In February 1945, the Georgian battalion was transferred to Texel, severing connections with the Zandvoort resistance. Nevertheless, contacts with the Texel resistance movement were soon established. Artemidze carried a letter of introduction written by Averink. However, the island's underground boasted few communists. Therefore, Averink's letter was of little help and the relationships remained distant. Besides, communication was bothersome; very few Georgians spoke any German, a language that most islanders were familiar with.

By this time, it was abundantly clear that the Third Reich's collapse was near. Parts of the Netherlands had already been liberated as of September 1944, although a large area still remained occupied. But by late March 1945, Canadian troops were advancing rapidly in the country's east, fighting their way westward. On 5 April, German commanders decided that 500 Georgians would be assigned to the front. The Georgian leaders, among them Lieutenant Shalva Loladze and political commissars Evgeni Artemidze and Sergei Gudzabidze, knew that their compatriots would surely defect. This would have serious repercussions for the almost 300 Georgians who were to remain on Texel. Moreover, even if they would survive, they feared that as 'defectors' they would not be welcome upon return in the Georgian Soviet Socialist Republic, while the allied forces would doubtlessly treat them as enemy soldiers. Hence they were caught between a rock and a hard place. Any chance of rehabilitation at home hinged on a last-minute attempt to fight their overseers. The Georgian leadership decided that the battalion would rise up against the Germans in a surprise operation with the Russian codename *Den' Rozjdenija* ('Birthday').

At 01:00 a.m. on 6 April 1945, the Georgians killed 180 German soldiers. In a concerted action, they slit their throats when the latter were sound asleep. However, they did not succeed in slaying all the Germans, and they failed to capture the German-staffed coastal batteries that could also aim inland. German officers immediately called in reinforcements from the mainland. Heavy combat ensued. Soon after the uprising began, Loladze sent for resistance leader W. N. Kelder. The Georgians had left him completely ignorant of their plans. Loladze told him that all able-bodied Texel men had to report to the Georgian headquarters in the main town of Den Burg to fight along with them. He also announced that not doing so would have grave repercussions. Some 200 Texelians showed up that night and were provided with weapons. However, when the Germans began shelling Den Burg, many Texel 'volunteers' decided to disarm again and go into hiding. In less than half an hour, the shelling caused 38 civilian casualties, while the Germans killed 10 innocent Texel men they had arrested. Within a couple of days, the Georgians were driven back and eventually cornered in the north of the island. For a week, they tenaciously held on to the airfield, hoping in vain that allied troops would come to their rescue. With their light weaponry, they stood little chance against the 2,000 German soldiers that had meanwhile been rushed to the island. On 21 April, the Georgians fought their last battle near the lighthouse on Texel's northernmost tip. Finding themselves in a hopeless situation, some committed suicide while a few managed to escape. Fifty-nine men surrendered. They were forced to undress and dig a grave, whereupon the Germans shot them. Under Nazi martial law, this was the usual penalty for defectors.

The Georgians who had survived the skirmishes—by then some 350—concealed themselves from the Germans, either on their own or with the help of islanders, who ran the risk of being summarily executed. The Germans conducted search parties and showed no mercy with any Georgian they discovered. They often set fire to farms or houses in the vicinity of which they were found. This was done to discourage islanders from extending aid to the 'turncoats'. Some Georgians continued to fight as partisans, but obviously ammunition and supplies were running out and most of them hid in the woods or dunes. The situation was grim. In the final stages of the war, German soldiers harassed civilians and pillaged and burned houses, which aroused deep hatred.

It was abundantly clear to the islanders, however, that the Germans' defeat and the return to freedom were imminent. The islanders' stance towards them had already deteriorated when 800 Texel men had to conduct forced labour in 1944–5. Moreover, the fresh German troops showed little compassion for the islanders. This had only increased the latter's sympathy for the Georgians and inflamed their hatred of the Germans. This explains to a large extent why the Georgians won the hearts and minds of many civilians almost overnight: they were the enemy of the islanders' enemy. Despite the Germans' systematic search parties, 228 Georgians managed to survive. Instead of being the island's glorious liberators, they owed their lives to the non-combatant islanders.

On 20 May 1945, a fortnight after the German occupiers had officially surrendered in the Netherlands, Canadian troops finally arrived on Texel. They liberated the island without combat and disarmed the Germans, who were expelled from the island the next day. The Canadians also attempted to disarm the Georgians, but they refused to do so voluntarily, unless the Canadians would speak on their behalf to the Soviet authorities. The Georgians only left the island on 17 June 1945. Their political commissar Artemidze had meanwhile paid several visits to the mainland, where he saw his communist contacts. He realized that for a safe return to the Soviet Union, it would be vital to avail himself of documents proving the Georgians' courageous comportment during the rebellion. He obtained a document from the Communist Party Netherlands (CPN), stating that he had been involved in the underground work of Dutch partisans and communists. Artemidze also visited the editorial office of the communist daily *De Waarheid*. He left a message for the Texelians, expressing gratefulness for all the help the Georgians had received: 'Once we have returned home, we will never forget the comradeship that emerged in our joint fight.'[1] Right after Texel's liberation, Annie Averink, who would soon become a communist figurehead, visited the island to see her friend Artemidze and other Georgians. These warm relationships would prove extremely valuable for the latter. Moreover, the commander of the Canadian unit on Texel, Lieutenant Colonel Lord Tweedsmuir, wrote a report on the Georgian rebellion. In a letter directed to the Soviet High Command, he stated that the rebels had been valiant Soviet allies who were 'entitled to all the honours of a brave and dauntless Ally' (quoted in van

Reeuwijk 2002, 58). Myth making thus started immediately: the rebellion was *not* a joint fight with the Texel population, and former Red Army POWs who served in the German army could hardly be considered allies.

When the Georgians assembled at the quay of Oudeschild harbour for their departure, many islanders showed up to bid them farewell and wish them Godspeed (Figure 5.2). Congladze and Artemidze thanked the Texelians and apologized for all the suffering the rebellion had caused. Their words prompted rounds of applause. It was only when boarding the ferry that the Georgians, except for their officers, disarmed. Tweedsmuir accompanied them to a displaced persons camp near Wilhelmshaven, where Soviet liaison officers were awaiting them. They arrived in Tbilisi, Georgia's capital, in early 1946, after having 'served' with the Soviet Army as construction workers for some time. Perhaps owing to the Dutch communists' and Tweedsmuir's praise, the majority returned home, unlike other Soviet defectors. It would take a long time, however, before this news reached the Netherlands. Soon after the Georgians had left Texel, the non-communist press began raising questions about their fate. It surmised that the Stalinist regime would have executed them or detained them in camps. The Cold War had begun. This also became clear in the memorialization of the uprising.

Figure 5.2 Georgians arriving with the Texel ferry in Den Helder on 17 June 1945 (Image collection Aerial and War Museum Texel).

COLD WAR COMMEMORATIONS: THE RUSSIAN CEMETERY AS A SITE OF REMEMBRANCE

Right after the Germans left, it was decided to create a cemetery for the fallen Georgians (van Dijk *et al.* 2005, 87*ff.*). A meadow on a scenic spot was selected for this purpose. With the assistance of Texelians who had been forced to witness executions, the Georgians located mass graves and individual field graves and identified their fallen compatriots before reburial. On 26 May 1945, 187 victims were put to rest in nine long rows, with their commander Shalva Loladze at the head. Solemn speeches and salutes accompanied the reburials at the so-called Russian cemetery (*Russenbegraafplaats*), as the islanders soon dubbed it.

Initially, the Georgian leaders decided that only those who had fought to the death deserved an honourable reburial. Those who had surrendered only to be subsequently executed were denied this ceremony. They were left in their mass graves. It was only in 1947, after consultation with the Soviet embassy, that these Georgians were also exhumed and reinterred at the cemetery, adding another three rows of graves. Eventually, there would be 485 graves.

Before their departure, the Georgians erected a memorial at the cemetery's premises. It consisted of a simple red-painted conical structure with a star, a hammer and a sickle atop. Behind glass were a photograph of Loladze and an epitaph in Georgian and Dutch: 'Here rests the brave Soviet-captain Loladse Schwalwa and his brothers-in-arms who fought for their freedom against German terror on the island of Texel. We are and will remain Georgians and we will never forget our brave fallen comrades.' The monument was inaugurated on 15 June 1945. Apparently, it was not to everyone's liking: more than once the glass was shattered. This in turn caused outrage in the local press. It reported that the perpetrator, whose identity and motives remained unknown, 'must have a twisted soul'.[2]

Possibly to prevent further vandalism, the municipal authorities put a barbed wire fence around the premises. This in turn soon triggered indignation. The local newspaper likened the site to a concentration camp and wrote that the fence was a 'horrible error'. 'The only solution is: tear it down, rectify this mistake!'[3] The report met with the approval of several islanders. Many deemed it an impious act or even sacrilege to have barbed wire around a cemetery. The national press also began devoting attention to the matter. The communist daily *De Waarheid* cried shame:

> It is one of the most disgraceful expressions of neglect and ingratitude vis-à-vis fallen combatants for Dutch liberty that we have experienced so far. What will the Russians think of our civilization, that we like to boast about so much?[4]

However, to consider the Georgian soldiers 'combatants for Dutch liberty' was to give them much more credit than they in fact deserved. Most Texelians had deemed their insurrection a disaster for the island. As a local minister pointed out:

> Texel civilians have been involved in the uprising partly of their own free will and partly through coercion. Since the Russians' rebellion aimed at their self-preservation, their battle was not our battle. . . . It is simply foolish to think that the mutineers became 'allied fighters' because they rose up.[5]

It was immediately clear that the Georgians did not fit a neat category of 'heroes'. But Dutch communists demanded that they be recognized as allies. It is important to bear in mind that in the immediate post-war era, *De Waarheid* was the largest daily newspaper in the Netherlands, and communists were an important political force. Owing to their over-representation in the resistance, they had earned a lot of respect. Their wartime comportment warranted that they meddled in matters of morality. Thus, remembrance had almost immediately become the subject of politics.

The pressure was on Texel's municipal authorities. They announced that a reputed sculptor would design a monument for the cemetery. However, its materialization would take considerable time. The municipality did make sure, however, that the cemetery was tidied up and gradually aestheticized. Trees and shrubs were planted, shell-covered paths laid out and the rows of graves that lacked headstones were provided with beds of red bush roses in a field of grass. A brick wall was constructed around the cemetery. In many respects, the site conformed to the order, tidiness and rationality characteristic of so many war cemeteries (Mosse 1990). In addition, annual rituals sacralized the Russian cemetery.

A committee organized an annual wreath-laying ceremony. Its board included Texel's former resistance leader Kelder and Cornelia Boon-Verberg, a communist and one of the local women who had helped the Georgians. On the board was also the commander of Texel's small naval base. Civil and military authorities and approximately 400 islanders attended the first anniversary of the uprising at the cemetery. Several representatives of *De Waarheid* and the CPN, including Annie Averink, were also there. Notably absent were members of the Dutch government or their representatives. The commander-in-chief of the Dutch Royal Navy arrived with the ambassador of the Soviet Union to the Netherlands and embassy staff. The ambassador donated a 3,000-guilder cheque for a permanent monument and laid a wreath. In a brief speech in Russian, he commemorated the islanders and the Georgians who as Soviet soldiers had died 'in a battle against an enemy that tried to conquer the world with the fascist terror system'.[6]

The phrasing speaks volumes. The Soviets claimed to have battled fascism, a claim that was extended to satellite states. The rehabilitation of the Georgians who had served in the German army demanded that they were reframed as anti-fascist freedom fighters. With the help of Dutch communists, they were portrayed as Texel's liberators. The Soviet Union sought recognition for its wartime role in the West and eagerly seized every opportunity to let its officials attend remembrances. For the islanders, however, the 'real' commemoration occurred later that day, in Den Burg. Here, civilians and allied soldiers who died as a consequence of the war stood in the limelight. According to *De Waarheid*, this commemoration was organized by 'Texel reactionaries', 'who attempted to avoid that the emphasis would be on the ceremony for the Russians'.[7]

As of 1947, the Dutch government decided that all remembrance meetings would be held on the evening of 4 May (the official Dutch National Remembrance Day of the war dead, *Dodenherdenking*). In order to commemorate the fallen Georgian soldiers, the Texel municipality facilitated brief afternoon meetings at the Russian cemetery. A delegation of the Soviet embassy was usually present, with Soviet diplomats in their commemorative speeches always referring to the rebels as 'POWs' who rose up against their German guards. The cemetery still lacked a permanent monument. The embassy agreed on a design only in November 1952. The memorial—a granite wall with engravings of the island's and the Soviet Union's coats of arms—was unveiled six months later (Figure 5.3).

Although the embassy had immediately put the Georgian rebels on a pedestal for political purposes, at home they were officially rehabilitated only in

Figure 5.3 Monument for the fallen Georgians (Image copyright Rob van Ginkel).

1956, under the Khrushchev regime. In the meantime, nothing was known on Texel about their situation. For almost a decade, attempts to get in touch had failed time and again. During the tenth commemoration of the rebellion, the Soviet embassy's military attaché read a statement that was supposedly written by the veterans. They expressed their gratitude to the Texel population. In the autumn of 1956, several Dutch dailies devoted attention to a news item in the Russian-language journal *Zary Vostoka*, which coined the Georgian rebels as 'keen Soviet patriots'.[8] A few Dutch newspapers believed this to be evidence of their rehabilitation, and subsequently *De Waarheid* occasionally published news about and photographs of Artemidze.

Nonetheless, non-communists usually dismissed news from behind the Iron Curtain as propaganda; it was the heyday of the Cold War. This also had consequences for the commemoration on Texel. The Dutch Intelligence Service photographed Texelians who joined the remembrance rituals at the cemetery. They were suspected of being socialists or at least entertaining communist sympathies. Rumours about mysterious mainland photographers soon percolated. The islanders began avoiding the commemoration because they feared being mistaken for communists (van Reeuwijk 2002, 67). The islanders feared that if they attended the ceremonies, they would experience disadvantages, including not being accepted for certain civil service jobs. There were never huge crowds of Texelians attending the commemoration at the cemetery anyway, but when islanders began avoiding the remembrance meetings, the communist presence became all the more visible.

GEORGIAN–TEXELIAN REUNIONS AND MEMORIAL INTENSIFICATION

On 7 April 1960, *De Waarheid* published an item by a reporter who met eleven veterans of the Georgian uprising in Tbilisi. They still vividly remembered Texel, and some exchanged letters with islanders, despite the language problems. 'Texel is our second home,' they claimed. The author emphasized that no harm had been done to any of the former soldiers upon return to the USSR and that they lived happy lives and had fine jobs. It was counter-propaganda against newspapers that still claimed that the Georgians who fought on Texel had all been executed, despite overwhelming evidence to the contrary.

The fact that the communist daily newspaper published stories about the Georgians' well-being was a prelude to closer relationships between Texelians and Georgian representatives. At long last, Cornelia Boon-Verberg, who had been trying to establish contact for years, began receiving letters from veterans. In 1963, three islanders—including Boon-Verberg—received an honorary membership of the Association of Soviet War Veterans. A Russian delegation, amongst whom was a Georgian veteran, attended the 4 May commemoration. In 1964, Artemidze, Congladze and three others once more expressed their gratefulness in a letter to Texel's mayor.[9]

The veterans had gained more leeway to get in touch with Texelians. Over the next few years, more Georgians, including veterans of the rebellion, visited the island. They brought soil from home to scatter on the graves of their comrades, so that the fallen soldiers would be buried in Georgian ground. In 1966, they also gave a copper plaque with a chased-work image of Shalva Loladze to serve as a memorial tombstone on his grave. There were several emotional reunions of Georgians and islanders. Contacts intensified, although it remained difficult to obtain visas. A Georgian cultural association invited Boon-Verberg, who was now in touch with several of the rebellion's veterans, to visit the country in 1966. She received a very warm welcome. On her return, she reported that she had had a banquet with 120 former rebels and talked to many of them. She was dubbed 'the mother of the Georgians', and she later received the Freedom of the City for Georgia's capital, Tbilisi. As she noted, the Dutch government had bestowed no such honours upon her or other islanders for helping the Georgians.

In the 1970s and early 1980s, the memorialization of WWII intensified, in the Netherlands as a whole and on Texel as well.[10] It was in this decade that some films, documentaries and books about the Georgian rebellion were launched, and several new monuments were unveiled. Georgians donated all of these memorials. It was belated reciprocal gift giving in return for the help received during the rebellion. The Western European première of the Soviet film *The Crucified Island* was on screen on Texel in April 1970. It presented the uprising as if the Georgians had been held prisoners of war by the Germans on the island and broke lose to fight for their freedom together with the islanders. Of course, the film had to suit the official Soviet version of the historical events, namely that the rebels fought fascism. Despite this historically rather peculiar representation, the first public performance turned into a commemorative event. Georgian film director Rezo Tabukashvili brought a copper name board that said in Dutch, Georgian and Russian, 'Joint grave of Soviet soldiers from Georgia'. It was fixed to the cemetery's entrance gate. A few years later, 'the Georgian people' gave a new gate of wrought iron. It was decorated with a copper relief—an image of Saint George defeating the dragon ('good' conquering 'evil'). Gifts continued to come, including a 4-metre-high copper Tree of Life memorial, symbolizing immortality. It was erected at Den Burg's general cemetery in 1975. By putting it at the entrance, it was ensured that the memorial would have an important place in the *Dodenherdenking*.

In addition to memorial intensification, a documentary and new or revised books about the Georgian uprising appeared (van der Vlis 1978 [1945]; Bartels and Kalkman 1980; Bartels 1986). The Dutch documentary '*Sondermeldung Texel*' ('*Special Message Texel*') was broadcast on national TV in October 1979, prompting renewed media and public interest in the history of the rebellion. Less than two weeks later, two MPs for the Dutch Labour Party addressed the government, stating that many islanders were still suffering from war trauma as a consequence of extending aid to the

Georgian rebels. In their view, this was particularly so because post-war governments had not recognized their support to the rebels. Some of them had, on the contrary, been painfully confronted with state obstructions due to their alleged sympathy for 'Russians'. Some civil servants conducted research into the matter. They found no evidence that Texelians suffered war traumas but also concluded that little was known about the rebellion beyond the island society. It was therefore decided to donate 75,000 guilders for a permanent exhibition in a local museum.

Despite all the attention for the rebellion, the Soviet ambassador and his staff still faced restrictions from the Dutch Foreign Office when travelling to the island. They had to avoid Den Helder's naval base. The communist presence at commemorations at the Russian cemetery had not declined, and some considered this to be a thorn in their side. On the face of it, most islanders did not seem to care much about all the fuss and bother. Very few of them actually attended the remembrance meetings at the cemetery.

It is therefore surprising that more than 200 Texelians travelled to Georgia in October 1983 to meet veterans of the rebellion. The local Texel newspaper organized the trip. Its editor-in-chief was deeply involved in the Foundation Texel–Georgia Contact. Owing to the cooperation of a travel agency specializing in Eastern Europe combined with Soviet sponsorship, the journey was a real bargain. Approximately eighty veterans were still alive then, and several older Texelians wanted to meet them again. Many islanders considered it an interesting adventure to have a look behind the Iron Curtain. A few thought they might have an opportunity to finally meet their genetic fathers, as a small number of Georgian soldiers had conducted relationships with local women. Be this as it may, the islanders received a hearty welcome, although the Soviet authorities tried to make sure that the few meetings between them and the veterans—twenty-nine of whom showed up—were tightly organized and supervised. The Dutch media covered the trip extensively. For quite different reasons, the Dutch secret service also showed an interest.

Not all Texelians appreciated the warm relationships with the Georgian veterans. Theo Witte, who lost his brother due to the uprising, maintained:

> Back then, the Georgians chose for themselves. At the very last moment, they wanted to clear their names because they knew that if they didn't do that, they couldn't expect much on their return to Russia. Stalin had ordered them to save the last bullet for themselves. Had they not risen up, we would have been spared from much misery.[11]

It went against the grain for him that so many islanders, most of whom 'hadn't lost anyone', visited Georgia. 'I had problems with the fact that the Georgians who survived never recognized . . . that they never told people back home what they had caused on the island of Texel'.[12] Witte was certainly not the only one who thought this way. Many others who had lost family did not want to have anything to do with visiting Georgians. Like Witte, they did not appreciate the fact that so many islanders went to Georgia. The matter led to some dissension in the island society, which was also expressed via letters to the editor of the local newspaper, but it was never pushed to extremes.

Still, while many islanders reckoned the Georgians to be reckless egotists, in 1983 many apparently deemed the time ripe to go and see for themselves how the veterans fared in Georgia. Why? Time had healed some wounds, while for others they were still raw. For islanders who had helped the Georgians, time was running out, as they were septuagenarians or octogenarians by then. They did not have to fear for their jobs if the secret service suspected them of entertaining communist sympathies. The Cold War was considered to be a matter for politicians while people were craving peace, which showed in the contemporary upsurge of the peace movement. This had brought about a political climate in which communism was no longer viewed as a demonic creed. As stated, many considered the trip a unique and cheap opportunity to visit a country behind the Iron Curtain. Last but certainly not least, there was the feeling that Texel was 'special' and that the bond with Georgians reinforced this singularity.

Two years after the large Texel delegation had visited Georgia, seventeen Georgians and seven Russians paid a return visit to the island to participate in the celebration of the fortieth anniversary of the liberation. Six of them were veterans of the rebellion. Again, the Georgians donated a memorial: a commemorative stone with an image of Cornelia Boon-Verberg. It was placed at a corner of the cemetery, near Loladze's grave and the granite monument. Apart from some minor additions, the Russian cemetery was now believed to be complete. The 1985 commemoration at the cemetery was the last of its kind in the sense that the veterans were there as *Soviet* heroes.

THE GEORGIAN APPROPRIATION OF THE CEMETERY

The breakdown of the Soviet empire from 1989 onwards also had consequences for the Russian cemetery, which would be turned into the cemetery of the Georgians. Until 1991, the Soviet ambassador usually went to the island to attend National Remembrance Day. That year, Georgia regained independence. The fallen soldiers were no longer revered as Soviet communists but as Georgian patriots. Obviously, the communist influence over the commemorations disappeared. This was due not just to the breakdown of the USSR but also because the Dutch communists disappeared from the national political stage. Their influence on the remembrance of the Georgian rebellion had already been abating for several years, but by now it had disappeared altogether. The Communist Party of the Netherlands merged with other left-wing parties in 1989, and it was officially dissolved in 1991. The communist daily *De Waarheid* ceased publication in 1990.

During most of the 1990s, very few people showed up to attend the annual commemoration at the Georgian cemetery, which most islanders still stubbornly dub the 'Russian cemetery'. In 1995, the municipal council decided that memorializing the rebellion was too controversial and had been a source of dissension for too long. For several years, no official representation had visited the cemetery anyway. But in 1999, the first ambassador of independent Georgia in the Benelux, Zurab Abashidze, and the patriarch of the Georgian Orthodox Church, Ilia II, appeared with a large company of Georgian religious and civil authorities. Abashidze extended apologies to the people of Texel, like many of his Soviet predecessors had done. Ilia II led a remembrance ceremony at the cemetery in the afternoon of 4 May. He said: 'The Georgian soldiers have defended the Dutch territory against the fascists. Many gave their lives.' Apparently, some of the old socialist rhetoric lingered, but the remark was certainly not intended as a nostalgic veneration of Soviet empire. Though the fallen soldiers were now no longer classified as *Soviet* heroes, they were considered to be *heroes* nonetheless. And whereas remembrance meetings had always been strictly secular, a religious dimension was added to the commemoration.

In the following years, Georgia's ambassador would invariably show up at the cemetery and later in Den Burg during Dutch National Remembrance Day. In 2005, Georgian president Mikheil Saakashvili and his Dutch wife, Sandra Roelofs, paid a 'private' visit to the site. Several delegates, including Patriarch Ilia II, members of the Georgian Parliament and two veterans of the uprising, accompanied them, with many Georgian and Dutch news reporters in their wake. In an earlier press release, Saakashvili contended that the ceremony would honour all Georgians 'as heroes who fought bravely in that war against more than one totalitarian regime', explaining that '[t]hey suffered simultaneously at the hands of Nazi Germany and Stalin's tyranny.'[13] The idea was that sixty years after the uprising, the Georgian soldiers of the Red Army who turned into the Germans' auxiliaries would at once be de-Nazified, de-socialized and reframed as *Georgian* heroes—even though some of them had been Stalinist diehards. However, shortly prior to the remembrance ceremony, Theo Witte told Saakashvili what had 'really' happened, whereupon the president refrained from using the word 'heroes', at least during his visit to the cemetery. It was obvious that some islanders no longer tolerated the fallen Georgian soldiers' heroization. If Saakashvili had intended to use his visit to the cemetery for national political purposes, he must have been disappointed.

At home in Georgia, however, the heroization went on. On 18 July 2007, Saakashvili began a government meeting by asking a minute of silence in honour of the memory of a veteran who had died a few days earlier as the last of Texel's rebels. He was buried with military honours. In him, Saakashvili paid tribute to all the Georgian 'heroes' who participated in the Texel rebellion. He 'expressed regret that in due time the heroism of the revolt's participants were [sic] not appraised in the adequate manner, many of them were arrested, exiled, many of them were depreciated'.[14] With tensions between

Georgia and Russia mounting, the president apparently deemed it a good moment to point out that the leading force of the Soviet empire had not dealt with its Georgian heroes properly. He carefully avoided mentioning that the Georgians fought on the side of the Germans before the rebellion.

Currently, around two score people gather at the Russian cemetery for Dutch Remembrance Day, usually around four o'clock in the afternoon (Figure 5.4). Among them are a delegation from the embassy, including the Georgian ambassador to the Netherlands, and several Georgian expats. Texel's mayor serves as the ambassador's host, but does not pay tribute to the dead. The meeting is low-key and almost devoid of ceremony. Usually, no more than twenty islanders show up. Following the laying of wreaths and flowers and the spilling of some wine or spirits on the graves, the visitors leave the premises to join for a drink to the dead on the adjacent green. Some subsequently head for Den Burg. In the evening, the official national remembrance of the war dead takes place there. After a memorial meeting in a local church, people silently walk to the town's cemetery, which includes a Commonwealth War Graves section. It is attended by a much larger crowd than the unofficial event in the afternoon. People join in silence for a tour of the cemetery premises, where they pay their respects at the headstones

Figure 5.4 Georgians and Texelians meet informally outside the war cemetery premises on Remembrance Day 2011 (Image copyright Rob van Ginkel).

marking the graves of allied air and naval personnel. Wreaths are laid at the Cross of Sacrifice and the Georgian Tree of Life memorial. It is obvious, however, that the focus of this official remembrance are the Texel civilian war victims and the fallen *allied* soldiers: officially, the fallen Georgian rebels are not commemorated.

CONCLUSIONS

Texelians remain divided and ambivalent about the Georgian uprising. Most of them feel sorry for the men who, through the whims of history but mostly on account of seeking their self-preservation, ended up being buried in foreign soil. A minority defend their actions or are apologetic, arguing that they had no alternative and fought bravely in an attempt to liberate the island. The majority, however, despite feeling compassion for these young men, are of the opinion that their desperate and inconsiderate action was utterly irresponsible and put the civilian population needlessly at risk. There is consensus that without the rebellion, the war would have been over in a matter of weeks anyway. Without a doubt, the Texelians would then have escaped from the war relatively unscathed, and there would be no need for a remembrance of the so-called Russian War. The belligerence cost scores of islanders their lives, caused huge material damage and protracted the war for at least a fortnight after the rest of the Netherlands enjoyed freedom again after five years of Nazi occupation. Still, in a 2011 opinion poll of a local website, more than 60 per cent of the islanders stated that the Georgians should be commemorated at National Remembrance Day, while almost 40 per cent disagreed.[15]

The island's war history was unique by sheer coincidence. Had the Georgian battalion remained in Zandvoort, nothing much out of the ordinary would have happened on Texel, except probably for the fact that it would have survived five years of occupation relatively unharmed. This is what happened on the other Dutch Wadden Islands (Vlieland, Terschelling, Ameland and Schiermonnikoog). At the same time, the fact that Texel *is* an island implied that the hostilities between Germans and Georgians were geographically bound to its territory. The latter had little opportunity to escape to the mainland. They were trapped. So were the civilian islanders. Due to the fierce combat that ensued, Texel was designated as 'Europe's last battlefield'. This historical contingency and the particular relationship with the Georgians provided the island with a special position that played an important role not only in memory and remembrance but also in identity politics. For better or worse, many islanders cherished and continue to cherish their connection with Georgia and the Georgians because it gives them and the island a uniqueness that is also emphasized in many other respects (van Ginkel 1995). The belligerence is but a footnote in the larger scheme of things, but due to the Georgian uprising,

WWII is also *their* war, providing the island and the islanders with a very particular history that is used in their identity formation of 'being different' in relation to the Dutch mainland and other islands.

Like all societies, small islands need to come to terms with the material and immaterial heritage of war and occupation. The options range from active remembrance, memorialization and commemoration on the one hand to negligence, suppression and oblivion on the other. The balance of this continuum is usually dynamic and may shift considerably in the course of time. For example, following liberation, remnants of the Atlantic Wall on Texel were initially put to practical use as summer homes, then demolished as 'the marks of Cain' and what little was left is currently carefully mapped and preserved as war heritage. However, the local balance may deviate significantly from dominant national narratives and myths. As peripheral and clearly bounded entities and social formations, most island societies tend to emphasize their singularity vis-à-vis the dominant mainland, nurturing special feats that are believed to be unique. For better or worse, most Texelians cherished the history of the Georgian rebellion and their special relationships with its veterans. Of course, it *was* a unique local history. At the same time, however, the *dynamic of its appropriation* is a more common feature of small island societies. Almost any historical fact that lends itself for this purpose may be singled out. For instance, unlike elsewhere in the Netherlands, friend and foe are buried side by side at Schiermonnikoog's war cemetery and commemorated together. The islanders are proud of this and never tire of pointing it out. Generally, bits and pieces of war history are selected and become integral parts of island identity politics. The Georgian insurrection and its morally ambivalent heritage provided Texelians ample opportunity to do so.

As a site of remembrance, the Russian cemetery continues to be a source of controversy and dissension. The fallen Georgian soldiers do not fit into any neat binary category of good and bad. Most local people deem them neither friend nor foe. They are, so to speak, 'dead out of place' (cf. Douglas 1966). Now it is true that all deceased persons or 'dead bodies' are ridden with ambiguity and 'are open to many different readings' (Verdery 1999, 28), but this is especially the case with the Georgians interred on Texel. Their ambiguity may have led to avoidance: islanders avoided the cemetery not only because they were afraid of being regarded as communists by the intelligence service but also because the Georgians who are buried there were and are ambiguous and hence taboo. Their ambivalent trajectory began when they were still alive: from being POWs they turned into the Germans' auxiliaries; from auxiliaries they turned into communist turncoats; from turncoats they turned into would-be liberators; and from would-be liberators they indirectly turned into a menace for the civilian population who nonetheless helped many of them survive the war. Following the liberation, the rather pitiful bunch of fallen and surviving soldiers were redefined—at least by Dutch and Soviet communists—as 'Soviet heroes' who fought fascism, and lastly, the Georgian authorities attempted to re-appropriate them as Georgian patriots. Very few Texelians would now or in the past glorify the Georgian rebels or grant them any noble motives. The fallen Georgian soldiers remain betwixt and between because their sacrifice was not in any way selfless but aimed solely at self-preservation and thus ultimately not a sacrifice at all. Alive and dead, they turned their coats so many times that they could be nothing other than ambiguous.

NOTES

1. *De Waarheid*, 1 June 1945. During the war, *De Waarheid* (*The Truth*, the Dutch equivalent of *Pravda*) was an underground journal.
2. *Texelse Courant*, 22 August 1945.
3. *Texelse Courant*, 10 October 1945.
4. *De Waarheid*, 11 October 1945.
5. *Texelse Courant*, 12 December 1945.
6. *Het Vrije Volk*, 8 April 1946.
7. *De Waarheid*, 8 April 1946.
8. *Zary Vostoka*, 6 October 1956.
9. *De Waarheid*, 21 May 1964.
10. There are many reasons for the memorial intensification in the Netherlands (see van Ginkel 2011). Importantly, a new generation began doubting the dominant myth of nationwide resistance; the Nazi persecution of Jews and widespread collaboration came into the limelight, and new publications and television broadcasts about the war brought about lively discussions, historical revisions and attention for 'forgotten' war episodes.
11. *NRC Handelsblad*, 30 August 2008.
12. Videotaped interview: http://getuigenverhalen.nl/interview/interview-14–0.
13. See www.president.gov.ge/en/PressOffice/News/MeetingsAndVisits?p=2211&i=1 (accessed on 6 March 2013). The press release is dated 5 February 2005.
14. See press release: www.government.gov.ge/index.php?lang_id=ENG&sec_id=148&info_id=1711 (accessed on 24 March 2013).
15. See www.texel-plaza.nl, 11 May 2011 (accessed 12 May 2011).

BIBLIOGRAPHY

Bartels, J.A.C. (1986) *Muiterij aan het Marsdiep. De tragedie van de Georgiërs op Texel, April 1945* [*Mutiny at the Marsdiep. The Tragedy of the Georgians on Texel, April 1945*], Amsterdam: De Bataafsche Leeuw.

Bartels, J.A.C. and Kalkman, W. (1980) *Texel: Nederlands laatste slagveld. De muiterij van de Georgiërs* [*Texel: Holland's Last Battlefield. The Mutiny of the Georgians*], Zutphen: De Walburg Pers.

Douglas, M. (1966) *Purity and Danger. An Analysis of Concepts of Pollution and Taboo*, London: Routledge.

Mosse, G. L. (1990) *Fallen Soldiers: Reshaping the Memory of the World Wars*, New York: Oxford University Press.

van der Vlis, J. A. (1978 [1945]) *Tragedie op Texel* [*Tragedy on Texel*], 8th ed, Den Burg: Langeveld & de Rooij.

van Dijk, A., van Es (jr.), J. A., Bakker, J. J. and Engelmoer, G. J. (2005) *Hun leven voor ons leven. Verzamelde werken ter gelegenheid van 60 jaar bevrijding* [*Their

Lives for Our Life. Collected Works on the Occasion of 60 Years Liberation], Texel: Comité 4/5 mei Texel.
van Ginkel, R. (1995) '"Texelian at heart": the articulation of identity in a Dutch island society', *Ethnos*, 60, 265–86.
van Ginkel, R. (2011) *Rondom de Stilte. Herdenkingscultuur in Nederland* [*Around the Silence. Commemorative Culture in the Netherlands*], Amsterdam: Bert Bakker.
van Reeuwijk, D. (2002) *'Sondermeldung Texel': The Georgian Rebellion on Texel*, Den Burg: Het Open Boek.
Verdery, K. (1999) *The Political Lives of Dead Bodies. Reburial and Postsocialist Change*, New York: Columbia University Press.

Section II

Islands of Tourism, Landscapes of War

6 The HMS *Royal Oak* and the 'Ownership of Tragedy' in Orkney

Daniel Travers

THE ORKNEY ISLANDS AT WAR

The Orkney Islands lie just a few miles north of the Scottish mainland, across the Pentland Firth from John O'Groats (Figure 6.1). Though officially a county of Scotland, Orkney has a unique sense of identity and culture, dating back to the time when the islands formed a significant outpost in a vast Norse naval empire. Because of the presence of a British naval base at Scapa Flow, Orkney was heavily militarized during WWII, and Orkney's defensive role is a large part of Orkney's domestic war experience. As a result, hundreds of derelict wartime buildings litter the landscape, such as turrets, blockhouses, barracks, hangars and sheds. Most of these have been in a serious state of disrepair since the late 1950s, only now finding heritage groups to champion their restoration. Orkney's involvement in two more tangential aspects of the British narrative have taken the forefront within Orkney's wartime commemoration: the sinking of the HMS *Royal Oak* in October 1939 and their association with Italian POWs after 1942. The Italian Chapel, an ornate Roman Catholic chapel saved from scrap after the war is now, unlike the many bunkers, lovingly maintained by local Orkney craftspersons. The HMS *Royal Oak* and the Italian POW stories form a large portion of the commemorative culture of Orkney. This is because they are something which Orcadians can claim as their own. Orcadians annually commemorate the *Royal Oak* tragedy, while diligently taking care of the Italian Chapel. These provide the concrete symbols of Orkney's collective remembrance. Possession of the wreck of the *Royal Oak* has meant that this now informs part of their commemorative experience; it has been appropriated as part of 'Orkney's War' and worthy of constant remembrance. Though the POW experience and the *Royal Oak* tragedy are tangential to Britain's finest-hour war narrative, they have been appropriated by Orcadians in order to show off the islands' own unique heritage and sense of difference. This represents an 'ownership of tragedy' in Orkney (Ugolini 2011, 228–30). These divisive aspects of the war are part of the landscape of remembrance there because Orcadians choose to maintain their associations with it—in essence to remember where the rest of Britain has chosen to forget.

It has been said that 'the Battle of Orkney was fought and won long before the Battle of Britain began' (Smith 1989, 4). Orkney holds the distinction of firing the first shots of the war from its battery at Stromness (Dorman 1996, 28), having the first bomb to make landfall on the British Isles at Lyness, and claiming the first German plane of the war to anti-aircraft fire, shot down by Orcadian gunners in October 1939 (Lamb 1991, 30). An Orcadian also holds the unfortunate distinction of being

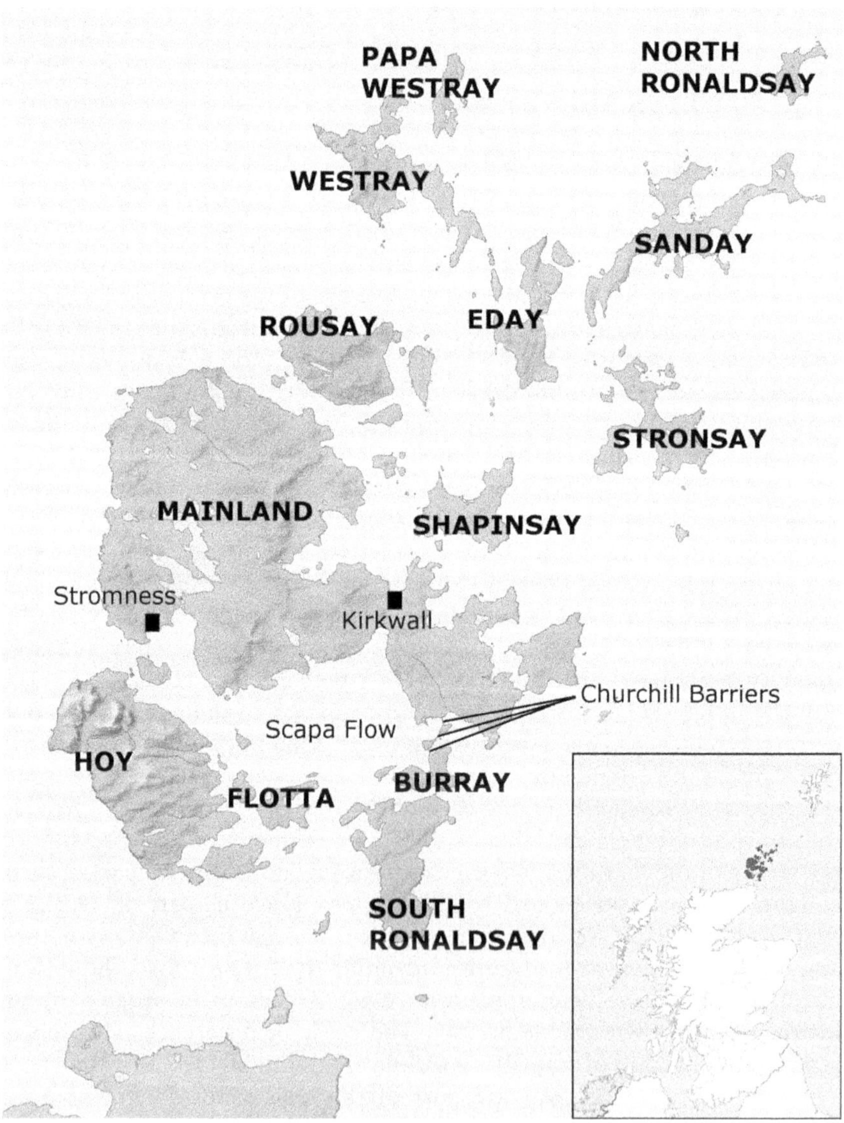

Figure 6.1 The Orkney Islands (Creative Commons).

the first British civilian killed by bombing; James Isbister, a resident of Loch Stenness, suffered fatal shrapnel wounds after a bomb landed near his house in March 1940 (Schei and Moberg 1985, 91). Though being beyond the support range for most German fighter aircraft, German bombers frequently attacked ships in harbour at Scapa Flow, and Orkney was bombed routinely in the lead-up to and shortly after the invasion of Norway in 1940. A total of 228 bombs made landfall in Orkney during the war, the vast majority of these within the first year, killing three civilians and seriously injuring sixteen others. Several houses were also damaged by bombs, and the town of Stromness had at one point been strafed up and down its main street (Hewison 2005, 320–1).

Orkney proved vulnerable from the sea as well as the air. On the 14 October, Gunther Prien, Commander of the German U-boat U47, was able to sneak through Kirk Sound on Scapa Flow's eastern approaches and fire a salvo of torpedoes at the HMS *Royal Oak*, a large but somewhat antiquated capital ship in anchor at the Flow (Miller 2000, 83–95). The result was the loss of 834 lives[1] and what has been called a 'national disaster' (Turner 2009, 7). It was a tragedy compounded by the fact that many of the crew were boys aged between 15 and 18 years old, an issue which would later cause the British Admiralty to revise its naval policy, disallowing youths from active service (Turner 2009, 96–9).

While the rest of Britain experienced a relatively quiet early war—the British Expeditionary Force patiently awaiting the invasion of France—Orkney saw significant action. Though wholly incomparable with what many British (or indeed European) cities experienced during the war, being bombed was a part of life for Orcadians in the early months, and many were concerned by the constant air raids. The result was that Orkney gained fearsome notoriety in late 1939 and early 1940, something that the Admiralty was well aware of, and the Navy found it difficult to recruit labour from the mainland to work in the islands. Large sums of money were offered to non-Orcadian labourers (mostly Irishmen hired from Liverpool and Glasgow) to provide a temporary workforce. Many of these workers proved unsatisfactory, and considerable concern was raised when anti-British propaganda was found in one of the accommodation huts (Hewison 2005, 241, 269–70). A solution to the labour shortage was found by employing prisoners of war, mostly Italians captured in the desert campaigns of 1940 and 1941, on essential building projects in Orkney (Cormack 1992, 5–11). Most famous of these projects were the large Churchill Barriers, lengthy causeways stretching across Orkney's south-eastern islands, closing the entrance of Scapa Flow to any further attack by U-boat from the east. The causeways had been ordered shortly after the *Royal Oak* sinking by Churchill himself (hence the name) to prevent another such attack, although the concept of closing the eastern approaches was not his alone.[2] The Italians caused considerable difficulty for the engineers at first: some protested at being forced to do war work, a task prohibited by the Geneva Convention (MacDonald 1992, 7–9); others were put off by the accommodation provided for them or found their

'Mediterranean temperament' unsuitable to Orkney's climate (Hewison 2005, 270–4). By the end of their internment, however, many Italian labourers had taken 'considerable pride' in their work on Orkney (Cormack 1992, 11). The Italian Chapel, constructed by Domenico Chiocchetti as a place of worship for the captives, stands to this day as a testament of Italian gasconade. Though Orkney's use of POW labour was small in comparison with rural mainland Britain, where interned combatants provided a significant part of the agricultural labour force (Hellen 1999, 191–217), Orkney's POWs left a lasting legacy etched into the very landscape of the islands. The Italian Chapel serves now as a cultural meeting place, whereas the Churchill Barriers provide a much needed transport link between the southernmost islands and the Orkney mainland.

WWII IN ORCADIAN MEMORY

The sinking of the HMS *Royal Oak* has also become something which Orcadians have absorbed into their war experience, largely due to the fact that the wreck still lies within Orkney waters. It has become something immensely important for the islanders and both literally and figuratively part of the Orkney landscape. Since 1945, commemorations for Britain's naval dead, including the HMS *Royal Oak*, have typically occurred at the Portsmouth Naval Memorial on Southsea Common. The WWII memorial there contains the names of 833 men who lost their lives in the *Royal Oak* disaster and for years provided the meeting place for survivors to commemorate the tragedy.

Though *Royal Oak* dead were included in the many Remembrance Day services held in St Magnus cathedral in Kirkwall, large-scale commemoration of the event on the anniversary of the sinking was limited to Portsmouth. Strangely enough, although a large service was held in Kirkwall for the ninth anniversary of the disaster in 1948, a ceremony which saw the commissioning of a permanent memorial to the disaster in St Magnus Cathedral,[3] there is no newspaper record of any service held at the site in 1964, the twenty-fifth anniversary or in 1984 for the forty-fifth. By 1989, however, Orkney had taken over from Portsmouth the role of commemorating the event, and it has now become a yearly event.[4] Since 1989, a memorial service for those who lost their lives has occurred every year in Orkney. This memorialization mirrors heritage trends in Britain, with a conspicuous lack of interest in performative commemoration in the 1960s and 1970s and a resurgence in the 1980s and 1990s, with particular emphasis on five- and ten-year anniversaries. The seventieth anniversary was attended by the Princess Royal, the first time a member of the Royal Family has attended such an event.[5] On most occasions, survivors themselves travel from England to attend the commemorations, although the youngest in 2009 was in his late eighties. A service in St Magnus Cathedral is held on the closest Sunday to the anniversary of the disaster, and a wreath-laying ceremony occurs at the site each year, usually from the deck of the HMS *Orkney*, on the morning of 14 October.

The *Royal Oak* has now found a place in Orkney's landscape of remembrance. It has been appropriated as part of Orkney's heritage. At the end of Old Scapa Road, in Scapa Bay, less than 4 kilometres from the wreck site, a memorial garden to the victims of the *Royal Oak* was created. Today, a modest wooden hut acts as a shrine for those who leave wreaths, poppies, crosses and even notes to family members who were lost in the disaster. Inside, pinned to the bulletin board is a notice about the *Royal Oak* commemorations planned for that year. A picture of the ship is flanked by two flags, the Union flag on the left and the Orkney flag on the right. St Magnus Cathedral still plays host to a permanent memorial to the ship's crew, complete with two naval ensigns and the ship's bell, recovered by divers in 1982. A page in the book of remembrance is turned every day by the cathedral's staff.

Orcadians, it seems, have now assumed the role of official commemorators of the *Royal Oak* tragedy. No longer are the services held in Portsmouth; annual events are organized for 14 October and are conducted on board a ship in Scapa Flow. The *Royal Oak* has become part of the Orkney landscape. There is now a Royal Oak Road, a Royal Oak Naval Hall in Longhope and a Royal Oak Guesthouse in Kirkwall. The site of the wreck itself can often be seen a few metres beneath the surface on a clear day, and a green buoy sits on the sea in Scapa to mark the spot. The most recent chapter in the *Royal Oak* story is the creation of a granite obelisk, erected by survivors, friends and family of the *Royal Oak* association on 14 October 2011 and unveiled in the garden of remembrance near the wreck site.[6]

The sinking of the *Royal Oak* is a unique part of Orkney's heritage which Orcadians can, with some justification, claim as their own. Since being proudly 'saved from scrap' by Orcadian residents in the 1950s,[7] Orkney's museums now contain photographs, ship statistics and mementos taken from the ship. The brass nameplate of the ship, pillaged by illegal divers in the 1970s and recovered by the Royal Navy in 1994, is now on display at the Scapa Flow Visitors Centre. For the sixtieth anniversary of the sinking in 1999, the Royal Navy philatelic society launched a limited edition commemorative cover postmarked at the Kirkwall post office. Signed by survivors and by families of the victims, the one hundred covers which went on sale were snapped up in minutes by Orkney residents.[8]

The Orcadian public has a vested interest in the history and commemoration of the *Royal Oak*. It is something which now forms part of Orkney's domestic wartime experience and its reputation. Publications about the *Royal Oak* have, on occasion, stirred discontent among the Orkney public. A recent example is David Turner's book *The Ultimate Sacrifice*, published in 2004 (later republished with the title *Last Dawn*), which created widespread disapproval by alleging that Orcadians were prevented from rescuing survivors on the night of 14 October and that an enemy spy stationed in Orkney had been responsible for guiding the U-boat through Scapa's eastern approaches.[9] Lately, current issues surrounding the wreck site, including the safety of armaments sunk within it, the leaking of oil from the wreck and illegal diving operations, have hit Orkney's local headlines on a monthly basis and help to ensure that the *Royal Oak* remains in the minds of the local people throughout the year. As one *Royal Oak* survivor stated: 'Orcadians think more about the *Royal Oak* because it is part of *their* history . . . it's always there!'[10]

The HMS *Royal Oak* and the story of its demise in Orcadian waters now espouse feelings of Orcadianness, at the same time creating a sense of shared history with the rest of the UK and the Royal Navy. Though only two of the 834 men who died on the ship were from Orkney, it is the physical possession of the wreck and constant memorialization that engrained it on the collective memory of Orcadians. It has now become part of the island which they love; the tragedy has now become a part of their identity.

By contrast, the rest of Orkney's WWII sites have been left, for the most part, unglorified. After the withdrawal of the Royal Navy in 1957, most of the naval buildings were left abandoned and unmaintained. Only a handful of the buildings now serve a more modern purpose; the vast majority are crumbling badly. Nowhere is this more prevalent than on the island of Hoy, where the former Lyness naval base stood. The area is now a ghost town of abandoned buildings, vehicles and poorly maintained roads. Only in the last few years have local agents banded together to stop what they see as a loss of Orcadian heritage. In November 2008, the *Orkney Defence Interest Network* (ODIN), headed by local administrator Anne Billing, met for the first time. Its mission is to record, categorize and preserve Orkney's wartime heritage.[11] Created from the community in response to the 'insufficient preservation' of Orkney's historic sites, the group now leads surveys to historic buildings, researches and provides materials and information about Orkney's defence heritage, and records Orkney's oral histories primarily from those remaining who experienced life during WWII.[12] Similarly, in October 2008, the Aviation Research Group Orkney and Shetland (ARGOS) was formed focusing on the documentation of Orkney's aviation heritage and crash sites and with the long-term goal of opening an aviation museum on the island.[13] Though primarily concerned with the aerial history of Orkney, ARGOS works closely with ODIN to achieve these preservation goals.[14] Lastly, in August 2009, Orkney Islands Council's own initiative, the Scapa Flow Landscape Project, championed many similar aims and the goal of promoting 'understanding and appreciation of the wartime heritage of the Scapa Flow area'. This project is now developing a trail of wartime sites, as well as doing restoration work to Ness Battery, a coastal defence battery near Stromness and one of the more salvageable sites in the islands.[15]

Wartime heritage in Orkney is still currently under development. Although the interest has perhaps always existed, initiatives to categorize and preserve such sites are just now gaining momentum. One building dating from WWII, however, has escaped the neglect that the other sites have been subject to over the last six decades. The Italian Chapel, an ornate Roman Catholic chapel constructed from Nissen huts and spare concrete by Italian Prisoners of War brought to Orkney in 1942–3, has been lovingly preserved and maintained.

THE ITALIAN CONNECTION

A fine example of Italian culture thousands of miles from Italy, the Italian Chapel defies the wind and rain of Lamb Holm, one of the smaller islands in Orkney, and seems completely out of place next to the large concrete causeways, the Churchill Barriers, which the POWs were brought to build (Figure 6.2). Many Orcadian cultural objects and celebrations now carry with them Italian elements. The islands' festivals and religious services include some song or dedication to an Italian theme, and the islands have a treaty of friendship with Moena, a village in Northern Italy. The Italian Chapel now stands as a testament to friendship between Orkney and Italy. Yet of the 19,000 or so inhabitants on the islands, very few if any consider themselves Italian. There is no Italian community on the island, nor has the island been subject to the same post-war mass migration as have some other places in Britain. Interest in preserving Orkney's WWII sites is just beginning to peak, whereas the Italian POW connection has been well maintained and nurtured by the Orkney community.

The Italian Chapel, saved from destruction in 1945, has been entirely maintained by the Orcadian residents themselves over the last sixty-five years. Initially retouched and repainted in 1960 by the original artist, Domenico Chiocchetti, the chapel has since needed constant work to keep it in its original condition—all of which has been undertaken by Orkney craftspersons (Paris 2010, 200–1). Regular visits to the Italian Chapel by the Italian POWs themselves and their families have been a feature of the chapel since the 1960s, and local newspapers report on the return of one of them almost every year. The Chapel also serves as an important tourist attraction, drawing over 92,000 visitors every year.[16] The Chapel is owned by Orkney Islands Council, but it is a group of amateur enthusiasts, the Italian Chapel Preservation Committee, who ensure that the building is maintained and accessible for visitors. The secretary, John Muir, a retiree from Holm, is well known to most Orcadians for his charitable work and for taking it upon himself to walk down and open the Italian Chapel early every morning.[17]

The Italian Chapel has become a cultural and spiritual building for the people of Orkney. Many local weddings are held there, and religious services have been regularly conducted inside it. A mass on the first Sunday of each month is still held between April and September, as well as one on the Sunday closest to the anniversary of the sinking of the HMS *Royal Oak* in October (Paris 2010, 197). The Churchill Barriers constructed by the POWs

Figure 6.2 The Italian Chapel on Lamb Holm (Copyright Daniel Travers).

serve today as causeways that provide a road link from the southernmost islands of Orkney to the mainland. The opening of the barriers so close to VE Day on 12 May 1945 has meant that commemorations of the end of the war have also been connected with the POW story. This is unusual in the sense that it combines a typically 'British' commemorative occasion with a unique display of Orcadian identity. The POW experience does not typically factor into VE Day celebrations on the mainland, as it highlights the work of non-Britons in the winning of the war. This tends to contradict the 'dig for victory' and 'make do and mend' wartime work ethic which the Churchillian paradigm[18] privileges. In Orkney, however, it is seen as an essential part of the war experience, one that can and should be connected with what has now become a celebration of Britain's role in the war. As a result, the 1995 VE Day commemorations in Orkney saw representatives from Balfour Beatty and the Admiralty together with a delegation of over thirty ex-POWs, local dignitaries, and members of the Orkney community participate in a celebration of the opening of the barriers.[19] To coincide with this, the Tankerness House Museum in Kirkwall opened a special exhibition on Italian Prisoners of War entitled 'Building the Barriers'.[20] In 2002, to mark the anniversary of the arrival of the POWs, Stromness Museum also held a special exhibition on the Italian POWs focusing on their arrival, camp life and finally their return to Orkney after the war.[21] Walking tours or coach tours of the barriers and the Chapel are regularly enjoyed by resident and visitor alike, often coinciding with significant anniversaries such as the sinking of the *Royal Oak*.[22] Associating this permanent symbol with the larger picture of WWII in Orkney has been encouraged by at least one local newspaper pundit, the significance of the construction project to modern-day Orkney being a 'sentiment to celebrate' on VE Day.[23] One of the concrete blocks used in the construction of the barriers now marks the entrance to the first barrier at Holm Sound (Figure 6.3). On it, a plaque makes reference to the barriers being constructed 'with the aid of Italian Prisoners of War'.

Many Orcadian residents still possess *objets d'art* given to them by the Italians during their captivity. Museums across the islands display a number of small items either made by the Italians or pertaining to the Italians. Ornate ashtrays, cigarette cases, coat hangers and vanity boxes can be found at the Orkney Museum, the Orkney Wireless Museum, and the Fossil and Heritage Centre. Some of these items were recognized as their own by a group of eight ex-POWs who returned to Orkney in 1992.[24] Italian influence has been allowed to permeate even the most fundamental aspects of Orcadian culture. The St Magnus festival, a decades-old music festival held in Orkney every year, often sees concerts performed in the chapel, and in 2000, a musical tribute entitled *Domenico* took place there,

Figure 6.3 The Churchill Barriers constructed '. . . with the aid of Italian Prisoners of War' (Copyright Daniel Travers).

attended by former POWs.[25] In 2002, the anniversary of the arrival of POWs in Orkney, a play entitled *Barriers*, written by Alan Plater and with music by Sir Peter Maxwell Davis, was performed as part of the festival at the Orkney Arts Theatre. The play focused on the interactions between the Italian POWs and the Orkney community and involved both a British brass band and an Italian accordion and wind band.[26] Highland Park Whisky, an Orkney distiller for over two centuries, commissioned a special edition bottle of Capella whisky, complete with a painting of the Italian Chapel painted by Chiochetti, for the sixtieth anniversary of the departure of the POWs in 2004.

Orkney also has strong contemporary associations with the *commune* (municipality) of Moena in Italy, the hometown of Domenico Chiocchetti. This created a very real link between Orkney and its wartime heritage, one which was heavily supported by the island community. Almost like Italian communities in Britain, which often have very strong links with areas of origin within Italy, even going as far as having relationships with specific villages, Orkney's relationship with Moena is distinctive. Local choirs have on occasion led a cultural exchange with the Italian village, and on one occasion the Moena Choir Canticum Novum, with Chiocchetti's daughter as a member, performed in Orkney.[27] Orkney enjoys a larger than average proportion of Italian visitors in response to possession of the Italian Chapel and the strong links the island has with Italy.

ORKNEY'S 'OWNERSHIP OF TRAGEDY'

Possession of the Italian Chapel, the most recognizable symbol of this type of remembrance, has meant that the Italian connection in Orkney goes beyond a simple romanticization. The possession and care for the Italian Chapel, as well as the deliberate and ongoing attempts to forge and maintain contemporary connections with Italy, are unique for a non-Italian community in the British Isles. Orkney's desire to give pride of place to its Italian war heritage, above more traditional themes, is unusual. Orcadians defy British war remembrance by highlighting this aspect of their war story, a more tangential narrative, above others. Although a prisoner-of-war story forms a part of British memory, it is almost exclusively that of British POWs in captivity rather than enemy combatants interned on British soil. Episodes such as those which occurred at Stalag Luft III have inspired films such as *The Wooden Horse* (1950) and *The Colditz Story* (1955), which privilege the narrative of Britons using their ingenuity and resolve. Films focusing on Axis prisoners of war, including *The One That Got Away* (1955) and *The MacKenzie Break* (1970) have similar themes, rather than directing their attention to the contribution of Axis POWs to the allied war effort. Sophie Jackson has argued that the enemy POW story has become 'almost ignored in the annals of Second World War history' (Jackson 2011, 172) and has lamented how most information about the camps is now 'based on archaeological work—ironic since the camps existed a mere 60 years ago'. The positive experience POWs had in Orkney can also be contrasted with what Wendy Ugolini has called the 'communal myth' shared by Italian-Scots after the war. It focuses on and highlights the contributions of Italians to the war effort, rather than focusing on their role as victims of racial mistreatment (Ugolini 2008, 421–36). Since 1945, the story of Italians in Britain, both as civilians and soldiers, has remained an embarrassing footnote to the greater war myth. Britain's Churchillian paradigm has tended to marginalize the story of enemy aliens, much like the story of internment, in order to favour conceptions of unity (Connelly 2004; Smith 2000). Though the British government saw little to fear in the Italians, both in their ideology and in their (perceived lack of) military ability, civilians did not always take the same view (Moore 2005, 25–40). The beatings, humiliation and destruction of personal property which occurred in England and Scotland after the declaration of war by Mussolini in 1940 was, as Terri Colpi has shown, 'no less than devastating' for the Italian community (Colpi 1991, 99). It is this experience which has manifested itself within the collective memory of the Scottish-Italian community, perhaps in part because some of the worst violence in the United Kingdom was directed against Scottish-Italians in Edinburgh (Ugolini 2004b, 149). The British government's handling of Italians in Britain during the war, both as POWs and as civilian internees, is a dark aspect of the British war experience. The collar-the-lot attitude of the government towards British nationals and the HMT *Dunera* and *Arandora Star* experiences, where many internees and POWs died, have come to represent the 'Italian story' in Scotland, a deeply negative aspect of the war.[28] Little was made of such incidents after the war by the government, and these experiences were very quickly buried under the weight of positive collective remembrance. Race riots and the mistreatment of POWs defy the notions of 'British decency and fairness', an aspect which is central to how Britain as a whole chooses to understand its wartime heritage (Sponza 2006, 57–74; Kushner 2004).

The use of POW labour in Orkney was small in comparison with mainland Britain, where interned combatants provided a significant part of the agricultural work force. Elsewhere in the United Kingdom, however, little lasting legacy of the POW story exists. Though chapels built by Italian prisoners of war were common in Britain, all except two were destroyed at war's end. Of these, Orkney's chapel is by far the best preserved, used and displayed (Paris 2010, 220–1).[29] Orkney's commemoration of the Italian experience is strong. While the rest of Britain saw the Italians come and go, the British national narrative has minimized the massive impact that prisoner-labourers had on the wartime economy. They have become mostly forgotten contributors to the war effort. Orcadians, on the other hand, nurture their association with the Italian POWs to the point that it has become both an important part of the island's wartime narrative and heritage, as well as of their own sense of Orcadianness. It appears that none of the POWs stayed on in Orkney directly after the war, as they did elsewhere in Britain, and Orkney has no ethnic Italian community which provides the driving force for memorializing this aspect of the war. It is the Orcadians themselves who have provided the momentum for such initiatives. Several decades after their departure, the Italian prisoners' experience of war has now established a place within the very landscape of Orkney and in the island's material and commemorative culture. The *Royal Oak* and the Italian story have become part of who Orcadians are, the Churchill Barriers providing a tangible link between Orkney's past and future. The Italian Chapel, constructed in traditional Italian style, has been carefully maintained by Orcadians over the decades to preserve the memory of this aspect of their historical experience.

Recognition of these tangential aspects of Britain's war experience says something important about the need for Orcadians to assert their independence over their own history and heritage. Over the last fifteen years, there has been an upsurge in interest in commemorating WWII across Britain, which has led to the erection of hundreds of plaques, monuments, obelisks and other sites of memory. The presence of new memorials such as the Civilian Memorial Park at Hermitage Wharf in Wapping, the memorial to Civilian Workers in Coventry Cathedral built in 1999 and the Bomb Disposal memorial at Eden Camp erected in 2001 all reflect what Angela Gaffney has called groups 'belatedly seeking recognition for their efforts' (Gaffney 2007, 196). On occasion, more than one community can claim ownership of a particular aspect of the war. Wendy Ugolini has shown a rise in need for memorialization of the 800 internees and crew killed on the SS *Arandora Star* when it was sunk by a U-boat in July 1940 (incidentally by the same U-boat and captain that sank the *Royal Oak* in 1939). In 2005, Colonsay islanders unveiled a granite tablet to victims of the sinking, while in 2008 Liverpool unveiled a memorial plaque at Pier Head, where the ship sailed from. In July 2009 and 2010, memorials were commissioned to the loss in Middlesbrough and in Cardiff, respectively. After sixty some years of silence on the issue, Ugolini argues, it now seems like multiple regions are competing for 'ownership' of the tragedy (Ugolini 2011, 228–30).

This 'ownership of tragedy', in part, explains why Orcadians prioritize the commemoration of the HMS *Royal Oak* and the Italian connection more than any other part of their domestic war experience. Orkney is perhaps one of the last places in Scotland where a regional identity takes precedent over Scottish national unity. It is a place where, as Schei and Moberg have observed, 'People are Orcadian first, then Scots or British . . .' (Schei and Moberg 1985, 10). Maintaining a sense of difference is a key feature of what makes Orcadians and Orcadian culture distinct. A sense of duty accompanies the creation of heritage which ensures that sites of memory have unique elements. Orkney prefers its distinct heritage, and islanders proudly display the objects and mementos left behind by visitors to the islands over many centuries in order to distinguish their own culture and identity as separate from Scotland or Britain. For Orcadians, the Italian connection is an emotional one and one which for the moment still resides in living memory. The deliberate maintenance of the Italian Chapel has served both to reflect and to reinforce this connection, and now stands as an icon, a physical reminder of the associations and friendships made between Italians and the local populace during their brief time in the islands. The inclusion of an Italian connection within Orkney's physical and mnemonic landscape also serves as yet another way to differentiate themselves from traditional 'British' or 'Scottish' identity, adding a significant dimension to our understanding of Orcadianness. For Orcadians, the HMS *Royal Oak* and the Italian POW stories form a large portion of their 'heritage culture', largely because they are something which Orcadians can claim as uniquely their own. By displaying their own unique story in the form of their POW and *Royal Oak* history, they exert their cultural sovereignty. Possession of the wreck of the *Royal Oak* has meant that this now forms part of their commemorative experience. It has been appropriated as part of Orkney's war, worthy of constant remembrance despite the tragic nature of the events which created it. Similarly, the Italian Chapel has become a signifier of Orkney's difference. The islands choose to make the POW experience something in which they show enormous pride, emotional attachment and a willingness to remember while the rest of Britain has chosen to forget. Perhaps spurred on by the sheer beauty of the Italian Chapel and what it represents, the POW story has been championed in Orkney. This is something which the Orcadians highlight to assert their distinctiveness from the rest of the British Isles. The Italian Chapel has become much more than the sum of its parts—of spare concrete and materials salvaged from blockships. It now serves as both a reminder of the friendships created during a difficult time in the islands' past, while serving to deepen Orkney's own heritage, making it distinct from their Scottish or British neighbours. The pride the islanders feel over their association with this aspect of the war experience is unique in the British Isles. Heritage in the islands reflects this. The HMS *Royal Oak* disaster and Orkney's POW story serve as devices with which Orcadians can show off their own identity, culture and values. It is this sense of difference which is integral to the maintenance of a distinct culture, identity and way of life for Orcadians. The wreck of the HMS *Royal Oak* and the Italian Chapel are the tangible remains of culture absorbed, appropriated not as symbols of war but of Orcadianness. The use of these historical events to deepen the islands' own heritage suggests that heritage can be used as a tool—both consciously and unconsciously—for small islands to differentiate themselves from a larger and perhaps more politically and culturally powerful neighbour. There are dozens, perhaps hundreds of island communities around the globe which possess an 'island identity' and share Orkney's need for uniqueness. By charting their own mnemonic course, each island's finite boundaries and geographical separation can be combined with a type of historical independence that allows them to escape absorption. As world nations continue to devolve and fragment in the twenty-first century, no doubt island societies seeking greater cultural, political and legal independence will begin (if they have not already) to look more deeply into their domestic past for elements which can strengthen their own sense of sovereignty.

NOTES

1. The official number recorded by the War office is 833, although research conducted by the Royal British Legion Orkney Branch suggests that there may have been 834 lives lost. One man, it has been argued, is neglected from this tally. As of 2011, the memorial obelisk at the garden of remembrance in Orkney makes reference to '834 men'.
2. Sir William Halcrow had first proposed such measures after a visit to Orkney in 1915; however, it was decided that such an undertaking was unnecessary at such a 'late' stage in the war, and the sinking of block ships was decided upon as the most cost-effective method to protect against attack from submarine. See Hewison 2005, 269.
3. Ninth anniversary of *Royal Oak* Disaster', 14 October 1948, *The Orcadian*, 5.
4. 'Remembering the HMS *Royal Oak* . . . 50 years on', 19 October 1989, *The Orcadian*, 1, 12–13.
5. 'Royal tribute to *Royal Oak* remembrance', 15 October 2009, *Orkney Today*, 1.
6. '*Royal Oak* 72nd Anniversary Memorial Services 2011', HMS *Royal Oak*. Online. Available HTTP: www.hmsroyaloak.co.uk/72nd.html (accessed 22 November 2011).
7. 'Orkney protests saved the *Royal Oak* from scrap', 24 January 2008, *The Orcadian*, 20.
8. 'Cover marks *Royal Oak* 60th anniversary', 14 October 1999, *The Orcadian*, 10; '*Royal Oak* remembered', 21 October 2009, *The Orcadian*, 1.
9. '*Oak* book stirs up hornet's nest', 6 April 2006, *Orkney Today,* 1, 3.
10. Interview with Herbert E. Pocock, HMS *Royal Oak* survivor, by the author, 11 October 2009.
11. 'ODIN group's first meeting', 27 November 2008, *The Orcadian*, 34.
12. Interview with Anne Billing, president of Orkney Defence Interest Network, by the author, 15 October 2009.
13. 'Planes that were lost . . . but not forgotten', September 2010, *Living Orkney*, 58, 34–8.
14. Personal correspondence with David Earl, Aviation Research Group Orkney and Shetland founder, 2 March 2011.
15. 'Scapa Flow Heritage Project in place', 23 August 2009, *Orkney Today*, 2. For more information, see www.orkneycommunities.co.uk/SCAPAFLOW/members.asp
16. Interview with John Muir, Chapel Preservation committee, by the author, 2 July 2010.
17. Muir was at one point nominated for Orkney Citizen of the year for his efforts in making Orkney's heritage available to the general public.

18. For more on the Churchillian paradigm, see Sanders (2005) *The British Channel Islands Under German Occupation 1940–1945*, Jersey: Société Jersiaise/ Jersey Heritage Trust.
19. 'Italians return for ceremony to mark barriers anniversary', 11 May 1995, *The Orcadian*, 21.
20. 'Building the Barriers—Tankerness House Museum', 4 May 1995, advertisement, *The Orcadian*, 21.
21. 'Orkney Tribute to POWs who became part of the family', 1 April 2002, *The Scotsman*. Online. Available HTTP: http://news.scotsman.com/inverness/ Orkney-tribute-to-Italian-PoWs.2314848.jp (accessed 12 June 2011).
22. A number of people have been responsible for these tours over the years. One large tour of over 100 people was conducted by Geoffrey Stell for the sixtieth anniversary in 2005. As of 2010, Anne Bignall had taken the reigns, conducting photographic excursions. See '60 years on: The Churchill barriers re-visited Excursion Guide', May 2005, Orkney Archive Reference: D1/959/1, and 'What's on in nature, archaeology and heritage', Orkney Community Environment Awareness Network, October/November 2009.
23. 'Blether by Bumpkin', 16 December 2004, *The Orcadian*, 18.
24. 'Captive memories at the chapel', 11 June 1992, *The Orcadian*, 2–3.
25. 'Chapel premier for "Domenico"', 20 July 2000, *The Orcadian, 5.*
26. 'No barriers in the way of pure brilliance at festival launch', 27 June 2002, *The Orcadian,* 23.
27. 'Moena link will be strengthened by choir visit', 7 March 2002, *The Orcadian*, 5.
28. The HMT *Dunera* from Liverpool to Australia was particularly well known for the cruelty that British guards could inflict on their Italian captives. The HMT *Arandora Star* was a ship bound for Canada full of Italian internees and German POWs which was sunk in July 1940 with the loss of over 700 lives. It was a tragedy for the Italian community in Britain, very few of whom were not personally affected by the loss. For more on the *Arandora Star*'s effects, see Colpi (1991, 115–23).
29. There is another very small chapel in Henllan, West Wales, which was not destroyed after the war. Abandoned until the 1970s, the Henllan chapel is not nearly as well known or as well visited as the one on Lamb Holm.

BIBLIOGRAPHY

Colpi, T. (1991) *The Italian Factor: The Italian Community in Britain*, Edinburgh: Mainstream.

Connelly, M. (2004) *We Can Take It! Britain and the Memory of the Second World War*, London: Longman.

Cormack, A. (1992) *Bolsters, Blocks, Barriers: The Story of the Building of the Churchill Barriers in Orkney*, Kirkwall: Orkney View.

Dorman, J. (1996) *Orkney Coastal Batteries 1914–1956*, Kirkwall: Twin Six Production.

Gaffney, A. (2007) '"The Second Armageddon:" remembering the Second World War in Wales', in M. Cragoe and C. Williams (eds), *Wales and War: Society, Politics and Religion in the Nineteenth and Twentieth Centuries*, Cardiff: University of Wales Press, 184–203.

Hellen, J. A. (1999) 'Temporary settlement and transient populations: the legacy of Britain's prisoner of war camps 1940–1948', *Erdkunde*, 53, 191–217.

Hewison, W. S. (2005 [1985]) *This Great Harbour: Scapa Flow*, Edinburgh: Birlinn.

Jackson, S. (2011) *Churchill's Unexpected Guests: Prisoners of War in Britain in World War II*, Stroud, Gloucestershire: History Press.
Kushner, T. (2004) *We Europeans? Mass-Observation, 'Race' and British Identity in the Twentieth Century*, Hampshire: Ashgate.
Lamb, G. (1991) *Sky over Scapa: 1939–45*, Kirkwall: Byrgisey.
MacDonald, J. (1992) *Churchill's Prisoners: The Italians in Orkney 1942–1944*, Kirkwall: Orkney Wireless Museum.
Miller, J. (2000) *Scapa: Britain's Famous Wartime Naval Base*, Edinburgh: Birlinn.
Moore, B. (2005) 'British Perceptions of Italian Prisoners of War, 1940–1947', in B. Moore and B. Hately-Borad (eds), *Prisoners of War, Prisoners of Peace: Captivity, Homecoming and Memory in World War II*, Oxford: Berg, 25–40.
Paris, P. (2010) *Orkney's Italian Chapel: The True Story of an Icon*, Edinburgh: Black and White.
Sanders, P. (2005) *The British Channel Islands Under German Occupation 1940–1945*, Jersey: Société Jersiaise/Jersey Heritage Trust.
Schei, L. K. and Moberg, G. (1985) *The Orkney Story*, New York: Hippocrene.
Smith, D. J. (1989) *Action Stations 7: Military Airfields of Scotland, the North East, and Northern Ireland*, Northamptonshire: Patrick Stephens.
Smith, M. (2000) *Britain and 1940: History, Myth, and Popular Memory*, London: Routledge.
Sponza, L. (2006) 'Italian immigrants in Britain: perceptions and self perceptions', in K. Burrell and P. Panayi (eds), *Histories and Memories: Migrants and Their History in Britain*, London: Tauris, 57–74.
Turner, D. (2009) *Last Dawn: The* Royal Oak *Tragedy at Scapa Flow*, Glendaruel: Argyll [first published as *The Ultimate Sacrifice* in 2004].
Ugolini, W. (2004a) 'Communal myth and silenced memories: the unremembered experience of Italians in Scotland during the Second World War', in W. Kidd and B. Murdoch (eds), *Memory and Memorials: The Commemorative Century*, Aldershot: Ashgate, 151–66.
Ugolini, W. (2004b) 'The internal enemy 'other': recovering the World War Two narratives of Italian Scottish women', *Journal of Scottish Historical Studies,* 24 (2), 137–58.
Ugolini, W. (2008) 'Memory, war and the Italians in Edinburgh: the role of communal myth', *National Identities,* 8 (4), 421–36.
Ugolini, W. (2011) *Experiencing War as the 'Enemy Other': Italian Scottish Experience in World War II*, Manchester: Manchester University Press.

7 *"Tingbaot Wol Wo II Long Pasifik Aelan"*

Managing Memories of WWII Heritage in the Pacific

Keir Reeves and Joseph Cheer

INTRODUCTION

> Regarding the native peoples, whether near the front or rear bases, there had never been anything before of such strangeness and magnitude.
>
> (Bennett 2009, 7)

As Judith Bennett's quotation observes, the impact of the Pacific War on Pacific Islanders was a unique event that permanently changed patterns of life throughout the region. This chapter investigates and analyses the present-day cultural heritage legacy issues associated with WWII. It emphasizes the importance of Islander perspectives of the conflict and how their heritage is remembered and commemorated at the present time. It also considers how official heritage dialogues commemorate the war. For many, the memory of the Pacific War, as well as in turn attention to the remaining heritage, is fading and in some instances is no longer relevant. Seemingly there are competing trends regarding commemoration of war heritage or, increasingly, forgetting it. Many Pacific Islands exhibit longevity in nurturing their war memory and heritage. An indicative example is the Solomon Islands, where there has been a recent spate of commemorative events and public sculptures to honour the coastwatchers. Yet other Pacific Island societies place less and less value on their war heritage as time goes by. This tension between remembering and forgetting is primarily associated with the Pacific War's intangible heritage.

However, there is also extensive physical, particularly built heritage of the war era that lies scattered throughout the former conflict sites of the Pacific (see Figure 7.1 for the key sites discussed in this chapter). Many sites continue to be uncovered by maritime and historical archaeologists in the opening decades of the twenty-first century. Vanuatu, Micronesia, and the Solomon Islands all have extensive remnants of war in various stages of ageing that provide an archaeological record of the Pacific War. This includes key sites such as the maritime warscape of Guadalcanal in the Solomon Islands, a site of some of the bloodiest fighting of WWII. The lesser known Truk lagoon in Micronesia, the site of Japan's main South Pacific base during the war, includes over fifty shipwrecks and extensive aircraft. These sites are emblematic of the enduring legacy of war. There, dozens of naval vessels belonging to both the Allied and Japanese naval forces, along with scores of unexploded ordinances, lie submerged. The wreckage of the troopship the USS *President Coolidge* at Espiritu Santo in Vanuatu is another remnant from the war that is an enduring reminder of the conflict, as is nearby Million Dollar Point. These subsea vestiges of WWII that serve as prominent reminders are matched on land by rusting tanks and aircraft bodies that lay alongside monuments that have since been erected as memorials to Allied and Japanese soldiers. Throughout the Pacific, this war heritage includes extensive military matériel and ordinance, as well as altered former warscapes of the conflict (for example, see the war era artefacts from the Solomon Islands in Figure 7.2). The physical legacy of the war in Melanesia is also demonstrated from the almost pristine artillery batteries at Momi Bay in Fiji, to the Kokoda Track in Papua New Guinea, Pearl Harbor and the veneration of John Frum in Tanna situated in the southern region of the Vanuatu archipelago.

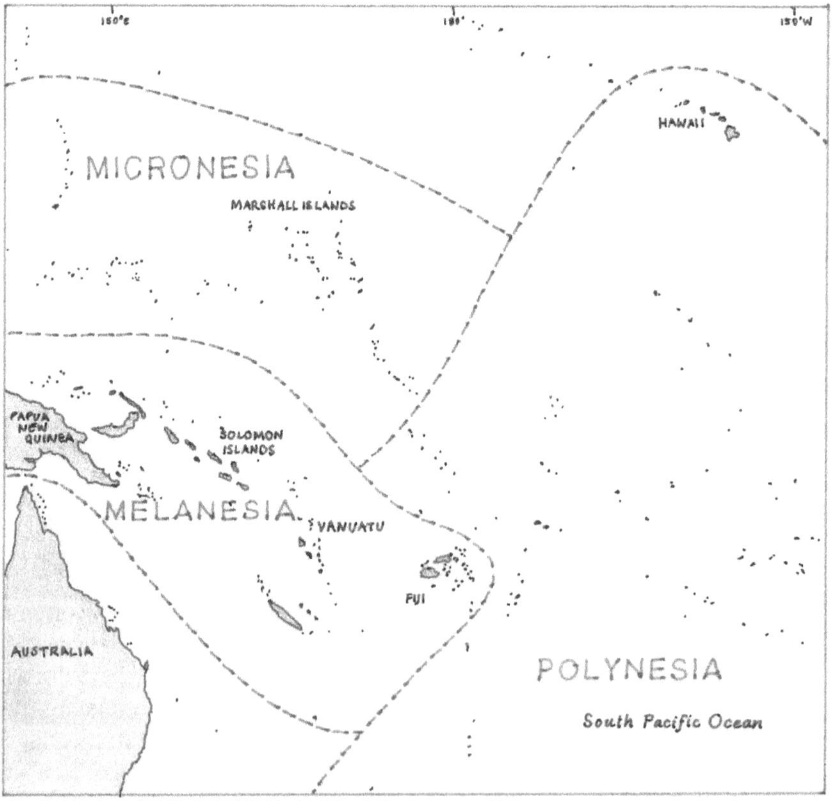

Figure 7.1 Map of Western Pacific Theatre of War (Map copyright Antoinette Dillon).

Figure 7.2 WWII-related ephemera, Solomon Islands (Image copyright Stefan Krasowski).

The focus of this chapter is on the present-day memory and heritage associated with WWII in the Pacific (also known as the Pacific War). In it we examine the extent to which the conflict continues to resonate and intersect with contemporary Pacific Island societies. The central question posed in Bislama is *Wannem yufala tingting long Wol Wo Two*?[1] Which, when translated into English, means, "What does World War Two mean to you?" This question gauges local-level attitudes towards the war in the present day. This is a critical and timely question about the politics of heritage and memory of the Pacific War given recent geopolitical tussles for regional authority amongst the former combatants (particularly Australia, America and Japan) and others (notably China). It also raises further the question about whether alliances forged during WWII hold any currency in present diplomatic dealings. Therefore, this research sets out to examine the nuanced, local-level personifications of war.

Vanuatu, Fiji and the Solomon Islands are part of the Melanesian cluster of archipelagic nations that share significant commonalities, histories and understandings. While the Solomon Islands group was by far the only one of the three to have directly experienced the full intensity of war, Vanuatu and Fiji, to a lesser extent, bore witness to the build-up of the war arsenal and construction of wartime infrastructure (housing, artillery batteries, bunkers and hangars). Most importantly, it was the presence of soldiers and the commerce that came with their arrival that was to have marked effects on the way Islanders perceived themselves and on the futures they came to conceptualize and desire. As Firth outlines, 'Islanders had been changed by the war' and 'colonies were now seen not as permanent additions to the imperial realms but as territories set upon paths of development that would ultimately lead to self-government or independence' (Firth 1997, 309).

In the small island countries of Micronesia, the impact of war was felt on a much larger scale than in Melanesia as Micronesia's land mass and population size were considerably smaller. Micronesia's proximity to Japan and Hawaii meant that it was at the epicentre of the conflict and in turn quickly felt the reverberations of war that emanated from the Japanese mainland and the American response following the bombing of Pearl Harbor in December 1941. According to Firth, Japan's determination and stubbornness to defend Micronesia were unflinching: 'the Japanese fought tenaciously, and the island sites of bloody battles have passed into American and Japanese folklore' (Firth 1997, 296). Much of the war activity in Micronesia stretched across the vast expanses of the region from the Marshall and Gilbert island groups in the east, to Palau in the west and to Saipan, Guam and Truk in the centre. For the Japanese Imperial Army, Micronesia quickly became the base from which their foray into the greater Pacific Islands region was launched. As Falgout, Poyer, and Carucci (2008, 17) point out, '[r]ealising the need for a defensive posture in their Micronesian territory, Japan poured in reinforcements and hastily bolstered island installations'. This naturally brought about a concerted focus by the Allied forces to destabilize Japanese control resulting in prolonged and widespread fighting and the forced internment of Islanders by the Japanese to aid the war effort (the physical legacy can still be readily observed in the present day at sites such at Truk Lagoon; see Figure 7.3).

We argue that both intangible and built heritage needs to be considered in conjunction. This is not just to gauge war heritage throughout the region but also to examine how this heritage is remembered in the present day. In doing so, we contend that war remembrance and cultural heritage management of conflict commemoration sites constitute an ongoing process that is best understood as part of a continuum of the experience of colonialism in the Pacific. For the purposes of this volume, the Islands of the Pacific region are all islands of war; however, increasingly the war memories and heritage associated with each one are understood through social, political and economic priorities of the present day.

This coupling of colonial heritage and war heritage is under increasing threat from the desire of contemporary Pacific Islanders to build on and emphasize the 'Melanesian way'. The Melanesian way of life is a multidimensional construct concerning the respect for customary lore and the epistemologies, cosmologies and cultural underpinnings that govern the way people should live. A key proponent of the Melanesian way, Bernard

Figure 7.3 Tank on the deck of the wreck of the *Nippo Maru*, Truk Lagoon, Micronesia (Image copyright Geoff Sparkes).

Narakobi (1980) argues that to demarcate what this is undermines its very nature given the multiplicity of meanings and presentations. In a sense, the Melanesian way is a reaction against all of the exogenous influences that have impinged on Melanesians and appeals for the reaffirmation of their agency. Colonization, Christianity and WWII stand as events that have had the most profound impact on Melanesians, and reaffirming autonomy, voice and assuredness underlines notions of the Melanesian way. This is a phenomenon that is explicitly geared towards promoting indigenous Pacific Islander cultural heritage and ideologies. It is in contrast to external interpretations of WWII heritage that continue to emphasize the history of the overarching international narrative of the Pacific Theatre as part of that war. In doing so, these Western views understandably examine the heritage sites of conflict and the commemoration of the war predominantly from an Allied historical viewpoint. Tellingly, the Japanese perspective on Pacific war heritage remains largely muted.

In *Made in Oceania: Social Movements, Cultural Heritage and the State in the Pacific*, Edvard Hviding and Knut Rio argued that 'throughout the Pacific, local cultural heritage has been a central element in political innovation in and beyond the local' (Hviding and Rio 2011, 325). There are many reasons for this apart from the fact that much war heritage is either

Japanese, Allied or indeed nontraditional heritage, all of which is of questionable cultural and heritage significance to Pacific Islanders. The proliferation of private museums for tourism in the Solomon Islands, whose collections predominantly feature random war ephemera (for instance 1940s-era Coca-Cola bottles), leftover military hardware (shell casings, helmets, boots, vehicles and other war-related matériel) and other collectables is emblematic of the commoditization of the remnants of war. This suggests that the value of war heritage lies in its commercial potential much more than in its historical significance. Kiko suggests that unless war heritage has commercial value, encouraging local-level compulsion to preserve and nurture its conservation will become increasingly problematic (Kiko 2014).

Any discussion of the historical legacy and site management issues of WWII Pacific Island locations invariably emphasizes the enduring significance of the Allied conflict with the Empire of Japan. As in many other Islands impacted by WWII, the conflict remains an unprecedented event in the Pacific. Indeed Lindstrom's discussion of war heritage and its present-day meanings emphasizes that war heritage 'convey[s] people's enthusiasm in cultivating these' (Lindstrom, Chapter 9 this volume). While Lindstrom's close reading of Efate raises important issues regarding cultural memory and practical implications, such as management of former sites and artefacts, it is important also to recognize that not all indigenous Pacific Islanders in the present day are aware of or indeed culturally engaged with the war in the Pacific. Therefore, for many, it is important to Pacific Islanders only if it confers some material recompense. Accordingly, war heritage is important for Pacific Island communities but not necessarily for the lofty anthropological or historiographical reasons that are conferred upon it by external researchers or for the heritage values ascribed by UNESCO or the respective national governments. This heritage significance may be the case to an extent, but ultimately the embrace of Pacific War heritage in island communities such as Vanuatu, the Solomon Islands, Papua New Guinea and in Micronesia is determined by potential financial opportunities. Hawaii enjoys relatively greater funding support for war heritage management and interpretation. Accordingly, the conservation and commemoration of sites such as Pearl Harbor are not subject to the same financial pressures.

WWII was unparalleled in the way it thrust the small island nations of the northern and south-western Pacific headlong into the global spotlight (Lindstrom and White 1994). Bennett's close examination of how the arrival of warring parties and their military paraphernalia was received by Islanders highlights the legacy of the transformative impact of the war in the present day. The appearance of aircraft, the divisions of men, the construction of bases such as Truk (a Japanese base colloquially known as the Gibraltar of the Pacific) and the Allies' logistical and supply base in Espiritu Santo in Vanuatu for the most part was received with a mix of wonderment, awe and apprehension (Lindstrom, Chapter 9 this volume; Lindemann 1982, 20–2). The onset of war magnified the sense of fear and bewilderment felt by Islanders who had little reason to comprehend an event of such magnitude. Bennett maintains that the war expedited processes of change that were already in process and that would come to have a profound impact on Islanders' everyday life and culture (Bennett 2009). In particular, Bennett argues that '[i]n Melanesia, a greater awareness of the cash economy emerged, partly as a result of widespread employment of native labour and sale of curios and local produce' (Bennett 2009, 293). One distinctive long-term legacy of WWII that remains prevalent throughout the Pacific Islands has been the monetization of previously custom-focused societies.[2]

Perhaps the war was the catalyst that provided fertile grounds for what ethnographer and anthropologist Marshall Sahlins termed 'developman', where Pacific Islanders unambiguously articulated their notions of development, monetization and integration into globalized systems (Sahlins 2005). Similarly, Firth has argued that 'the war heralded the coming of developmental colonization' suggesting that the most enduring impression was that of the presence of soldiers and of the extensive physical damage that was wrought on some islands, as well as the way this enforced a sudden and sharp departure from the placidity of Islanders' everyday life (Firth 1997, 322).

> The war fought over their beaches, islets, lagoons, mountains and swamps had a profound impact on many people. Armies brought destruction, disorder and massive loss of life.
>
> (Firth 1997, 323)

Firth's visualization of the way war was manifested in the island landscape is evocative of the massive disjuncture that was created. The question of Islander agency, sovereignty and determination over whether such a conflict was welcomed is to a large extent a moot point given that 'Islanders were caught in a vast conflict that was not of their making' (Firth 1997, 323). For Islanders throughout the Pacific, it was they who bore the full brunt of this conflict, and indeed its impact was unprecedented and catastrophic. Although spared the experience of direct bombardment and fighting and instead hosting support bases, Islanders in Fiji and Vanuatu were still awed by the magnitude of troops, equipment and supplies which they had never before encountered. For the combatants, war was ostensibly linked to the enlargement and securitization of colonization, empire building and influence in the region. All of these developments have had a considerable influence on Pacific Island nations and their peoples. As Nicholas Thomas points out, 'Empire was powerfully suggestive, it prompted people to rethink themselves, their lives and their future' (Thomas 2010, 297). While Thomas was writing about the colonial encounter, many legacies of WWII can be understood in the context of Thomas's observations of an earlier imperial era.

The manner in which war memories proliferate alongside unfolding modern modalities of life in Melanesia complements Mary Patterson and

Martha Macintyre's argument that 'while the production of modernity is multiple, its invariable set of ruptures in relation to personhood, property, exchange and so on belie heterogeneity at the source of modernity itself' (Patterson and Macintyre 2011, 3). Indeed, while war left immediate and marked impressions in some locales, in others the war was just another event in a line of milestones, none better or worse than the other—just different and sometimes forgotten.

THE TREE AND THE CANOE: METAPHORS OF WAR

The twin notions of the tree and the canoe, drawn from the francophone insights of Joel Bonnemaison's time spent in Tanna, Vanuatu, characterize the two pillars of Islander socialities that have influenced the manner by which the war was experienced and also the way it has been interpreted and memorialized (Bonnemaison 1984). For Bonnemaison, the tree represents rootedness, connection with country and a sense of permanence. The canoe, on the other hand, is indicative of mobility and adaptiveness to changing circumstances (a metaphorical feature of all Pacific Island societies). We apply the metaphors that Bonnemaison identified to a war heritage and memory analysis because when considered together, the tree and the canoe characterize the way war upended the social solidity that was emblematic of the tree and rendered Islanders immobile for the duration of the conflict. The cultural and economic linkages that were driven by inter-island trade and movement were curtailed. In turn, this meant that the impact of the Pacific War rendered both the tree and the canoe, as social reference points and cultural motifs, redundant.

The metaphors also speak of the way tradition and modernity can benignly coexist, clash or simply diminish over time. For Bonnemaison, 'the tree symbolises connectedness to place: the man who lives within his place and who stands straight will take root along with the tree', while the canoe embodies community; its 'destiny is to circulate, to go beyond the tree, to move from place to place and island to island, wherever its roads lead it' (Bonnemaison 1984, 321–2). Community memories emphasizing WWII dismembered the tree and disoriented the canoe. As Firth observes, '[F]or some people the coming of Japan or America fulfilled prophecies', while for others it brought deliverance 'from the oppression of colonial governments' and set in train future moves towards independence (Firth 1997, 316). Similarly, Evelyn Colbert regards the war as having a long-term impact:

> When the war ended for the rest of the world, the islands returned to the obscurity from which they had so briefly emerged. But the war's impact on the islands helped to shape their future.
>
> (Colbert 1997, 33)

In a sense, postcolonial independence developments had their nascent origins during this era and can be understood as another subtle legacy of the conflict.

Colbert argues that the historical implications of this era are marked and that the impact of the war was paradoxical in that although it generated a 'new prosperity'; at the same time, the island 'populations were uprooted' (Colbert 1997, 34). WWII, or *Bikfala Faet* (literally translated from Solomon Islands Pijin or Vanuatu Bislama meaning 'Big Fight') has in many ways left lasting legacies on Islanders and their domain (White, Gegeo, Akin, and Watson-Gegeo, 1988). This is redolent in the ongoing memorialization of the war, especially in transmuted, old and new oral histories and the presence of war tourists and pilgrims such as those along the Kokoda Track. It is visible in the widespread evidence of wartime, including the remains of military armaments strewn across the island landscape—the rusting remains of gunneries, tanks, aircraft and ships. It can also be observed through the permanent alterations to island topographies as a result of bombardment and conflict. Entire plantations were razed, coral reefs destroyed and hills levelled.

The heritage, historical narratives and most of the commemoration practices of WWII throughout the Pacific are ostensibly removed from Islanders in that they have originated from elsewhere and belong to a conflict that they hosted but for the most part have come to emphasize non-Islander contributions. There is little doubt that victory in the Solomon Islands, particularly the battle for Guadalcanal, was pivotal to Allied war efforts in the Pacific Theatre. As Belshaw observed, 'The Solomon Islands, although small in area and economically unimportant, proved to be of considerable strategic value to the United States, Australia, and New Zealand during the last war' (Belshaw 1948, 95). The intensity of war in the Solomon Islands was most evident in Sealark Channel, the body of water in which clusters a number of islands including Guadalcanal, Savo and Florida Islands and which has since become known as Iron Bottom Sound on account of the dozens of ships that lay on the sea floor. To characterize the experience of war in the Solomon Islands is to speak of the intensity of fear and uncertainty. Solomon Islander Jan Sanga enunciated the sense of apprehension and nervousness that developed as war took hold:

> There was much bombing and shelling at the time. The Americans fired shells at the Japanese at Lunga and along the shore, and the Japanese fired back at the Americans. Aeroplanes dropped bombs which exploded and caused much confusion and fear among the people.
>
> (Alasia and Laracy 1989, 24)

The extent to which Pacific Islander deaths ensued as a result of WWII is unclear. Equivocation of absolute numbers is likely to be inexact at best and misleading at worst. The reason for this is that unlike Allied and Japanese casualties, for whom a census of combatants was accessible to base assumptions on, for many Islander communities no such census was available. Indeed, many fled the scene of combat while others who remained were engaged in assisting the efforts of either the Japanese or the Allied Forces. Inevitably and perhaps unavoidably, Islanders were 'absorbed by the war effort, participating in its martial economy as scouts, soldiers, coastwatchers and workers' (Lindstrom and White 1989, 30). As Lindstrom and White explained, 'Many were caught up and transformed by supporting the Japanese or the Allies as workers and fighters' (Lindstrom and White 1989, 30). That 'labour was essential to the war effort' is widely acknowledged, and this point is made ever more important given that 'the native troops had the advantage of environmental knowledge' (Bennett 2009, 133, 137).

The recollections of George Maelalo, a Solomon Islander in service with the Allies during the war, capture the sense of loss that was felt: 'After the war was over I sadly thought about the many young men who died'. Indeed, while specific numbers are difficult to ascertain, anecdotally it seems the case that 'hundreds if not thousands of casualties resulted from the bombing of villages' (Lindstrom and White 1989, 24; Bennett 2009; Firth 1997). In many other cases, Islanders enlisted in the war effort as labourers with zeal, keen for adventure and cognizant of the financial, material and status-related rewards of doing so. Thus, it seems that the best estimate of Islander casualties lies in examining the numbers who had enlisted to fight as soldiers or as de facto soldiers. Both Bennett and Firth also draw links to wartime deaths that were the result of retaliatory actions or vendettas against local rivals or of confrontations with fighting Islanders who had enlisted to fight with the Japanese (Bennett 2009; Firth 1997). Further, many Islanders forcibly evacuated or fled their villages and islands, in many cases never to return. These should be acknowledged in the tally of Islander fatalities. The challenge of substantiating the Islander war deaths is outlined by Coulter who argued soon after the war that 'the populations of hundreds of villages in the Solomons and New Guinea are decimated and the survivors dispersed' (Bennett 2009; Firth 1997). Perhaps Coulter's statement offers the best guidance to the gauging the extent of Islander deaths: 'At this time one can only generalize on the effects of the war on the people of the Pacific islands' (Bennett 2009; Firth 1997.) With the passing of time almost seventy-five years after the conflict, this remains the case.

Establishing how pronounced WWII has been in terms of its influence over life in the present day is difficult. Notwithstanding, it is clear that in some cases war has left an enduring mark. For example, Bennett argues that '[t]he war catalyzed a range of changes, many relating to the use of environment' and 'expanded demand for exogenous manufactured goods, creating dependency' (Bennett 2009, 293). These changes imply that the war enforced a reframing of how natural and cultural resource usage was viewed. This included the commercialization of aspects of the traditional economy and how customary and non-customary groups sought to maximize their use. Bennett also contends that war intensified local-level contests for resources use in places such as Bougainville. Interestingly this is a trend that continues in the opening decades of the twenty-first century (Bennett 2009).

For the people of Vanuatu, their involvement with the war emphasizes their role as providing a logistical and supply conduit for the assault on the Japanese Imperial Army. This was principally concentrated on the northern island, Espiritu Santo, which served as a launching pad for American air and sea assaults on the established Japanese strongholds in New Guinea and the Solomon Islands and as support for the effort elsewhere in the region, including Truk and Hawaii. As a consequence of its proximity to the Australian mainland, the New Hebrides (now the Republic of Vanuatu) stood as the final staging post for the Japanese naval assault that paved the way for advancement to Australia. As Earl Hinz outlines, the New Hebrides were in the 'right place, at the right time', and 'the South Pacific war's major supply depot and staging base was to be hacked out of a jungle and made into an instant supermarket of war goods and services in an area most of the world had never heard of' (Hinz 1995, 54).

IMPACT OF WWII IN THE PACIFIC

War literally transformed the physical environs of islands that experienced heavy combat, and in places such as Iron Bottom Sound, the remains of unexploded ordinances continue to be found. Coulter observed after the war that 'coconut trees have been reduced to gaunt trunks and tattered leaves by artillery fire and bombing, and plantation lands are riddled with shell holes' (Coulter 1946, 409). To characterize the legacy of WWII in the Pacific Islands region as a homogeneous encounter elides the reality that the experiences across Melanesia, Micronesia and Polynesia differed considerably. For the most part, the legacy of war was largely dependent on geographic proximity to the zones of conflict. The countries of Polynesia with the exception of Hawaii were largely spared because of their remote location in the South Pacific in relation to key zones of conflict. However, the close proximity of Melanesia and Micronesia, respectively, to the Pacific Theatre meant that for them, there was no escaping active participation and first-hand impressions (Lindstrom and White, 1989). As Morrell explained, WWII 'had a more unsettling effect on island life than any event since the first coming of the European; and this unsettlement by no means ceased when the Japanese were driven out and the British and American forces moved away' (Morrell 1960, 440). To suggest that the consequences of the war was in some ways lessened or heightened in either Melanesia or Micronesia is fraught given that in both cases, the tangible and intangible consequences were monumental and unprecedented.

In the Melanesian case, New Guinea and the Solomon Islands shouldered much of the burden of WWII. It was here that the carnage and disturbance to island life was at its zenith—especially on the mainland and coastal areas of the north-eastern flanks of New Guinea and at Guadalcanal in the Solomons. In New Guinea, the township of Rabaul on New Britain Island encountered much of the fierce fighting. As Hinz points out, 'Rabaul offered great strategic value to Japan's World War Two drive into the southwest Pacific' (Morrell 1960, 440). Central to the Allied forces winning the Pacific campaign was the capture of Rabaul because so much of the Japanese effort was dependent on maintaining control over the township that provided an essential supply line to the Japanese front line. Consequently the township encountered severe bombardment from Allied forces, stoking the fears of island residents and resulting in marked changes to and long-term scarring of the island landscape. This war heritage is now insignificant as the eruption of the Rabaul caldera destroyed the old precinct of the town in 1994. Accordingly, war memories, not physical heritage, are the only ongoing legacies associated with WWII.

Intense fighting also occurred on the New Guinea mainland especially following attempts to stave off Japanese advances toward Port Moresby that may have had disastrous effects for the Allied push. Firth outlines that the Japanese New Guinea campaign was driven by 300,000 troops, of which only 127,000 survived (Firth 1997, 296). Such large numbers stretched local resources across the Pacific Islands region, especially food and water resources. Firth suggests that the Palau case was emblematic of how Islanders struggled during the war years, when '50,000 Japanese competed with 5,000 Palauans for food and survival' (Firth 1997, 297). Preventing the penetration of the Owen Stanley Range by Japanese forces proved to be critical in not only securing New Guinea for the Allies but also in preventing the serious breaching of Australian borders.

Wol Wo II: hemi gud or no gud?[3]

Colbert made the critical point that 'islanders were not merely the victims of hostilities; many became active participants', perhaps drawn to the possibility of adventure and reward, and mostly unacquainted to conflict of such magnitude and complexity (Colbert 1997). Additionally, Colbert observed that 'out of these new experiences ideas developed about future relations between islanders and Europeans' (Colbert 1997, 35). We concur and argue also that Islander agency during the war and memories about the war are more ambiguous than commonly held views that depict Islanders as inanimate and disengaged or as loyal and compliant to Allied or Japanese agendas. So in discussing the heritage and memory of the Pacific War, it is important to be wary of Western academic discourses of heritage and memory of the war that can render Islander views on their role in the Pacific War, and in turn their war heritage, marginal.

CONCLUSION

This chapter has drawn on long-term ethnographic, archival and observational fieldwork in the Pacific Islands with an emphasis on Melanesia and Micronesia. While some observations and insights were gleaned from throughout the Western Pacific region, this chapter has emphasized the Ni-Vanuatuan and Solomon Islander experience.[4] This has concentrated on the islands of Efate, Espiritu Santo and Tanna where the presence of soldiers and military hardware was most apparent and on Guadalcanal where the fighting was the heaviest in the Pacific War and has left an extensive archaeological legacy of the conflict.

In this chapter we have explored the ambiguous and fluid nature of how WWII heritage is remembered, managed and indeed understood in the present day. Pacific Islands are often typified by tranquil settings of idyllic tropical beauty. However, many of the former conflict sites such as Guadalcanal, Truk and Palau are sites of heavy conflict and ultraviolence where some of the bloodiest conflicts of WWII occurred. Interestingly these locations continue to remain the most socially and economically vulnerable in the Pacific Islands region. A minor but related point that must be raised is that in the early twenty-first century, there is now a clear sense that the United States and Australia no longer dominate Islander regional agendas and that accordingly there has been a diminution of interest in war heritage throughout the Pacific region. For some, war commemoration is no longer as relevant as it once was throughout the region, and the bulk of the war debris from the Allied effort is not curated and remains strewn across the warscapes and jungles where it is souvenired and utilized as required. The question that arises for Islanders is why should they conserve and curate this war matériel? Ultimately there is ambivalence amongst locals, who largely feel that unless war heritage has a commercial imperative, it is unlikely that there is the capacity or interest to preserve it. This is irrespective of the high level of heritage significance attributed to WWII war heritage by former combatant countries such as America, Japan, Australia and New Zealand.

ACKNOWLEDGEMENTS

The authors acknowledge the assistance of Dr Elizabeth Cory-Pearce, Merian Numake, Antoinette Dillon, Kirk Huffman, Geoffrey White, Marcelin Abong, Peter Harvey, Geoff Sparkes, Thomas Bong and the Vanuatu Cultural Centre, the McDonald Institute for Archaeological Research Cambridge and the Cambridge Heritage Research Group. Fieldwork for this chapter was supported by the Australian Research Council Linkage Projects Scheme, the Monash Fellowship Scheme and the Faculty of Arts, Monash University.

NOTES

1. Bislama is the local Creole used widely in Vanuatu. It is akin to Tok Pisin in Papua New Guinea and Solomon Island Pijin and has its roots in the colonial era from the interactions between islanders and 'Europeans' or '*waet man*' (white man).
2. Custom, or *kastom*, lifestyle is a traditional cultural life and practice throughout Melanesia. It forms a cornerstone of Melanesian, indeed Pacific Islander, cultural identity. Custom social practice is often presented as conflicting with modern, Western-oriented modalities of life throughout the Pacific.
3. Translates from Bislama into English as, 'World War Two, was it good or bad?'
4. Ni-Vanuatu is a demonym that refers to people who come from the Pacific Island nation of the Republic of Vanuatu. It strictly refers to people of Melanesian extraction.

BIBLIOGRAPHY

Alasia, S. and Laracy, H. (1989) *Ples blong iumi: Solomon Islands, the Past Four Thousand Years*, Suva, Fiji: Institute of Pacific Studies, University of the South Pacific.

Belshaw, C. (1948) 'The postwar Solomon Islands', *Far Eastern Survey*, 17 (8), 95–8.

Bennett, J. A. (2009) *Natives and Exotics: World War II and Environment in the Southern Pacific*, Honolulu: University of Hawaii Press.

Bonnemaison, J. (1984) 'Social and cultural aspects of land tenure', in P. Larmour (ed), *Land Tenure in Vanuatu*, Suva: Institute of Pacific Studies, University of the South Pacific.

Colbert, E. (1997) *The Pacific Islands: Paths to the Present*, Boulder, CO: Westview Press.

Coulter, J. W. (1946) 'Impact of the war on South Sea islands', *Geographical Review*, 36 (3), 409–19.

Denoon, D., with Firth, S., Linnekin, J., Melieisea, M. and Nero K. (eds). (1997) *The Cambridge History of the Pacific Islanders*, Cambridge: Cambridge University Press.

Falgout, S., Poyer, L. and Carucci, L. (2008) *Memories of War: Micronesians in the Pacific War*, Honolulu: University of Hawai'i Press.

Firth, S. (1997) 'The war in the Pacific', in D. Denoon, with Firth, S., Linnekin, J., Melieisea, M. and Nero K. (eds), *The Cambridge History of the Pacific Islanders*, Cambridge: Cambridge University Press, 291–323.

Hinz, E. (1995) *Pacific Island Battleground of World War II: Then and Now*, Honolulu: Bess Press.

Hviding, E. and Rio, K. (eds). (2011) *Made in Oceania: Social Movements, Cultural Heritage and the State in the Pacific*, Wantage, UK: Sean Kingston.

Kiko, L. (2014) 'Preservation of WW2 sites and WW2 materials in Solomon Islands—a crisis,' paper presented at Oceanscapes: Cooperation Across the Pacific, Australian Association for Pacific Studies (AAPS) conference 2014, Sydney University, 22–6 April.

Lindemann, K. P. (1982) *Hailstorm over Truk Lagoon*, Singapore: Maruzen Asia.

Lindstrom, L. and White, G. M. (1989) 'War stories', in G. M. White and L. Lindstrom (eds), *Pacific Theater: Island Representations of World War II*, Honolulu: University of Hawai'i Press, 3–40.

Lindstrom, L. and White, G. (eds), (1994). *Culture, Kastom, Tradition: Developing Cultural Policy in Melanesia*, Suva: Institute for Pacific Studies, University of the South Pacific.

Morrell, W. P. (1960) *Britain in the Pacific Islands*, Oxford: Clarendon.
Narakobi, B. (1980) *The Melanesian Way*, Boroko: Institute of Papua New Guinea Studies.
Patterson, M. and Macintyre, M. (2011) *Managing Modernity in the Western Pacific*, St Lucia: University of Queensland Press.
Sahlins, M. (2005) 'The economics of developman in the Pacific', in J. Robbins and H. Wardlow (eds), *The Making of Global and Local Modernities in Melanesia: Humiliation, Transformation and the Nature of Cultural Change*, Aldershot, Hampshire: Ashgate, 23–42.
Thomas, N. (2010) *Islanders: The Pacific in the Age of Empire*, New Haven, CT: Yale University Press.
White, G. and Lindstrom, L. (eds). (1989) *The Pacific Theater: Island Representations of World War II, Pacific Islands Monograph Series, No. 8*, Honolulu: Centre for Pacific Island Studies, School of Hawaiian, Asian and Pacific Studies, University of Hawai'i Press.
White, G.M., Gegeo, D., Akin, D. and Watson-Gegeo, K. (eds). (1988) *The Big Death: Solomon Islanders Remember World War II*, Suva: Institute of Pacific Studies, University of the South Pacific.

8 Malta G.C.

War Memories and Cultural Narratives of a Mediterranean Island

Sandro Debono

INTRODUCTION

On 15 April 1942, King George VI awarded the George Cross to Malta. The brief, albeit significant letter sent to the then governor of Malta articulates the King's decision to personally bestow this honour:

> To honour her brave people I award the George Cross to the Island Fortress of Malta to bear witness to a heroism and devotion that will long be famous in history.

A visual image of the citation is inscribed in marble and placed on the main façade of the former Governor General's Palace in Valletta (now the President's Palace), alongside a similar marble plaque featuring a copy of Franklin D. Roosevelt's illuminated scroll presented to the 'people and defenders of Malta' in 1943, as well as commemorative plaques for Malta's Independence (1964), for the Republic Constitution (1974) and for the closure of the British military base (1979). The prominence given to the George Cross citation letter, next to other marble plaques commemorating key episodes in Malta's twentieth-century history, is unquestionably commensurate with the relevance of the award of the George Cross as a public recognition. The public presence of the citation letter provided the island's inhabitants with direct access to the King's words, a constant reminder of why the George Cross was awarded to them. President Roosevelt's scroll citation, dated 7 December 1943, almost a year and a half after the George Cross award, is less succinctly written and articulates further the Allied tribute to Malta's war effort:

> In the name of the people of the United States of America, I salute the Island of Malta, its people and defenders, who, in the cause of freedom and justice and decency throughout the world, have rendered valorous service far above and beyond the call of duty. Under repeated fire from the skies, Malta stood alone, but unafraid in the center of the sea, one tiny bright flame in the darkness—a beacon of hope for the clearer days

> which have come. Malta's bright story of human fortitude and courage will be read by posterity with wonder and with gratitude through all the ages. What was done in this Island maintains the highest traditions of gallant men and women who from the beginning of time have lived and died to preserve civilization for all mankind.

In 1942, Malta was Allied frontier territory and, in consequence, under heavy siege by fascist Italy and the Luftwaffe based in Sicily. This was anything but a rare circumstance. Malta had been frontier territory to the advancing Ottoman Empire for centuries, particularly during the times of the Order of St John of Jerusalem. This military and hospitaller chivalric order, ruling Malta between 1530 and 1798, had effectively transformed the outpost island into a powerful fortress. The Great Siege of 1565, a four-month onslaught by a powerful Ottoman force which the knights and the islanders resisted heroically, stood as a heroic deed to emulate and celebrate. That was the event for which Malta was recognized as the bulwark of Europe (*Propugnaculum Europea*), and imagery, particularly maps, celebrated this heroic legacy (Freller 2009, 30). A frontier has been understood to be a place under threat (Brogini 2013), in need of standing armies and a strictly imposed reactionary political framework promoting otherness. Malta was such a territory on the European frontier, particularly during the sixteenth and seventeenth centuries when the Ottoman Empire had a dominant presence in the Mediterranean (Brogini 2006). Military installations were an ever pressing necessity, and no expenses were spared to keep the island's military installations ready to repel potential attacks. Symbols of chivalry, including heraldic escutcheons, clearly promoted allegiances and authority (Figure 8.1). The eight-pointed cross of the Order of St John still remains the symbol par excellence of this time.

WWII gave Malta a frontier status similar to that of the times of the Order of St John of Jerusalem, albeit not as protracted. This second frontier status was telescoped and condensed into a brief historic time frame during which time both colonizer and colonized had to fight a common enemy. Absolute dedication to the British cause was then considered a fundamental necessity, and the values which had stood for the frontier during the times of the Order of St John were now evoked in the collective subconscious. Much like the Great Siege of 1565, Malta then stood alone, surrounded by Axis-held territory, and endured constant aerial bombing. This frontier mindset explains the dialectic between Malta and WWII and how memory is perceived, acknowledged or otherwise in this strategic island.

King George VI's citation letter provides the reasons why the George Cross, a recently instituted insignia, was awarded to Malta. The letter, a handwritten holograph, suggests a directness which is not always the case with protocol, especially when compared to Roosevelt's more articulated citation. Rather than a token piece of rhetoric, this letter reflects what the British understood Malta to stand for at such a crucial point in history and

Figure 8.1 Map by S. J. Turner (From Ritchie [1942, 7] copyright expired).

its significance as an island fortress. It also defines a British view of the Maltese, whom the letter recognizes as devout and brave heroes. It does not distinguish between population and defenders as Roosevelt's citation letter does, even though both refer to universal values. George VI cites heroism and devotion as the values for which Malta stood; Roosevelt cites human fortitude and courage. Beyond the generic references to universal values, meanings can also be read into the careful selection of words and sentence structure. This would rethink the citation into a brief yet focused summary of Malta's Britishness. The meanings of the chosen words and phrases have since changed, and the appropriate reading cannot ignore context. The citation holds layered meanings that may ironically provide the conflicting evidence that continues to fuel the debate about the significance, status and memory which the George Cross stands for.

'HER BRAVE PEOPLE . . .'

The King unquestionably singles out the Maltese as the recipients of the George Cross, unlike Roosevelt's clear distinction between population and defenders. They were the ones who had weathered constant aerial bombing—a definite rethinking of the traditional battlefield beyond the strict meaning of a geographic terrain hosting two opposing armies. Anywhere within the flight range of bomb-loaded airplanes could now become a battlefield. In consequence, the distinction between population and defenders was blurred, given that the role of civilians in the war effort became as important as that of the soldier on the traditional battlefield. The George Cross was intended from the onset as a highly prestigious civilian decoration in response to these recent developments in warfare and, significantly, instituted in time for Malta's war efforts to be duly recognized.

Britain's recognition of a 'brave people' gives due recognition to the support and perseverance of the Maltese and their vital contribution towards the defence of the island fortress. What the British understood the brave people of Malta to be is what the citation's wording seems to miss. Images from the years preceding the outbreak of hostilities and those produced as war propaganda are indicative of what the correct interpretation of the citation's reference to 'brave people' might be. A picture book entitled *Malta*, published in 1927, features reproductions of watercolours by the pro-British Maltese artist Edward Caruana Dingli (1876–1950). These illustrate the Maltese wearing traditional clothes and inhabiting a landscape which might have looked North African or Middle Eastern in the eyes of the British (Debono 2012). There are no city dwellers wearing fashionable clothes, although they are depicted by Caruana Dingli in some of his other paintings. *The Epic of Malta*, another picture book published probably shortly after the award of the George Cross in April 1942 (Ritchie 1943), describes Maltese farmers during harvest time and other indigenous activities, lace workers and fishermen, in much the same way as Caruana Dingli's picture book. Paintings by Leslie Cole (1910–76), the British war artist sent to Malta by the War Artists' Advisory Committee Board, also depict a similar choice of subject matter (Foss 2007, 196). Cole's images, painted in Malta between May and November 1943, describe the war effort and related military activity with images of the local inhabitants, the brave people just recognized as war heroes by the George Cross award. Cole depicts the Maltese as they struggle to survive in heavily bombed housing or in underground church crypts, rock-hewn shelters and what is now considered Malta's prime heritage site, the Hal Saflieni Hypogeum. These images picture the Maltese as natives of an island colonized by the British years before. Indeed, the image which the visuals pose require the George Cross to be understood in terms of a perceived redress recognizing the Maltese natives, who have now become anti-aircraft gunners, soldiers and committed British subjects in the service of Empire. This was by far an exceptional rise to the occasion by the Maltese, duly recognized by the British.

'THE ISLAND FORTRESS OF MALTA . . .'

The citation also refers to the island fortress of Malta, the strategic outpost of the Empire which stood alone against all odds in tough times even though the garrison had come close to surrender. Military and domestic supplies were running low, and sustained Axis aerial bombing was having a field day. The King wrote his citation when the fortress was feared close to capitulation and soon to be lost. By the time of the official presentation of the George Cross to the Maltese five months later, on 13 September 1942, the Allies had successfully tipped the scales in their favour. In mid-August 1942, five of the original fourteen ships forming part of a relief convoy code-named Operation Pedestal had sailed into Grand Harbour. The severe shortage of food and military supplies was partly relieved, although far from definitely resolved. The oil tanker SS *Ohio* also managed to deliver its much needed cargo of fuel in spite of heavy damage so that the island fortress could soldier on. The official presentation of the George Cross may have already been understood to be a gesture of recognition for what had happened in the immediate past—a heroic memory which had already been made 'famous in history' in spite of the fact that the war in the Mediterranean was far from over.

The island fortress on the frontier had yet again stood firm, and the heroism of its current defenders could be vindicated in the island's historic past. President Roosevelt's citation refers to the gallant men and women who could claim to have lived and died to preserve civilization. By contrast, the British monarch's citation does not connect to the island's historic past. The connection is nonetheless implied. Indeed, the island fortress paradigm is indebted to the history and military quests of the Order of St John of which the Great Siege is the most historically significant military achievement. This historic legacy was also claimed to legitimize the British garrison in Malta and, for that matter, British colonial authority.

This British colonial narrative was given tangible form through images and heritage displays. In 1851, Governor Sir William Reed transformed the historic armoury of the knights at the then British Governor's Palace, formerly used by the Grandmasters of the Order of St John, into Malta's first public museum (Spiteri 2003).

Guidebooks describe the suits of armour therein displayed as relics from the past once belonging to medieval warriors akin to those featured in Sir Walter Scott's novels. The display of military heritage was complemented with majolica pharmaceutical jars once in use at the hospital of the Order of St John in Valletta. The combination of suits of armour and pharmaceutical majolica jars may come across as disjointed and incompatible. But this may have been a carefully scripted narrative. The Order of St John was both military and hospitaller; a complementing display of pharmacy jars may have been intended to historically legitimize colonial Malta's role as the 'Nurse of the Mediterranean' (Mackinnon 1916). Malta's medical facilities were then a necessity of Empire and had oftentimes proved to be strategically effective in times of conflict. The main corridor leading to the armoury was also decked with a display of historic suits of armour made to carry shields featuring the coat of arms of the Grandmasters of the Order of St John from the earlier ones in Jerusalem down to the latest British governor in office (Townsend 1869, 30). This narrative of display intentionally promoted the British as the moral and legitimate successors of the Knights of the Order of St John, sustaining the strategic military values of the island fortress.

This politically rethought historic display served its purpose well. It legitimized Empire and colonial rule in Malta by scripting a hereditary continuity with the Knights of St John. This political narrative of display was by far not occasional. It was scripted to go on display at the venue to stand for British colonial authority in Malta, and an abridged version of this narrative version was reproposed for the Malta Pavilion at the Wembley British Empire Exhibition of 1924 (Bonello 2001). The three-storey pavilion looked like an early modern fortified tower. The entry points leading into the structure were built as replicas of baroque gateways to the medieval city of Mdina, Malta's capital city before the building of Valletta, and the town of Vittoriosa, the headquarters of the Order of St John between 1530 and the Great Siege of 1565. The exhibition hall of the Order of St John in the pavilion, one of three spaces earmarked for display purposes, included a medley of paintings, mirrors, furniture, arms and armour in what could be best described as an antiquarian-styled display of historic relics in support of the historical legitimacy of the island-fortress paradigm. Contents and container mimicked the island-fortress paradigm within the remit of a colonial exhibition celebrating Empire.

The values listed by King George VI in his George Cross citation letter evoke the concept of the monument commemorating the fallen in the Great

Siege of 1565 (Figure 8.2), commissioned fifteen years before in 1927 by the Maltese colonial government under the first self-government constitution granted to the island fortress in 1921. The project was commissioned to the Maltese sculptor Antonio Sciortino (1879–1947), the then director of the British Academy in Rome (Vella 2000, 99–101). The chosen site for the monument, right next to the former Conventual Church of the Order of St John, was immediately pinpointed as the most suitable given that it may have been intended right from the start to connect with the Order of St John's most iconic building in Valletta. The monument stands close to where a good number of knights who fell during the Great Siege, including the hero of the siege himself, the Frenchman Grandmaster Jean de La Vallette, are buried. Their values, enshrined in the achievements of the Great Siege, were allegorized in the monument to bridge with the heroes buried close by. Sciortino's monument features three allegorical figures. A central male figure representing Courage stands tall behind his shield, bare chested, dressed in a stylized helmet and wearing armour only from the waist down. Two female complementing allegories, Religion and Civilization, stand one on each side. To his left, Religion holds the papal tiara in her hand and, to the right, Civilization holds a crowned portrait bust representing Liberty. The values for which Sciortino's three allegorical figures stand were also highlighted and articulated in the various speeches delivered during the monument's inauguration ceremony (Galea 1970, 8–11).

The values cited in King George VI's and Roosevelt's citation letters are anticipated in Sciortino's commemorative monument by fifteen years even though these were then also valid for an anticolonial narrative highlighting the Order of St John as the defender of Catholicism and civilization, which colonialism was perceived to be usurping. The anticolonial movement in Malta was essentially pro-Italian and indirectly sanctioned Italian historic claims to Maltese territory. The British island fortress of Malta was constantly pressured by Italian irredentist claims, particularly after the rise to power of the fascist dictator Benito Mussolini in 1922. Pro-Italian propaganda can be summarized in the claims put forward by Guido Puccio, an academic and correspondent for the Italian newspaper *La Tribuna*, in his book *Il Conflitto Anglo Maltese* (1932). The author delineates geographic claims, race, language and culture as strong values with an overtly strong Italian imprint. The British colonial authorities hit back by abolishing the use of the Italian language in official records, promoting the Maltese language instead. By doing so, they strategically sought to obliterate the strongest identity construct, language, that the pro-Italian elites could lay claim to (Frendo 2012, 219–68). Attempts at wiping out identity claims based on language were followed by more concerted efforts. Less than five years later, following the Abyssinian crisis and escalating tensions between Britain and Italy, forty-two Maltese pro-Italian elites, perceived to be critical of British colonial politics, were deported to Uganda, where they were interned (Frendo 2012, 677–98).

Figure 8.2 The gift presented to Malta by Great Britain on Independence in 1964 (Copyright Daniel Cilia).

The values which the pro-Italian nationalists understood the Great Siege of 1565 to stand for did not concur with the anglophile definition of the island fortress paradigm as described in King George VI's citation letter, even though the values recognized by both sides were, and for that matter still are, universal. The monument to the Great Siege soon acquired the trappings of a national shrine (*Monumento Nazionale*) where patriotic speeches took place, including those in Italian, which were anything but favourable to the British colonial authorities (Vella 2000, 102). In spite of this, the British acknowledged the values for which this monument stood and wreath-laying ceremonies were recognized as a token of admiration to the heroism of the Maltese forefathers (Schembri Bonaci, 2012, 57). This gesture confirms yet again that the British narrative of the island fortress paradigm also drew on the heroism of the defenders as much as the pro-Italian narrative did, albeit in different ways. Victory over fascist Italy undoubtedly consolidated the British narrative. The famous message sent by the commander-in-chief of the British Mediterranean fleet to Admiralty on 11 September 1943 sums up this achievement: 'Be pleased to inform their Lordships that the Italian battle fleet now lies at anchor under the guns of the fortress of Malta' (Vella 1985, 192).

The message implied complete and total submission, which reflected the unconditional surrender of the Italian fascist state. It also evoked the island fortress image, now rebranded as the George Cross island, which King George VI had evoked in his citation letter awarding the George Cross. More importantly, it signified a turn in Malta's colonial identity, now unhindered by Italian irredentist claims; the George Cross became the ideal identity symbol of post-war colonial Malta.

At the end of WWII, the British could claim the island fortress paradigm independently from the historic past of the Order of St John. In 1945, the Art Fund of the United Kingdom presented to Malta five early nineteenth-century views of Valletta and Grand Harbour painted by the Swiss painter Abraham Louis Ducros (1748–1810) in the first decade of the nineteenth century (Debono 2013, 40–3). These had been purposely acquired to be 'presented to the people of Malta as a tribute to their courage and fortitude during WWII'.[1] The reason for this donation, as registered in the Art Fund records, leaves no doubt as to the reasons guiding this donation—which echo the recognition of the same values as the George Cross. The views stood for a measure of nostalgia, but what stands out in some of the paintings is the presence of the British soldiers roaming around the city within the same spaces inhabited by the Maltese at a time when most could not understand, let alone speak, English. Even though all five paintings date back to the early years of the nineteenth century, subjects and composition complement the same values which *The Epic of Malta* sought to underpin.

ALIGNING NARRATIVES

The narrative which the citation broadly refers to—the perceived selective reading of the past to sustain it and its contested meanings—belonged to circumstances which WWII completely rethought. From June 1940, when Italy went to war against Britain and the Allies, Malta became subject to constant bombing from nearby Sicily, which blurred the distinction between colonizers and colonized. In consequence, the values which Sciortino's monument to the fallen heroes of the Great Siege of 1565 stood for were evoked once again into the present to become shared values for both sieges.

Pairing the 1565 Great Siege of Malta with the WWII Siege also suited the purpose of the British colonial narrative. Malta's National Day was then still celebrated on 8 September in commemoration of victory at the Great

Siege of 1565, which historically coincided with the Catholic festivities of the nativity of the Virgin Mary. The eighth of September was rethought to also stand for Allied victory in the Mediterranean, and both began to be jointly commemorated every year in front of Sciortino's monument. Malta's national feast now stood for victory in the two sieges. The common values which each event stood for and which the monument represented could also be recognized by the George Cross, a symbol akin to the eight-pointed cross of the Order of St John. Indeed, the British had by then acquired a new understanding of Malta's significance as a regional fortress and took advantage of circumstances to consolidate the narrative of that same identity which they had sought so much to articulate in their favour.

The recognized narratives of the Great Siege of 1565 and the WWII Siege of Malta are indeed similar (Debono 2012b, 483). The relief force sent by the Spanish Viceroy in Sicily to assist the knights at the eleventh hour in 1565 has its counterpart in Operation Pedestal, which delivered the crucial supply of fuel for the war effort on SS *Ohio* and five other merchant ships which successfully sailed through the Axis blockade of the island fortress. Operation Pedestal is also recognized as the event which turned the tide of the siege much as the relief force from Sicily helped liberate the islands from the Ottoman invaders, even though the threat continued unabated. Indeed, the fascists were now the new Ottomans invading Malta, usurping the values of civilization, and the twinning of the two sieges may have been deliberate although, as yet, no documentary evidence confirms this. The armistice with Italy, signed in secret on 3 September 1943, was made public on 8 September by General Dwight Eisenhower on Radio Algier. The award ceremony of the George Cross happened on 13 September, the Sunday following Malta's national day (8 September) which in that year fell on a Tuesday.

The Great Siege of 1565 and the Siege of Malta during WWII are consequently woven into one narrative of identity, and the symbols that stood for either are also jointly promoted. Maltese stamps displayed both symbols regularly, and the state gift presented by Great Britain to the Maltese people on Independence in 1964 is also a combination of the two (Figure 8.3). The gift features a George Cross in silver with an eight-pointed gold-plated Malta cross mounted on top. The dominant shape, indeed the shape of the gift itself, is undoubtedly the George Cross and the complementing eight-pointed cross suggests the intertwining legacies of narratives which each symbol stands for. The historic past of the island, decisively shaped by the Order of St John of Jerusalem, is defined and sustained by the heroism and devotion of its British subjects duly recognized with the award of the George Cross. The intended twinning of these two icons, which the Independence gift of the United Kingdom to Malta physically fused into one, stands for a historic interface between identity narratives which Said refers to: '[T]he power to narrate, or to block other narratives from forming and emerging is very important to culture and imperialism and constitutes one of the main connections between them' (Said 1993, xiii).

Figure 8.3 Monument to the Fallen of the Great Siege of 1565 (Copyright Daniel Cilia).

CONTEMPORARY SIGNIFICANCE AND WAR MEMORY

Malta's shift from Crown Colony, following Independence in 1964, to full membership of the European Union in 2004, led to a gradual reconsideration of how Maltese society perceived the George Cross and its relevance as an identity symbol, what it stood for and its historic significance. The George Cross still features prominently on Malta's national flag, but this eminent status does not carry the unanimous approval of the Maltese.

The still ongoing debate concerning the George Cross and its place on Malta's national flag occasionally flares up in the local media from time to time. The most recent of these happened in the last months of 2013 (Borg, Jacob 2013). The objections raised over these past decades have been surprisingly consistent. Complementing or swapping the George Cross for the eight-pointed cross of the Order of St John is a valid option for some (e.g. Borg, Julius 2013; Calleja 2013; Oliva 2007). Indeed, both stand for the same values and may also subconsciously refer to the twinning of the George and Malta crosses, promoted on local postage stamps prior to Independence and featured as the Independence gift of Britain to Malta in 1964. The George Cross is, however, perceived to be of no relevance to Maltese culture and identity, which is not the case of the Malta cross—the symbol of the Order of St John which also stands for a rich historic and cultural legacy now recognized as the backbone of Malta's cultural tourism industry

> The George Cross has been part of Malta's history for 67 years and was introduced to the nation because of the three-year siege of Malta. . . . Other than encourage a triumphant attitude by some Maltese, the George Cross has done nothing for the Maltese culture and way of life.
> (Calleja 2009)

The arguments against the removal of the George Cross are oftentimes emotive and categorical: 'The George Cross is our heritage and we, the people of Malta, should be proud to have received it' (Grech 2013). Those in favour of keeping the George Cross on the national flag recognize it to stand for values other than a mere British colonial recognition. Solidarity, courage, fortitude and determination in times of difficulty are recognized as the values which define the very fabric of the Maltese nation (Zammit Marmara 2013).

Indeed, the George Cross is, to all intents and purposes, a recognition awarded in times of war and hardship, albeit for the heroism and devotion then shown. For some, 'G.C.' could also stand for *Guh Cbir*, meaning widespread hunger as written in old, Italianized Maltese (Vassallo 2014). Besides this psychological and social impact, Malta's losses included the destruction of historic buildings and the loss of works of art. This widely recognized narrative of WWII is the one which guides the politics of display at Malta's National War Museum where the George Cross resides today. It is purposefully displayed within a chronological narrative akin to a war diary, and this deliberate curatorial choice underpins its reading as a war medal beyond the communal values highlighted in the King's address to the Maltese, although citation and medal are exhibited next to each other. Within this narrative of display, the George Cross stands for loss of life and heritage rather than for culture and heritage, as with the case of the eight-pointed Malta cross. British colonial heritage was, until recently, also considered to be unworthy of preservation; this was particularly felt in the years leading to the closure of the British military base on the island, completed by the end of March 1979.

Most of what has survived is now part and parcel of the island's cultural tourism package and promoted alongside other major attractions. It concerns military installations, including gun turrets and pillboxes built at strategic locations all over the island. The most prominent and surviving WWII ruin is the Royal Opera, originally built by the British in 1866 and now recognized as a ruin proper, transformed into an open air theatre following designs prepared by Italian architect Renzo Piano (Buhagiar, Dreyfuss and Mifsud, 2013, 308–13). This major WWII ruin standing at the main entrance to the capital city was and still is a hotly contested project with many advocates of rebuilding. The debate has also featured in the British press (Squires 2010). Indeed, the war memory for which the George Cross stands for has been re-engineered and commodified to suit the purpose of Malta's tourism industry, thanks to the universal values which define the island fortress paradigm in the name of which wars are perceived to have been fought.

The recent debate concerning the introduction of the Maltese euro currency explains the significance of the George Cross to Maltese contemporary society. In 2006, the Maltese were asked, through public consultation, to choose the images to feature on the Maltese euro coins. A pre-established selection was proposed, and the debate featured prominently in the local press (Micallef 2005). Given the wide circulation which these coins enjoy, the final selection was intended from the outset as a triad of identity symbols for each member state chosen as unique symbols of a nation's particular identity.

The original three shortlisted images, from the pre-established selection proposed, included the Mnajdra prehistoric temple and the Republic's coat of arms. The choice stands for recognized symbols which the Maltese endorse and acknowledge as meaningful. Indeed, both are recognized icons of identity, and the choice was not hard to make. Malta's prehistoric past stands for a culture that goes back centuries and that is reputed to have constructed the oldest free-standing buildings in the world. The Republic's coat-of-arms is Malta's official state symbol. Both concern Malta's contemporary narrative of identity, a constitutional democracy and an island with evidence of a past which goes back centuries.

The large marble group of the Baptism of Christ at the former Conventual Church of the Order of St John (now co-cathedral in Valletta) was to be the third chosen image and had indeed garnered the largest number of votes from the shortlisted three (Grech 2006). This would have stood for the inextricable link between politics and religion, a strong characteristic of Malta's early modern history. The high number of votes garnered was interpreted as a choice for Malta's Christian faith, although its appropriateness as a religious symbol to feature on coinage was nonetheless questioned. A second round of votes overwhelmingly singled out the Malta cross instead. The large marble group of the Baptism of Christ was consequently dropped in favour of the Malta cross, in spite of the fact that this had not featured in the pre-established selection proposed for discussion and choice (Balzan 2006).

In spite of its privileged position on Malta's national flag, the George Cross did not feature either in the pre-established selection or in the public debate, which confirmed the Malta cross as a powerful symbol of identity in spite of the fact that the George Cross had been featured on Maltese low-denomination coinage in circulation during the 1970s–80s. The respondents to the survey, who also voiced their opinion in the local press, confirmed outright that the eight-pointed cross was the emblem or symbol which distinguished the Maltese as a nation within the international community (Borg, J. G. 2006). The complementing narratives of the two sieges and the consistent twinning of both symbols have done little to sustain the constructed significance of the George Cross as an identity icon for modern Malta, as the final choice of symbols for Malta's euro coins suggests, and relegates it to the remit of war legacy as scripted within the narrative of the display of the Malta War Museum.

The presence of the George Cross on the national flag is caught between two meanings. Contemporary Maltese society acknowledges the award within the remit of a war memory belonging to the island's colonial past but also recognizes the universal values for which it stands and which King George VI's and Roosevelt's citation letters refer. These values are still recognized for their military significance and can perhaps be rethought as a legacy of the island-fortress paradigm. It is no mere coincidence that the Cross also features on the air squadron roundel and the naval jack of the maritime squadron of the Armed Forces of Malta. The national flag does not stand for an island fortress anymore, but there is no question that the history of Malta is the story of an island fortress. The George Cross has historical significance but can no longer be recognized as an identity icon. Beyond the universal values which the George Cross stands for, this historic significance concerns the war memory of the island's communities. In the eyes of the local population, the George Cross does not stand for the plight of WWII and the heavy blitz sustained by the islands. For local communities, war memory has to do with deceased relatives—personal connections and recollections which evoke the plight of the island. Most local communities have their war memorials. Some are complex multifigure monuments, others just a marble plaque. Most of these monuments were commissioned in the 1990s to coincide with the fiftieth anniversary of the end of WWII, and the phenomenon merits further study. Common elements can nonetheless be singled out. First, all monuments feature a list of those who died during the war and who hailed from that particular community. Second, the George Cross rarely features in the monument, and what is given prominence instead is the institutional symbol or identity icon of the community. Third, these monuments rarely feature British servicemen in the lists. Malta's war memory is of an essentially ethnic kind and based on kinship relations which personalize war memory. This may be symptomatic of the close-knit character of island communities, of which Malta is no exception. By their very nature, Maltese communities have kept a personal memory of the war, in contrast to the official memory now touted to tourists within the broader remit of the island fortress paradigm.

Malta's war memory concerns personal recollections and plights, the effort and sacrifice of the individual which contributed to the victory of Empire. The war museum itself is a collection of personalized memories, a narrative of themes and related objects which concerns items whose owners are known and who would have donated these with the intent of preserving personal war memories. The individual has a far greater share in scripting Malta's war memory in ways which empower grounded narratives. Such memories are far from being romanticized, as in the case of the war memory of the Great Siege of 1565. As a guardian of memory, the individual would be prone to define his or her personalized war memory correctly, although a measure of hagiography cannot be entirely ruled out. This memory is but a fragment from a bigger collective recognized by communities and institutions. The George Cross stands only for the universal values to which these personalized memories refer. It is the invisible thread which binds them all together and which stands for the universal values that can tinge personal recollections with hagiographic additions.

NOTE

1. Art Fund Records (United Kingdom), Acquisitions 1945.

BIBLIOGRAPHY

Balzan, A. (2006) 'Maltese Cross makes eurocoin shortlist', Euronews.com, 12 June. Online Available HTTP: https://euobserver.com/economic/21825 (accessed 18 February 2013).

Bonello, G. (2001) 'Malta in the Wembley Exhibition—1924', in G. Bonello (ed), *Histories of Malta—Figments and Fragments*, Vol. 2, Malta: Patrimonju Publishing, 215–9.

Borg, Jacob. (2013) 'Maltese cross on Facebook over flag change issue', *The Malta Independent*, 25 September. Online. Available HTTP: www.independent.com.mt/articles/2013–09–25/news/maltese-getting-cross-on-facebook-over-flag-change-issue-2716368898/ (accessed 12 November 2013).

Borg, J. G. (2006) 'The George Cross', *The Malta Independent on Sunday*, 12 February. Online. Available HTTP: www.independent.com.mt (accessed 25 September 2013).

Borg, Julius. (2013) 'Crosses on the flag', *Times of Malta*, 9 October. Online. Available HTTP: www.timesofmalta.com (accessed 12 November 2013).

Brogini, A. (2006) *Malte—Frontiere de Chretiente (1530–1675)*, Rome: Ecole Francaise de Rome.

Brogini, A. (2013) 'Diventare Una Citta-Frontiera: Nizza nella prima eta "moderna"', in *Mediterranea—ricerche storiche*, X (28) (August), 367–80.

Buhagiar, K., Dreyfuss, G. and Mifsud, M. (2013) 'The Royal Opera House in Malta: from ruin towards a new reality', in F. Castagneto and V. Fiore (eds), *Recupero, Valorizzazione, Manutenzione nei Centri Storici. Un tavolo di confronto interdisciplinare*, Siracusa: Lettera Ventidue Edizioni, 300–13.

Calleja, E. (2013) 'Both crosses on our flag', *Times of Malta*, 28 October. Online. Available HTTP: www.timesofmalta.com (accessed 12 November 2013).

Calleja, P. (2009) 'The right cross for the flag', *Times of Malta*, 4 August. Online. Available HTTP: www.timesofmalta.com (accessed 18 February 2013).

Debono, S. (2012a) 'Għonnella—a British styled allegory for 20th century Malta', in K. Gambin (ed), *Peasant Costume—Insights into Rural Life and Society*, Malta: Heritage Malta, 67–71.

Debono, S. (2012b) 'Malta L-Ewwel u Qabel Kollox—L-Arti u l-Identita Politiko-Kulturali Maltija tas-Seklu Ghoxrin [Malta first and foremost—the art and politico-cultural identity of 20th century Malta]', in M. Cutajar *Mintoff—il-bniedem u l-istorja* [*Mintoff—the Man and History*], Malta: Sensiela Kotba Socjalisti, 478–92.

Debono, S. (2013) 'British donations to the national collection: the case of the five views of Valletta and Grand Harbour by Louis Ducros (1758–1810) at the National Museum of Fine Arts', *Treasures of Malta*, XX (2), 40–3.

Foss, B. (2007) *War Paint: Art, War, State and Identity in Britain—1939–1945*, New Haven, CT: Yale University Press.

Freller, T. (2009) *Malta and the Grand Tour*, Malta: Midsea Books.

Frendo, H. (2012) *Europe and Empire—Culture, Politics and Identity in Malta and the Mediterranean*, Sta Venera, Malta: Midsea Books.

Galea, M. (1970) *Malta—Historical Sketches*, Malta: M. Galea.

Grech, H. (2006) 'New poll on euro coin images', *Times of Malta*, 30 May. Online. Available HTTP: www.timesofmalta.com (accessed 13 February 2013).

Grech, J. F. (2013) 'George Cross should make us all proud', *Times of Malta*, 13 November, Online. Available HTTP: www.timesofmalta.com (accessed 12 March 2014).

Mackinnon, M. A. (1916) *Malta—The Nurse of the Mediterranean*, London: Hodder and Stoughton.

Micallef, M. (2005) 'Popular say in choice of image for Malta euro coin', *Times of Malta*, 3 December. Online. Available HTTP: www.timesofmalta.com (accessed 10 January 2013).

Oliva, M. (2007) 'Two crosses, one flag', *Times of Malta*, 9 April. Online. Available HTTP: www.timesofmalta.com (accessed 13 February 2013).

Ritchie, L. (1943) *The Epic of Malta*, London: Odhams Press.

Said, E. W. (1993) *Culture and Imperialism*, London: Chatto and Windus.

Schembri Bonaci, G. and Moulden, S. (2012) *Antonio Sciortino and the British Academy of Arts in Rome*, Malta: Horizons.

Spiteri, S. (2003) *Armoury of the Knights*, Sta Venera, Malta: Midsea Books.

Squires, N. (2010) 'Maltese anger at plans to rebuild Valletta', *Daily Telegraph*, 8 May. Online. (accessed 12 September 2013).

Townsend, G. F. (1869) *English Handbook of Malta*, Malta: n.p.

Vassallo, R. (2014) 'National identity: the quest goes on | Charles Xuereb', *Maltatoday*, 14 January, Online. Available HTTP: www.maltatoday.com.mt/news/interview/33099/national-identity-the-quest-goes-on-charles-xuereb-20140113#.VLpq5sklkmw (accessed 18 February 2013).

Vella, P. (1985) *Malta: Blitzed but Not Beaten*, Mrieħel, Malta: Progress Press for the National War Museum Association.

Vella, D. (2000) *Antonio Sciortino—Monuments and Public Sculpture*. Valletta, Malta: National Museum of Fine Arts.

Zammit Marmara, D. (2013) 'George Cross Controversy', *Times of Malta*, 11 October. Online. Available HTTP: www.timesofmalta.com (accessed 12 February 2014).

9 Scraps of Memory

Pacific War Tourism on Efate Island (Vanuatu)

Lamont Lindstrom

Ernest Kalkoa, the one-man owner, curator and docent of a jumbled, local Pacific War museum located at Tanoliu village, along the shore of Efate's Havannah Harbour, makes sure visitors appreciate the corny pun 'Rust in Peace' (Figure 9.1). Or, in view of Kalkoa's collection of Coca-Cola bottles and fragments of other war relics, his museum's motto might also be 'Rest in Pieces'. Peace has followed war, and pieces of military scraps and trash today command attention as *relic*—this word a recent addition to modern Bislama, Vanuatu's Pidgin/Creole *lingua franca* (Figure 9.2).[1] *Artifak* is a second neo-Bislama term for another sort of material culture that tourism has also called forth—in this case, a portfolio of traditional carvings and other objects, versions of which are today sold to visitors.

Kalkoa competes with a second wayside war museum, run by the family of Erick Tom (Erick Kalman Songoman Tom), located a few miles along the road. These two local efforts curate and display Pacific War remains and, by this, materially recall war history for passing tourists. They comprise the few accessible venues in which war memories survive today in Vanuatu. I follow the history of war memory in these islands (formerly the colonial New Hebrides) as this has largely evaporated from personal recollection and drained away from scattered military installations, wreckage, ruins and installations to become bottled up instead in two home-grown museums. War memories, seventy years or so since the US military decamped in 1946, resonate less among local audiences as key events in people's life stories and community histories and more with escalating numbers of overseas tourists who come in search of *relik* and *artifak*.

Like once traditional carvings and artefacts, tourism has increased the value of relics, thus keeping war memory alive and circulating. WWII blew into the colonial Pacific from elsewhere. As many have noted, this was someone else's fight (e.g. Poyer, Falgout and Carucci 2001, 8; White, Chapter 11 in this volume). And today it is mostly someone else's memory. Apart from passing personal experience, independent island states have little impetus to celebrate the war within their nationalist history making or state memorial tributes. Vanuatu's national holidays commemorate first Prime Minister Walter Lini, custom chiefs, Independence Day, children, families, unity and scattered Christian holy days—apart from the latter, nothing, in fact, from

Figure 9.1 Kalkoa makes a point, 2013 (Copyright Lamont Lindstrom).

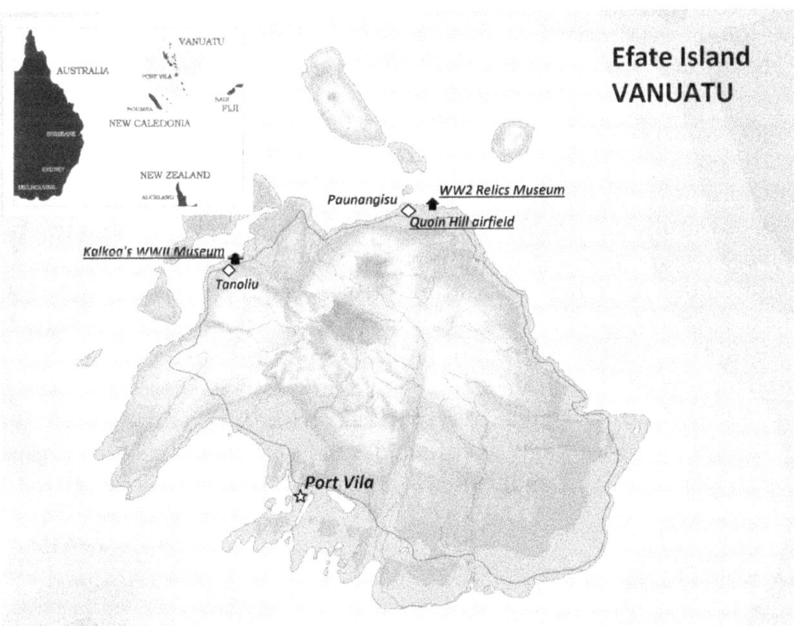

Figure 9.2 Map of Efate Island, Vanuatu, showing local war museums along America Road (Adapted from Hk kng, Creative Commons CC-BY-SA).

before Vanuatu's independence in 1980. Pacific War experience, including the service of several hundred ni-Vanuatu[2] in the New Hebrides Defence Force (NHDF) and thousands more in native labour corps, goes unobserved.

Unlike the Solomon Islands, New Guinea or Micronesia, no battles raged amidst island homes, and no villagers were caught in deadly crossfire. The New Hebrides (like the Ellice Islands (McQuarrie 1994), as well as Fiji, Tonga, and Samoa) merely hosted unexpected occupying forces that constructed bases from which to supply and stage battles elsewhere. Island war experience on Efate was more 'Big Work' (Lindstrom and Gwero 1998) and less 'Big Death', as in the Solomons (White, Gegeo, Akin and Watson-Gegeo 1988). In Vanuatu, unlike bloodier islands where many more people died, there is little need to worry over the ethics of balancing tourist attraction with memorial veneration (cf. Bennett 2012). Because of this, however, local war relics and places lack the strong sentiments of 'nostalgia and voyeurism' (Cooper 2006, 214) that attract battlefield thanatourism.

If memory of that first wave of intrusive military outsiders, over time, has easily washed away, a second touristic wave now uncovers renewed recollection. War heritage 'is not natural but contingent, and dependent on the work of those agents who bring it into being and sustain it' (Gegner and Ziino 2012, 2). Such agents in Vanuatu today are tourist entrepreneurs. Tourist arrivals to the archipelago have boomed in the last decades, and some visitors are keen to look beyond the beach. Museum entrepreneurs are digging up the war's relics, and these relics both demand and evoke storytelling (see Cappelletto 2005, 23–5). Vanuatu war memory today adheres in relics and in the stories they revitalize.

The impact of war on local communities and on shared memory is often severe—particularly so in compact places, on small islands like Efate. Even though located behind the front lines, people throughout the New Hebrides lived through four years of vibrant, unsettling, exciting and sometimes dangerous events, and they mixed for the first time with new and powerful kinds of outsiders. Shared war memory in small island communities may remain intense, circulated by stories told by one generation to the next. But small islands also suffer the continued vicissitudes of regional and global influence which can otherwise distract and direct people's attention. In Vanuatu, these have included a troubled independence in 1980, contentious national elections, a difficult economy and now increasing flows of overseas tourists themselves attracted to an island experience. The small island allure of a tropical holiday, in fact, is reviving island war memory. Given touristic interest, war relics have supplanted war stories as media for recollection.

REMEMBERING

There was a lot to remember about wartime events on Efate (Lindstrom 1996). Initial US units arrived there in March 1942 joining a small contingent of Australians already in place (Geslin 1956). A larger force came ashore in May, and the US military began construction of a large, extensive joint Army/Navy advanced operating base (Wallin 1967; Garrison 1983). Sentries patrolled Port Vila. Lights were blacked out at night. More than thirty-three Japanese nationals were arrested and eventually deported and imprisoned in Australia (Discombe 1979, 4). Batteries of coastal artillery were installed around Efate. Vila Bay was mined and antisubmarine nets protected the inner harbour. Arriving Seabees and Army engineers continued work clearing a 6,500-foot runway north of Vila. A large naval hospital was built on Bellevue plantation overlooking the town. Engineers extended colonial dirt roads up over Klehm Hill to Havannah Harbour and eventually around the island (Heinl 1944, 250). At Port Havannah, the Navy established a fleet anchorage, seaplane base, recreational facilities and another airstrip. Seabees built a third strip near Quoin Hill on north Efate for Marine flyers. Coastwatchers were landed on islands from Efate up to the Santa Cruz group. In July, American forces moved north to establish even larger installations on Espiritu Santo.

At the peak of base operations in 1945, 50,000 military personnel were stationed on Santo, another 15,000 were based on Efate, and tens of thousands more passed through on naval and supply ships or heading up to the front lines to the north-west (Kralovec 1945, 95, 109–12, 263). Military occupiers swamped Efate's pre-war population of some 3,000 islanders and colonists. The majority of American servicemen on Efate, moreover, were African-American—many members of the Army's segregated, Indian-fighting 24th Infantry Regiment put to work, in the New Hebrides, as stevedores, quartermasters and truck drivers. About 360 islanders, mostly from Malakula, joined the New Hebrides Defence Force (Discombe 1979). Even more would serve in native labour units. With a voracious need for local assistance, the military recruited and sometimes conscripted as many islander workers as it could (Lindstrom 1989).

When I moved in the late 1970s into a small Tanna village (on an island 200 kilometres south of Efate), men (and some women) still liked to talk about their Pacific War experience, sharing all sorts of personal memories. These friends were in their fifties and older, and almost all of the men had worked for at least three months at Allied military installations on Efate between 1942 and 1946 (see Lindstrom 1989). Islanders recalled their own war experiences to stake a variety of claims. These ranged from assertions of enduring (although typically inactive) relationships with military personnel, to proud recollections of personal agency and bravery, to invidious comparison of Allied generosity with Anglo/French colonial meanness, to complaints about wartime loss and suffering and more. Inspired by these memories, James Gwero and I, assisted by the Vanuatu Cultural Centre's local fieldworkers, collected and eventually published sample war stories from throughout the country (Lindstrom and Gwero 1998; see also Moon and Moon 1998). We recognized that many of those with war experiences and vivid memories of them had already passed on and that those remaining were elders. Most, though, were still vigorous and still had great stories to tell. In these stories, we followed a number of common themes that had been impressed upon local memory: American arrival; arresting Japanese settlers; all sorts of labour contributions; the New Hebrides Defence Force, coastwatching; road and airfield construction; military weaponry; equipment and cargo; new medical treatments; acts of violence, including rape; African-Americans; sports; music and cinema; and much more.

In the main, war stories in the 1980s remained personal, although some of the best and most gripping of these had also converted into family anecdote. A few younger folk told stories of events that had happened to their parents—particularly of wartime violence or other substantial community disruptions, including the rape and murder of a village woman on Santo, the labour recruitment of most of the male population of Tanna, and the removal of all residents of Tutuba, a small island off south-east Santo, so that the American military could use the islet for troop recreation and later for target practice.

Alongside story and anecdote, people also remembered and shared war experiences in a burst of recollected song. Some of these were mid-twentieth-century American standards that islanders had learned from servicemen; others were new compositions that celebrated a variety of war events, including the service of north Efate villagers' beginning the clearing of Bauerfield airport, the sinking of a submarine off north Efate's Emao Island, and the final departure of American troops (Lindstrom and White 1993). Narrative and song, in fact, are linked in that traditional stories in Vanuatu commonly climax in song, and war storytellers, too, liked to mix personal story or anecdote with a pertinent song.

Beyond human recollection, war memory also embedded itself in souvenirs, relics, abandoned military installations, new personal and place names and even a few children. Many in the 1970s kept various sorts of war salvage: Marsden matting reused for pigpens; mess trays and silverware; carefully hoarded scraps of uniform; and even pillowcases full of American coinage. Author James Michener, who revisited Santo in 1949, noted this residue: 'Stone-age families now have gasoline drums for water, army blankets, and cartridge cases to be used as trunks' (1951, 214; see Spennemann 1992). Salvaged relics evoked personal experience and connection, and people sometimes happily illustrated their stories with them but sometimes concealed them as too valuable for others' prying eyes.

Some people claimed fresh nicknames during the war which also generated new place names. Among the most resilient of these have been the names of two Port Vila neighbourhoods, Nambatu and Nambatri (Number Two, Number Three)—locations where Australian forces set up radio and signal stations—and Freswota (Freshwater). Some people still use other wartime locatives, including D-dock (one of the main American wharfs in Vila Harbour), Amerika (or Tanna) Road—the original highway circling Efate—and Million Dollar Point on Santo, one of the places where the US military dumped excess supplies at war's end (Lindstrom 2011).[3] The war had some impact on Bislama itself as islanders adopted new words like *samarin* (submarine), *bom* (bomb), *trak* (truck), *oke* (okay), *hedkwata* (headquarters) and more (Crowley 1990, 141).

Besides Efate's circular Amerika Road, wartime installations still in use include Bauerfield, now Vanuatu's international airfield, and Espiritu Santo's main airport, Pekoa, which the Americans cleared and named Bomber Two. The SS *President Coolidge*, submerged under 20–70 metres of water at the mouth of Segond Canal near Luganville, is perhaps the war's most spectacular relic—a favourite of tourist scuba divers. Originally a 654-foot long luxury liner converted into a troop carrier, the ship ran into an American mine and sank on 26 October 1942 (King 1991, 467; Stone 1997).[4] Submersion has engendered rust and decay but at least has preserved the ship from post-war scrap dealers or the ongoing souveniring that has carried away most other war material.

Back on Tanna, returnee labour corps recruits swiftly incorporated American martial elements into the liturgy and symbolic array of the John Frum movement (Lindstrom 1993). This anticolonial organization had emerged in the late 1930s, and one of its prophets had accurately predicted the subsequent arrival of American forces (although these disembarked to confront the Japanese, not the colonial system). Drilling and marching, bamboo rifles, oddments of army uniform, red crosses, model aeroplanes and identificatory dog tags all became part of cult paraphernalia (Figure 9.3). Still today, on cult holidays, John Frum supporters raise the American flag alongside a number of other movement banners.

Figure 9.3 Military mementos in a John Frum shrine overlooking Iasur volcano, 1978 (Copyright Lamont Lindstrom).

Finally, the war left behind genetic traces of passing romance and other sexual encounters. Although fewer than at other Pacific military installations, perhaps fifty or so children were born to women living near American bases on Efate, Santo, and the string of coastwatcher stations stretching to the north (see Wanhalla and Buxton 2013 for comment on marriages between American servicemen and women elsewhere in the Pacific).[5] In the 1980s, we were able to locate only a handful of these 'half caste' military offspring or their mothers; none had enjoyed any post-war association or much memory of their absentee fathers. Still, the presence of scattered half-American children along with personal war stories and anecdote, song repertories, place names, nicknames, wrecks and relics, and John Frum activism all worked for several decades to keep war memory fresh.

FORGETTING

Some seventy years beyond the war's end, war memories, along with the objects in which these were embedded, have now worn away (see Bennett 2012, 99–101). Since our oral history project in the late 1980s, those still living with personal war memories are elderly indeed. Most alive today would have been children in the 1940s. Ernest Kalkoa, for instance, guesses that he was born circa 1936 on Nguna (a small island off Efate's northwest coast), and he would have been about six years old when the first American forces landed in March 1942. Personal reminiscence has disappeared along with persons. Vanuatu war memory, nowadays, is 'vicarious' (Young 2000, 1–2) not only because the war mostly belonged to invasive outsiders but also because nearly all of its local survivors, too, have now perished. Reece Discombe and NHDF commander Ernie Reid, two Vanuatu military history buffs who did much through the years to stoke local memory, have also both passed on. Contemporary recollection and deployment of war stories in personal and community identity construction have shifted down from active adult reminiscence to passing childhood memory.

War relics, too, have seriously eroded over the years. Much indeed has 'rusted in peace.' Entrepreneurs like Reece Discombe, moreover, in the 1950s and 1960s, salvaged and exported to Southeast Asia and Japan many tons of war scrap metal—much of this retrieved from seabed military dumps (Discombe 1979, 16–7). Archaeologists complain about this obliteration of war heritage:

> These scrap metal drives continued the destruction of the historical resources at an unprecedented rate . . . In retrospect, the scrap metal collectors, as well as the well-intentioned cleanups and the removal of unexploded World War II ammunition during the same period caused more structural damage to the World War II heritage than the war itself.
> (Spenneman 1992, 15)

Over the years, collectors have continued attempts to strip the country of its accessible plane wrecks and other war remains. (Discombe estimated over 330 Allied planes crashed in these islands in the 1940s.) Vanuatu, along with the other Melanesian states, has, since its independence, passed laws restricting, legally at least, such export of wrecks and relics.

What scrap dealers left behind has now decayed. Luganville's former avenues and docks, once featuring row upon row of Quonset and Dallas storage huts, have mostly crumpled, and clusters of ammunition bunkers are now overgrown with forest and bush or visited only by the island's cattle herds. Some war material or remnant relic does still survive underwater, including sunken scuba dive favourites, *President Coolidge*, the *Tucker* and miscellaneous military surplus and trash dumped off Santo's Million Dollar Point. And tourists arrive and depart on what were originally wartime airfields, and they drive on roads first cleared by military bulldozers. At Bauerfield, however, many discover an opportunity to learn something of the airport's history only when *leaving* Vanuatu. Next to the door leading into customs and the departure lounge is a glass case displaying a propeller and engine from a wrecked Scout Bomber Douglas (SBD4) plane (salvaged from the end of Bauerfield) alongside a photograph of its pilot with information also about Marine Lieutenant Colonel Harold Bauer, commander of the 212th Fighter Squadron, after whom the airfield was named. Bauer ditched his plane at sea and disappeared, on 14 November 1942, during a skirmish over the Solomon Islands. The display also recounts basic Vanuatu war history.[6] Only when in the departure lounge itself may tourists learn more about Harold Bauer from several plaques and photographs hanging there, donated in 2008 by American veterans groups and the National Museum of Naval Aviation History in Pensacola, Florida.

Having no battlefields, Vanuatu also has no war cemeteries—or at least not any longer. Elsewhere, battlefields and cemeteries serve as foci of postwar memory and war tourism alike. During the war, the Americans commandeered Vila's small municipal cemetery (now Freshwater Cemetery) to bury more than 1,000 dead servicemen there. Most of these had been wounded up in the Solomon Islands and died in transit back to Vila or in the large naval base hospital built at Bellevue plantation. Others died of disease or road accidents on Efate. The military also buried its war dead on Santo's Surunda plantation. At war's end, however, the Americans sent a ship with a team of morticians to exhume these graves and return the remains to the United States. Many were reburied in Hawaii's (Punchbowl) National Memorial Cemetery of the Pacific. In the 1980s, local men who had helped in body retrieval remembered this in grisly detail. Today, only two lonely war graves remain: one army and one navy veteran from New Zealand, both of whom died in late 1942. Sextons gladly guide visitors to these graves, although only a very few come to pay respects. The bodies lie 25 metres or so apart, surrounded by local Vanuatu dead. Unlike New Zealand, which leaves its war dead to lie where they fall, the United States continues to send body collection teams from Hawaii's Joint POW/MIA Accounting Command Central Identification Laboratory to scrape together bone fragments scattered about occasionally rediscovered plane wrecks, repatriating these remains too.

War memory in Vanuatu has likewise faded with recent developments within Tanna's John Frum movement. In 2000, the movement split into three factions, the largest of which followed the new prophet Fred Nase, several of whose predictions had wonderfully come true. Nase replaced much of the group's once intense war symbology with a new focus on island unity, trance and healing, and he distributed colour-coded responsibilities for the organization's programmes and ritual events. He also abandoned much of the movement's Americophilia, offering a more critical appraisal of US military endeavour, past and present (Tabani 2008). Nase died in 2011, and remnant orthodox John Frum supporters at movement headquarters in Sulphur Bay have continued to display American-style uniforms and flags, notably during the annual movement holiday on 15 February. They attract far less governmental and touristic interest than they used to, however.

Nonetheless, a few surviving elders with war memory are still alive. In July 2013, I chatted with one old fellow from north Efate who, as had many earlier, recalled his former military friendships in song, serenading me with a version of 'You Are My Sunshine', one of the war's signature pop tunes. Scuba fans continue to visit in order to dive Santo's wrecks. And war residue continues to pop up out of land and sea, about which new stories can be told. Local entrepreneurs on Efate have established two roadside museums that display some of these relics to attract passing tourists. As elders fall away and local recollection fades, this growing touristic interest is in fact reviving, recollecting and now largely sustaining Vanuatu's war memory.

TOURISTIC RECOLLECTIONS

Tourism in Vanuatu increased after the country's independence in 1980 and, in recent years, has boomed. In 1995, 82,019 tourists arrived in Vanuatu. By 2011, visitor numbers had surged to 225,493. Of these, 124,818 were day visitors, passengers on cruise ships (Vanuatu National Statistics Office n.d.). Local entrepreneurs have responded to this evolving market with offerings ranging from local artefact sales, duty-free shopping, food and drink, and a variety of excursions featuring cultural, beach and nature tours. Several operators market Native Round Island Tours or Grand Efate Tours, made easier recently by the old America Road's new tarmac. Even day-trippers from cruise ships can find time for round island sightseeing. Among the road's new signage is one identifying Havannah Harbour's American Pool—a cement catchment that the US Navy built to water its ships.

Vanuatu's Tourism Office operates a storefront in downtown Port Vila. Staff there explain that while they have no specific war history brochure among the many offered, occasional tourists do come along with questions about war history.[7] They direct these mostly to Erick Tom's Second Lieutenant T. Vittitoe Museum near Quoin Hill, north Efate, mostly I suspect because Tom's mobile telephone is likely to work there. Tour operators, too, have incorporated the two local war museums into their itineraries, featuring these on their maps and brochures. Evergreen Vanuatu's and Jasons Travel Media's maps, for instance, plot Kalkoa's museum at Tanoliu as 'WW2 Memorabilia'; Tom's museum, east of Paonangisu village, as a 'WWII Museum'; and Quoin Hill as a 'disused American WW2 airstrip'. Jasons' accompanying pamphlet features tidbits of island war history (Jasons Travel Media n.d., 13, 30, 34, 45). Native Tours' brochure likewise invites tourists to stop at the Quoin Hill WW2 airstrip and Kalkoa's 'WWII Coca Cola bottle shop'. Bellevue Ranch's leaflet offers tourists a 4,000-vatu (a little more than US$40) horse trek that surveys the foundations and remains of the former American hospital located there. The Neil-Jones family's Secret Garden, a touristic 'cultural centre' located near Mele village, also displays several wartime photographs alongside potted war history on placards.[8]

For tourists who circle Efate, Kalkoa's and Tom's family museums curate Efate's most impressive relic collections. Kalkoa and Tom both established their collections about ten years ago, as Vanuatu's tourist numbers began to climb. Both museums, located 30 and 56 kilometres around Efate from Port Vila, house jumbled assortments of miscellaneous, portable relics: a helmet, propellers and other aeroplane and helicopter parts, mess trays, canteens, silverware, cups, jars, bullets, a partial bomb case and grenade, Japanese binoculars, mooring chains, pieces of water pumps and other sheet metal. Kalkoa's museum is a small (3- by 5-metre) wooden shack with a parking area in front over which flap Vanuatu, Australian, British and American national flags. Kalkoa especially displays hundreds of empty glass bottles: for battery acid, medicine, beer and particularly Coca-Cola (Figure 9.4). In the 1940s, plants across the United States bottled the drink, their locations embossed on bottle bottoms, and Kalkoa delights in reading off the names of these American cities and towns. He charges visitors 200 vatu (US$2.10) to enter his museum, but also makes money selling giant clam shells and salvaged Coca-Cola bottles for 1,000 vatu apiece. Such relabeling of war remains as 'relics' increases 'their spiritual value and thus the collector's justification for taking them in the first place' (Spennemann 1992, 15; see Sturken 2007, 12–13). War souveniring also puts a little cash into Kalkoa's pocket.

Both Kalkoa and Tom assert their interest in Vanuatu's history, particularly its war history, and a mission to share this with tourists but also with ni-Vanuatu youth who know little about wartime events in their country.

Figure 9.4 Kalkoa's relic bottles, 2013 (Copyright Lamont Lindstrom).

Simple labels identify Kalkoa's relics, although he provides most historical information through engaged storytelling. Acting as museum docent, he has polished an entertaining and rapid spiel, rattled off in his version of English, which includes testimony of his personal mission to preserve the war's history:

> Now lastly, ladies and gentlemen, girls and boys, I'd like to tell you why I do this. To me I can say many many times, many many people in this world, they don't like to know. They don't want to know their history. I can say, and I believe, I can say history is very very important to know. I mean, I have my own history, you all have your own history. This is Second World War history, and all our children will read about the Second World War and will know about the Second World War but they will not see or touch anything real. Also, why I like to keep this is because . . . if you read about Vanuatu, in the New Hebrides time . . . New Hebrides became one of the major supply points for the central Pacific campaign. Australia, New Zealand, US, they were together here. This is their base . . . So this is why, boys and girls and ladies and gentlemen, I like to collect and keep the history.
>
> (Kalkoa presentation)

A local museum keeper's essential prerequisite is to control access to a military dump. Kalkoa and his family salvage what they can from Havannah Harbour's seafloor, but they also mine a swamp at the north end of the harbour that is still full of American scrap. Here lies buried that cache of empty Coke bottles that Kalkoa is transforming into souvenir war relics. Further along the road, just east of Paunangisu village, Erick Tom likewise controls an old dumpsite close to where American Marine flyers once bivouacked near Quoin Hill airfield. His museum, moreover, is located under the approach slope of the runway, and, in addition to his salvaged war relics, he canoes tourists out to three wrecked planes that ditched in the sea while attempting to land. The closest of these is a submerged Marine F4U Corsair. Tom's museum is located near the seafront, 100 metres off the road. A sign attracts passing tourists to visit his WWII Relics Museum: 'fascinating, remarkable, unforgettable'. He charges 500 vatu to enter the museum, which is constructed of local bush material, and an additional 1,500 vatu if tourists want to visit or dive the underwater plane wreck sites. He also sells war-era Quoin Hill airfield maps for another 1,000 vatu.

Tom is too young to have experienced the war himself, but he explained that he is guided by his father Tasaruru Tom's profound wartime experiences and personal commitment to memorialize the war. Tasaruru, who died in April 2013, worked with and befriended American servicemen in the 1940s. Tom, in fact, has named the museum after the American pilot of the downed Corsair just offshore, Second Lieutenant James Vittitoe who crash-landed during a training mission in May 1944 (Vittitoe 1990). Vanuatuan and American flags frame his museum's doorway. Displayed war relics, somewhat less jumbled than Kalkoa's, include more canteens, mess trays, aluminium plates, silverware, helmets, a broken Corsair propeller blade and other airplane parts. Unlike Kalkoa's minimalist labels and active storytelling, Tom's museum features lengthier information placards featuring testimonials from Vittitoe (2004) and several other American servicemen. War museum storytelling here includes US veteran recollections (or those of their descendants) of their military service on Efate and also of subsequent returns to Vanuatu as tourists themselves. An excerpt from Reece Discombe's (1979) New Hebrides war history and various wartime maps and photographs are also displayed on the museum's walls.[9]

These two local war museums, stuffed with assorted relics, attract growing numbers of tourists (mostly from Australia and New Zealand, not the US). Each draws at least several groups of passing tourists daily. Local Vanuatu entrepreneurs are managing the salvage, selection, display and interpretation of war relics and the stories they want them to tell today. If seventy years after war's end, memory of personal experience has faded, recollected war relics and touristic appreciation of these (including the museum entrance fees they are willing to pay) bring memory back into play. Salvaged relics salvage history.

INDIGENOUS MUSEUMS

Pacific museums and cultural centres are commonly hybrid. Founded mostly in colonial times they mix metropolitan with island impulses to collect, curate and display objects; they blend diverse appreciations of nature, history and culture, and they address assorted audiences of tourists and locals (Geismar 2013, 128–9). Stanley asks of Melanesia, '[I]s there any cultural space left for a genuine autochthonous museum in any shape or form?' (2007, 4; see White 1996, 56). Efate's two local Pacific War collections suggest that there is indeed. Stanley, in fact, wondered whether this could be one of the war's outcomes: 'The detritus left behind by Europeans, particularly on battlefields, could either become pathetic relics or a source for future local display' (2007, 4). People have already taken up this challenge. On Guadalcanal, for example, Fred Kona and family collected war relics and displayed these at his homemade Vilu War Museum from 1975 until Kona passed away.[10]

As White notes for the Solomon Islands, apart from Fred Kona's curation efforts, Japanese and Allied veteran associations have designed and erected the majority of Guadalcanal's Pacific War memorials and mementos to tell their own stories (1996, 53). In Vanuatu, too, plans have been afoot for some years to develop a larger, sophisticated yet hybrid museum in Luganville, where war relics, fertile military dumpsites and remnant base installations much outnumber those extant on Efate. This, with galleries built to recall Quonset huts, would also serve a tourist clientele and any long-lived veterans still spry enough to return to Vanuatu.[11] Yet Kalkoa's and Tom's indigenous efforts to tell war stories and display relics in local terms are more significant and impressive than this international scheme. Even if indigenous museum visitors consist mostly of tourists, Kalkoa, Tom and their families control the selection and display of war relics and the stories they choose to tell about these.

As home-grown endeavours, Efate's indigenous war museums communicate more than war memory itself. They also convey people's enthusiasm for making new relationships and cultivating these—this interest in enduring cross-cultural friendship was one of the main themes we discovered when collecting war stories in the 1980s. As White notes, '[W]hen people who remember World War II tell their stories, they frequently do so by focusing on personal connections they developed with the foreigners who flooded through their islands' (1996, 56). Kalkoa spoke with pride about his contacts with a war historian in Canberra, with the Coca-Cola museum in Atlanta, Georgia, and with a visitor he identified as an American congresswoman. Tom, in addition to naming his museum after the lucky American pilot Vittitoe, also exhibits letters therein from Taylor Green, grandson of Marine meteorological officer William Green, and from Seabee veteran G. Robert Craig (for similar international connections invigorated by Kona's Vilu War Museum on Guadalcanal, see White 1996, 53–4).

War relic displays, moreover, are often less problematic in Melanesia than marketing artefacts or dance performances to tourists. Although local entrepreneurs over the years have transformed once sacred objects into suitcase-sized tourist artefacts and have founded dance teams or cultural villages to display the more spectacular aspects of island tradition, disputes frequently arise over rights to exhibit and profit from them. There are strong systems of *kastom* copyright throughout Melanesia that make problematic such displays of any sort of intellectual property, be this song, dance or object—in a museum, cultural centre, or hotel lobby (Geismar 2013, 68–71). The war and its surviving scraps, however, arrived from elsewhere. Foreign war relics, unlike local artefacts, at least partly escape traditional systems of knowledge control even if family access to a military dump site determines who might establish war museums. Indigenous war museums thus do more than keep alive selected war memories and war stories to be shared with tourists and the occasional interested local resident. They also figure in the emergence of new sorts of shared national culture in Vanuatu and an enlarging public domain (White 1996, 56). No resting in peace for the relics, they still have work to do.

ACKNOWLEDGEMENTS

I thank Ernest Kalkoa and Erick Tom for guiding me about their collections, as well as Vanuatu scholars Howard Van Trease and Augustron Asial for joining me on museum tours.

NOTES

1. Previously, in Bislama, trash salvaged from military dumps was *doti* ('dirty,' 'rubbish').
2. That is, citizens of Vanuatu.
3. In 2010, this road was rebuilt and paved with American aid funnelled through the Millennium Challenge Corporation.
4. The destroyer USS *Tucker* lies several hundred yards up the canal—also a scuba divers' favourite, although, unlike the *Coolidge*, it is scattered in pieces.
5. Judith Bennett (faculty of history, University of Otago) has an ongoing project on mixed-race children in the Pacific with American fathers and has helped set up a website for children to trace these men: www.otago.ac.nz/usfathers/ (accessed 20 August 2013).
6. Reece Discombe and Captain Robert Wyllie, with support from several Port Vila businesses and associations, prepared this exhibit, and Discombe continued to curate it while he lived.
7. Tourist Office publications regularly feature war history, e.g. 'SS *President Coolidge* puts Vanuatu at top 10 wreck dives world wide' (Vanuatu Tourism Office 2013, 5).
8. Local tour and cruise operators also market WWII history on Espiritu Santo, although scuba diving in Segond Canal and off Million Dollar Point is here the main attraction. American installations on Santo dwarfed those on Efate, and miscellaneous remains survive on land as well, including rusted Quonset huts

and ammunition bunkers near Matevulu College. Tourist arrivals on Santo, however, lag far behind those on Efate. In 2012, of 108,161 non-cruise tourists visiting Vanuatu, only 2,550 of these disembarked at Luganville where touristic infrastructure and celebrations of war history are relatively undeveloped (Vanuatu National Statistics Office 2013, 6).

9. Tom consulted former Vanuatu Cultural Centre Director Ralph Regenvanu and former Vanuatu US Peace Corps administrator Sara Lightner for museum display advice and to explore renewing connections with American veterans.
10. A second war collection was housed at Betikama Adventist College outside of Honiara, although the impetus for this may have come from the school's expatriate teachers.
11. Proposed museum maquettes are featured on a South Pacific WWII Museum Facebook page: www.facebook.com/SouthPacificWWllMuseum (accessed 22 August 2013).

BIBLIOGRAPHY

Bennett, J.A. (2012) '"Inevitable erosion of heroes and landmarks": an end to the politics of Allied war memorials in Tarawa?', in M. Gegner and B. Ziino (eds), *The Heritage of War*, London and New York: Routledge.

Cappelletto, F. (ed). (2005) 'Introduction', in F. Cappelletto (ed), *Memory and World War II: An Ethnographic Approach*, Oxford and New York: Berg.

Cooper, M. (2006) 'The Pacific War battlefields: tourist attractions or war memorials?', *International Journal of Tourism Research*, 8, 213–22.

Crowley, T. (1990) *Beach-la-Mar to Bislama: The Emergence of a National Language in Vanuatu*, Oxford: Clarendon Press.

Discombe, R. (1979) 'The New Hebrides at war: the part played by the New Hebrides during the following wars: Boer War 1902, Great War 1914–1918, World War II 1939–1945', unpublished manuscript, Port Vila, Vanuatu National Museum archives.

Garrison, R. (1983) *Task Force 9156 and III Island Command: A Story of a South Pacific Advanced Base During World War II Efate, New Hebrides*, Boston: Nimrod Press.

Gegner, M. and Ziino, B. (2012) 'Introduction: the heritage of war: agency, contingency, identity', in M. Gegner and B. Ziino (eds), *The Heritage of War*, London and New York: Routledge.

Geismar, H. (2013) *Treasured Possessions: Indigenous Interventions into Cultural and Intellectual Property*, Durham, NC: Duke University Press.

Geslin, Y. (1956) 'Les Américans aux Nouvelles-Hébrides', *Journal de la Société des Océanistes*, 12, 245–85.

Heinl, Jr., R.D. (1944) 'Palms and planes in the New Hebrides,' *National Geographic*, 86, 229–56.

Jasons Travel Media. (n.d.) *Vanuatu Visitor Guide 2012–2013*, Auckland: Jasons Travel Media.

King, A. G. (1991) *Vignettes of the South Pacific: The Lighter Side of World War II*, Cincinnati, OH: Self-published.

Kralovec, D. W. (compiler). (1945) 'A naval history of Espiritu Santo, New Hebrides', unpublished manuscript, Washington, DC, Command Files, World War II, US Naval Historical Center archives.

Lindstrom, L. (1989) 'Working encounters: oral histories of World War II labour corps from Tanna, Vanuatu', in G. M. White and L. Lindstrom (eds), *The Pacific Theater: Island Recollections of World War II*, Honolulu: University of Hawai'i Press.

Lindstrom, L. (1993) *Cargo Cult: Strange Stories of Desire from Melanesia and Beyond*, Honolulu: University of Hawai'i Press.

Lindstrom, L. (1996) *The American Occupation of the New Hebrides (Vanuatu)*, Macmillan Brown Centre for Pacific Studies working paper 4, Christchurch, Macmillan Brown Centre.

Lindstrom, L. (2011) 'Million dollar point', in C. Terrier, M. Abong and D. Tryon (eds), *101 Mots pour Comprendre le Vanuatu*, Nouméa: Éditions de GRHOC, 139–40.

Lindstrom, L. and Gwero, J. (eds). (1998) *Big Wok: Storian Blong Wol Wo Tu long Vanuatu*, Suva: Institute of Pacific Studies, University of the South Pacific; Christchurch: Macmillan Brown Centre for Pacific Studies, University of Canterbury.

Lindstrom, L. and White G.M. (1993) 'Singing history: island songs from the Pacific War', in P. Dark and R. Rose (eds), *Artistic Heritage in a Changing Pacific*, Bathurst, NSW: Crawford House Press, 185–96.

McQuarrie, P. (1994) *Strategic Atolls: Tuvalu and the Second World War*, Christchurch: Macmillan Brown Centre for Pacific Studies; Suva: Institute of Pacific Studies, University of the South Pacific.

Michener, J. A. (1951) *Return to Paradise*, New York: Random House.

Moon, M. and Moon, B. (1998) *Ni-Vanuatu Memories of World War II*, Diamond Harbour: Privately published.

Poyer, L., Falgout, S. and Carucci L. (2001) *The Typhoon of War: Micronesian Experiences of the Pacific War*, Honolulu: University of Hawai'i Press.

Spennemann, D.H.R. (1992) 'Apocalypse now? the fate of World War II sites on the central Pacific islands', *CRM Bulletin*, 15 (2), 15–6.

Stanley, N. (2007) 'Introduction: indigeneity and museum practice in the Southwest Pacific', in N. Stanley (ed), *The Future of Indigenous Museums*, New York and Oxford: Berghahn Books, 1–20.

Stone, P. (1997) *The Lady and the President: The Life and Loss of the SS* President Coolidge, Yarram, Victoria: Oceans Enterprises.

Sturken, M. (2007) *Tourists of History: Memory, Kitsch, and Consumerism from Oklahoma City to Ground Zero*, Durham, NC: Duke University Press.

Tabani, M. (2008) *Une Pirogue pour le Pardis,* Paris: Éditions de la Maison des Sciences de l'Homme.

Vanuatu National Statistics Office. (n.d.) 'Tourism summary'. Online. Available HTTP: www.vnso.gov.vu/index.php/tourism-stats (accessed 21 August 2013).

Vanuatu National Statistics Office. (2013) 'Statistics update: international migration' (May), Highlights, Port Vila: VNSO.

Vanuatu Tourism Office. (2013) 'One of the world's largest most accessible shipwrecks', *Vanuatu Tourism Island Nius*, 25 (12 August), 5.

Vittitoe, J. A. (1990) 'Passing thru . . .', *Korean War Educator, Veterans' Memories*. Online. Available HTTP: www.koreanwar-educator.org/memoirs/vittitoe_james/ (accessed 21 August 2013).

Vittitoe, J. A. (2004) 'Return to Efate', *Follow Me* (Official Publication of the Second Marine Division Association), 41 (5), 11–12.

Wallin, H.N. (1967) 'The project was roses', *The Navy Civil Engineer*, May/June, 16–19; July, 28–31; August, 26–28.

Wanhalla, A. and Buxton, E. (2013) 'Pacific brides: US forces and interracial marriage during the Pacific War', *Journal of New Zealand Studies*, 14, 138–51.

White, G. M. (1996) 'War remains: the culture of preservation in the Southwest Pacific', *CRM*, 19 (3), 52–6.

White, G., Gegeo, D. W., Akin, D. and Watson-Gegeo, K. (eds). (1988) *The Big Death: Solomon Islanders Remember World War II,* Honiara: Solomon Islands College of Higher Education; Suva: Institute of Pacific Studies, University of the South Pacific.

Young, J. E. (2000) *At Memory's Edge: After-Images of the Holocaust in Contemporary Art and Architecture*, New Haven, CT: Yale University Press.

10 Islands of No Return

Memory, Materiality and the Falklands War

Tony Pollard

Digging all day and digging all night
To keep my foxhole out of sight
Digging into dinner on a plate on my knees
The smell of damp webbing in the morning breeze
Fear in my stomach, fear in the sky
I eat my dinner with a weary eye
After all this it won't be the same
Messing around on Salisbury Plain

(From the song 'Island of No Return' by Billy Bragg, 1984)

INTRODUCTION

There is an increasing interest in the physical remains of recent wars, particularly those fought in the twentieth century. Archaeologists and anthropologists have offered meaningful insights into the human experience of the First and Second World Wars and their aftermaths through a close-grained analysis of monuments created not only as memorials but also as platforms for acts of commemoration (e.g. Tarlow 1999; Winter 1999, 55; Wall and Williams 2010; Carr 2012) and the material culture utilized in the fighting of those wars. Good examples of the latter are the examination of First World War objects that have been repurposed to create trench art (e.g. Saunders 2002). These approaches have informed the current study of memorials on the Falkland Islands, which considers among other things their diversity of form and location within the landscape. It is also suggested that debris left behind after the 1982 war plays a part in the process of remembrance (which does not necessarily involve the creation of memorials) and, along with the memorials, populates a landscape of memory.

This chapter is largely based on observations made during a visit to the islands by the author in December 2012, with the purpose of assessing the nature and extent of physical remains relating to the 1982 war. This discussion will focus on the land campaign and in the main, though not exclusively, on the British perspective.

HISTORICAL BACKGROUND

It is not the remit of this chapter to discuss in depth the history of the Falklands War or to rehearse the various debates surrounding the legitimacy of one side over the other on the thorny issue of sovereignty. It would, however, be disingenuous of the writer not to state that he believes that the feelings and aspirations of the islanders have primacy in any debate over sovereignty. It should also be added that, despite the term *Falklands War* being used throughout this chapter, neither side officially declared war on the other. Although the *Falklands Conflict* and the *Falklands Campaign* might be more accurate terms as far the intricacies of the laws of war are concerned, the events of May 1982 are most commonly remembered both locally and in the UK as the Falklands War. These caveats notwithstanding, it is clear that some basic historical background is required, as this is essential to an appreciation of the discussion that follows.

The Falklands War marked the climax of a long-standing disagreement between Argentina and Great Britain over the sovereignty of the islands, which are 8,000 miles away from the UK and 320 miles to the east of southern Argentina.

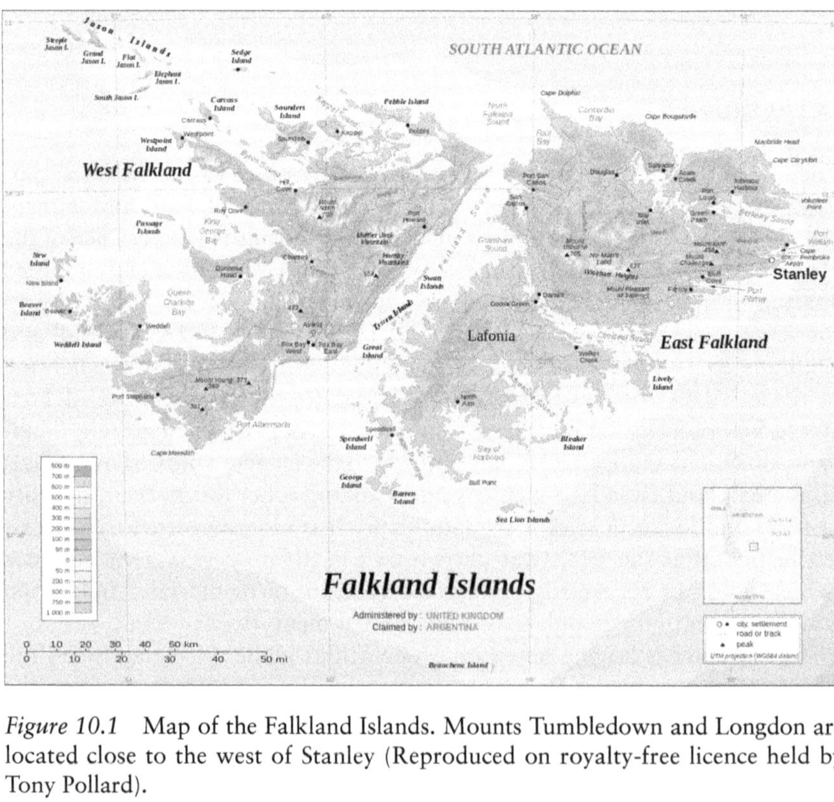

Figure 10.1 Map of the Falkland Islands. Mounts Tumbledown and Longdon are located close to the west of Stanley (Reproduced on royalty-free licence held by Tony Pollard).

The archipelago, which constitutes a British Overseas Territory, consists of East and West Falkland and over 700 smaller islands, with the overall landmass being around 4,700 square miles (slightly smaller than Northern Ireland). The capital, Stanley, is located in a sheltered bay on the eastern side of East Falkland (Figure 10.1). In 1982 the population was 1,820; in 2012 it was 2,932. The economy is traditionally based on sheep farming and fishing. In recent years, tourism has gained in importance, with cruise ships regularly berthing in Stanley during the summer months. In part because of this, battlefield tourism has also begun to make an impact on the islands' economy, though not all battlefield tourists originate on cruise ships as regular flights from the UK and South America bring in other visitors. The prospect of oil in the surrounding seas may see a further dramatic change, and this has also heightened tensions once again between the UK and Argentina.

The invasion of the islands by Argentina in April 1982 was an attempt by a fascist military junta in Buenos Aires to distract the populace from a free-falling economy and the inevitable backlash of having 'disappeared' thousands of Argentines with left-wing sympathies. For as long as there had been an Argentina (since 1816), the Malvinas (as they are known to Argentines) has been a place close to the average citizen's heart. As was hoped, the 1982 invasion fostered nationalist fervour, bringing Argentines of all political persuasions together for at least as long as the islands were flying their country's flag.

The British response to the invasion, which took the Argentines as much by surprise as their actions had the British, was to send a maritime task force 8,000 miles to retake the islands, with some of the 127 ships departing within just a few days of the Argentine landings. Ascension Island, situated at the halfway mark, was used as a way station and for a while accommodated the busiest airport in the world. Although negotiations, primarily involving the United States as a go-between, continued as the task force steamed south, it became clear that conflict was inevitable. The British had established a 250-mile exclusion zone, which, as the task force approached, became a total exclusion zone (TEZ) around the islands, and threatened to attack any Argentine ship or aircraft within it. The first casualties were suffered by the Argentines when, controversially, the Argentine cruiser ARA *Belgrano* was sunk with the loss of 323 lives by a torpedo launched by the British submarine HMS *Conqueror*, while she was outside the TEZ on 2 May. The first British casualties came when HMS *Sheffield* was badly damaged on 4 May (sinking on 10 May), with 20 men killed, by an Exocet missile fired from an Argentine aircraft.

British special forces were the first members of the task force to step foot on the islands. Prior to the main landings at San Carlos on 21 May, the Special Air Service (SAS) carried out a successful raid on an Argentine airbase on Pebble Island, from which aircraft could pose a serious danger. The landings were a relatively but not entirely bloodless affair for the British, and a beachhead was secured without much in the way of combat. However, ships anchored in San Carlos Water, which became known as Bomb Alley, were exposed to attack from Argentine aircraft, and several vessels were hit. The first major engagement took place on 28–9 May, when men of the 2nd Parachute Regiment captured the airfield and settlement of Goose Green, from where the garrison could have threatened the right flank of a British overland advance on Stanley to the east, where the main body of the occupying force was stationed. Among those killed in the battle, which like all of those to come cost the Argentines more casualties than the British, was the commander of 2nd Parachute Regiment, Colonel 'H' Jones, who was awarded a posthumous Victoria Cross for his actions.

An overland advance on foot was unavoidable as most of the British heavy lifting helicopter fleet had been destroyed in an Exocet attack on the MV *Atlantic Conveyor*, which had been transporting them to the islands. After taking just three days to negotiate over 50 miles of rough terrain carrying heavy packs, the men of 3 Commando Brigade prepared for attacks on the main Argentine defensive positions on the mountains to the west of Stanley. Meanwhile, British reinforcements in the form of 5 Infantry Brigade had arrived at San Carlos. Rather than have these men march across the islands, they were put on troop carriers and sailed to Bluff Cove, to the southwest of Stanley, where they could be disembarked to take part in the final push on the islands' capital. On 8 June, while the ships were at anchor with many men still on board, they were attacked by Argentine aircraft and suffered hits from bombs dropped at low altitude. Three aircraft were downed during the attack. The result was the worst single loss suffered by the British, with forty-eight men killed on RFA *Sir Galahad* and two on RFA *Sir Tristram*. In addition, six Royal Marines were killed when their landing craft was bombed in Choiseul Sound on its way to Bluff Cove. However, it was the Royal Navy's Sea Harriers and the RAF's Harriers flying from the aircraft carriers HMS *Invincible* and *Hermes* that won the war in the air, while also serving a vital ground support role.

Despite these losses British troops made their final approaches on the mountains. One by one the craggy peaks were assaulted and the Argentine defenders thrown back on Stanley, with the heaviest fighting taking place on Mount Longdon (11–12 June) and Mount Tumbledown (13–14 June). The final assault on Wireless Ridge (13–14 June) saw the bulk of the Argentine army stream back into the town, and the British braced themselves for a potentially costly assault, where the risk of casualties among the civilian population was great (three residents of Stanley were killed when Royal Navy gunfire accidently hit a house). Fortunately, the Argentines chose to surrender rather than to fight, and on 14 June British forces entered the town.

The Falklands War was a swift but costly conflict for the combatants, with 258 British and 649 Argentine servicemen killed and many more wounded on both sides in the seventy-four days between the Argentine landings on 2 April and the surrender in Stanley on 14 June. British success secured another term in office for Margaret Thatcher's Conservative government but doomed the junta in Argentina, with a return to democracy ushered in by elections in 1983. Relations between Argentina continue to be strained, as was evident on a number of occasions during the thirtieth anniversary in 2012 (see example related to 2012 Olympics later in the chapter). The presence of a strong British garrison airbase on the islands, however, makes it unlikely that the events of 1982 could be repeated.

MONUMENTAL ISLANDS—MEMORIALIZING THE WAR

The 1982 conflict created a physical legacy in the form of monuments erected for the purpose of memorialisation, mainly dedicated to the dead but also to the living[1]. The focus for the present discussion will be the memorials located on the islands (those in the UK will be covered in another paper). Memorials on the islands are so numerous and widely distributed across the entire theatre of the war that it is impossible not to regard them as a component of the landscape (cf. Trigg 2007, 295). Indeed the overall impression is very much in keeping with the idea of a *memorialscape* (Carr 2012), in which monuments punctuate narratives woven around the memory of the war and play an active role in interpreting past events (Pollard 2002; Carr 2012). It will, however, be argued here that the presence of these monuments within well preserved *battlescapes* (Pollard 2003), composed of topographic features, military structures and scatters of material culture (largely thanks to the isolated nature of the islands and the location of most battlefields away from areas of population), creates a mnemonic palimpsest, which today is mediated and manipulated through the interaction of battlefield guides and visitors.

The most immediate memorials to any war are the graves resulting from it, and an abiding image from 1982 is television footage of men from the 2nd Parachute Regiment standing over a mass grave containing a row of corpses in body bags as a padre reads out the names of the dead. These men died in the Battle of Goose Green and were given temporary burial close to the field hospital at Ajax. The Argentine dead were also subject to makeshift burial, often by the British, again in mass graves but usually close to where they died.

Only with the final Argentine surrender could thought be given to the permanent burial of the dead. Throughout the twentieth century, it was normal practice for British war dead to be buried close to where they fell. This was an official policy followed by the British government during the First World War (Sledge 2005, 204), and one that became entrenched in 1920 when it was decided, following debate in the House of Commons, that no programme of post-war repatriation would be instituted (Dyer 1994, 13). Since 1917, British and Commonwealth war cemeteries, with their uniform white headstones, have been the responsibility of the Commonwealth War

Graves Commission (CWGC, initially the Imperial War Graves Commission). Following the Falklands War, the decision was made to establish a war cemetery close to Blue Beach, which played a vital role during the landings. There is also an Argentine cemetery on the island, containing the graves of 247 men marked by stone crosses. Today, the British cemetery, contained within a circular enclosure incorporating a war memorial within this wall, accommodates just fourteen graves, including that of Colonel 'H' Jones, who was killed at Goose Green. This small number reflects a sea change in British attitudes towards the burial of the dead brought about by the Falklands War.

Even before the war was over there were pleas by bereaved families for the return of their dead and by early June, Margaret Thatcher's government was considering the issue (Anon 1982, 5). The result was a break with tradition. For the first time since the First World War, the bereaved were given the choice to have their deceased loved ones buried close to the battlefields or have them brought home. Perhaps not surprisingly, a large proportion of families opted to bring their men home, the distant and isolated location of the islands probably being a factor in some, if not all, of these decisions. In total, sixty-four of the dead were returned to the UK by ship in November 1982. With a precedent set, it became normal practice for the bodies of British personnel killed on active service to be taken back to the UK.

In addition to the graves, a wide range of other memorials are scattered across the battlefields. The usual type of unit memorial is the cross, normally set into a cairn. Examples include the 2nd Parachute Regiment memorials at Darwin/Goose Green and on Wireless Ridge, and those on the mountains including the Scots Guards cross on the summit of Mount Tumbledown and the 3rd Parachute Regiment cross on Mount Longdon (Figure 10.2). These provide a focus for smaller personal memorials, such as the wooden crosses and offerings that include wooden plaques bearing inscriptions and regimental badges and buttons.

As time passes, more monuments appear, with anniversaries such as the twenty-fifth and thirtieth marking important milestones for commemoration. Military pilgrimages to the islands also increase at these times. A notable example on Mount Longdon takes the form of a large granite cube, from the fractured corner of which steel poppies protrude, one each for the twenty-three British deaths during the battle (the Argentines lost thirty-one). The overall memorial, however, is dedicated to Sergeant Ian McKay of the 3rd Parachute Regiment, who won the Victoria Cross posthumously for storming an Argentine position in the crags, an act that cost him his life. This is one of two memorials to MacKay on Longdon, the other being a simple wooden cross, which, although modest, is larger than the small crosses that sometimes bear poppies and are found in greater numbers on the islands, while being ubiquitous in Britain on Remembrance Sunday. This is located at the base of the crag further to the east. This duplication reflects disagreement, even among the men who were fighting alongside him, as to

Figure 10.2 Scots Guards memorial on Mount Tumbledown and accumulation of personal memorials. Stanley is in the distance (Copyright Tony Pollard).

where MacKay was actually killed.[2] When viewed by the present author at the end of the thirtieth anniversary year, MacKay's cross, which bears a brass plaque and a parachute regiment badge, was accompanied by several plastic poppy wreaths and a red beret held in place under a stone. Also nearby was an ammunition box containing a tin of brass polish and polishing cloths intended for use on the plaque. This is a feature common to most of the memorials with brass plaques, for example the one dedicated to 'H' Jones and the main unit memorial crosses, and represents a good example of an unusual object biography, which features the reuse of military material culture in the memorialization process.

The steady increase in the number of individual memorials—be they stone or brass plaques, some mounted or resting on rocks, some on small stone cairns, or wooden crosses (again, larger than the 'poppy' crosses), placed by former comrades and/or family—is particularly apparent on Longdon and Tumbledown. Here, some of them cluster around the base of the regimental crosses, the plinths of which carry the names of all those killed. Others among these personal markers purport to mark the spot where an individual fell, but, as is the case with MacKay, certainty is difficult, especially as much of the fighting occurred in the dark. This 'repopulation' of the mountains is only possible due to the small numbers of men killed (compared to battles of earlier wars), but on another level it may be a practice in part born out of the removal of the dead from the theatre of war. When it comes to remembering them on the islands, the repatriation of these men (taking them home to the UK and away from the battlefields) has ironically created a new category of 'missing' men (Dyer 1994, 14), and these personal memorials serve to stand in for gravestones now located 8,000 miles away. The names of these men also appear on the back wall of the Blue Beach cemetery and on the memorial wall at the liberation monument in Stanley. There is also an arboretum in Stanley, where every British casualty is represented by a tree or shrub.

Other than the military cemetery to the north of the Darwin/Goose Green isthmus, there are no officially sanctioned Argentine memorials on the islands. Only one memorial to the dead of both sides was seen while exploring the mountains: a small marble plaque on Mount Longdon bearing a bilingual prayer to the dead of both sides, left by Jimmy O'Connell, a veteran of the 3rd Parachute Regiment. Argentine veterans do, however, visit their old battlefields, at which time they may, like their British counterparts, choose to erect personal memorials. Often, these are small white crosses, some of which bear the name of an individual. The intention, again as with the British examples, is to place these crosses where the men fell, and a number of them were encountered during site visits in 2012, including several on Mount Longdon. These are not welcomed by all, and it is not unusual for them to be removed, as shown by online visitor photographs which feature these crosses where there are now none. Also observed were four spots of epoxy glue on a rock, which according to a battlefield guide had secured an Argentine plaque prior to its removal. It is not entirely certain whether these are removed by islanders, British veterans or members of the present garrison, but it comes as no surprise as battlefields can continue to serve as a focus for symbolic conflict long after the real fighting is over (Pollard 2007, 121). The war might have ended on 14 June 1982, but there can be no doubt that relations between Argentina and Britain continue to be strained when it comes to sovereignty.

The Argentines have also participated in this war of symbols, and in May 2012, in the midst of the thirtieth anniversary, they released a television advert for their London Olympic team, which portrayed an Argentine hockey player exercising on the First World War memorial in Stanley and running up steps, at the bottom of which British Royal Marines were forced to lie face down after the surrender of Government House in April 1982. The advert closes with the provocative statement, 'To compete on English soil we train on Argentine soil' (Gilbert 2012a). Possible retaliation came in July 2012, when a glass case at the rear of the Argentine cemetery, containing a statue of the Virgin of Luján, was smashed, possibly by gunshot. The blue-and-white robes adorning the small figurine are said to be the only flag-like expression of Argentine national pride permitted on the islands (Gilbert 2012b).

British war memorials on the Falklands take a wide variety of forms, ranging from the formal and officially sanctioned to the totally personal and individual. Examples of the former include the CWGC cemetery at Blue Beach and the liberation monument in Stanley. These can be found not far from Thatcher Drive, the sign for which became a focus for remembrance at the time of her death in 2012 (since first writing of this chapter, a statue of Margaret Thatcher has been placed at this location—January 2015). These locations unintentionally serve to monumentalize the start and end points of the British land campaign. Less grandiose are the unit memorials, the most obvious form of which is the monumental cross, but in many cases these have been erected with military or other institutional support, such as that provided by the South Atlantic Medal Association.

METAL AND MEMORY—WEAVING NARRATIVES WITH MATERIAL CULTURE

Having discussed the wide variety of memorials to those killed on the islands in 1982, this section will consider how material remains have been incorporated into acts of remembrance. The mnemonic function of memorials and the commemorative acts centred on them, which stimulates and focuses memory, has been much discussed (e.g. Holtorf 1997, 50; Jarman 2001, 172; Walls 2010). The same can be said for objects associated with war modified into trench art and removed from the battlefield (e.g. Saunders 2002, 2003; Whittingham 2008). What has escaped attention is the part played by objects still situated within their original battlefield context, which in some cases include structures such as fortifications, in the creation and maintenance of the various and at times competing narratives that play an essential role in commemoration.

In order to examine the potential for a wide spectrum of remains situated within the landscape to serve as an *aide-memoire*, it is first necessary to look at the processes of deposition and attrition that have created the battlescapes that took the author to the islands in 2012. The Falklands War impacted the landscape of the islands in numerous ways. The Argentine occupation and preparation for a British military response ushered in an intensive programme of construction. The main objective of a British assault was always going to be the islands' capital, Stanley. Accordingly, large numbers of Argentine soldiers were placed on the slopes and summits of the mountains that provided a natural barrier to any advance on the town from the west. Numerous bunkers and sangars (defensive barricades) were constructed, and in places virtually every nook and cranny provided by these dramatic geological features was utilized. The vast majority of these structures were built from loose rocks, which exist in abundance in the screes and stone runs characteristic of this glaciated landscape. These structures served both as living spaces, many of the enclosures accommodating tents (as shown by fragments of tent fabric and brass eyelets in some), and as fighting positions. The onset of winter and the exposed conditions on the mountains provided very unfavourable conditions for soldiers, who were by and large poorly equipped to withstand extreme weather. Today, the presence on the mountains of numerous insoles from gym shoes is a result of conscripts thinking they were being sent on a routine exercise in warmer Argentina rather than to a potential combat zone in the Falklands (Middlebrook 1989, 221).

Many of these structures still survive today, despite some of them being damaged by explosive impacts during the battles which raged over the mountains in the final days of the conflict. Although a basic typology for these structures has been suggested (Pollard 2014), they are essentially makeshift in character and in part reflect the rapid advance of the task force. There can be no doubt that had the Argentines had longer to establish themselves, they would have built concrete pillboxes and other more permanent installations. The rock-built structures on the mountains, where even caves provided shelter, were also a response to the peculiarities of that environment. Elsewhere on the islands, where peaty soil predominates, trenches and foxholes were dug, with those on Darwin Ridge (guarding the northern approach to Goose Green) still being visible as negative features (some of which contain blankets and other objects). No matter what the basic nature of the structures, other materials were often used to enhance their protective qualities, be that protection from the elements or incoming fire. These other materials included timber, corrugated iron and iron rails. All of these materials were sourced locally or, more accurately, scavenged. For instance, the timber used on the mountains largely appears to originate from wooden pallets used by the islanders to stack drying peat cut from bogs on the lower ground, while the rails were removed from a light railway servicing a nearby quarry.

Unfortunately for the Argentine troops occupying these structures, they did not prove effective in halting the enemy advance, as demonstrated by British successes in taking Goose Green and storming the mountains. The occupation of these sites, before and after the British landings, as well as the fighting which raged around and in them, resulted in the deposition of numerous objects, including food containers, eating utensils, water bottles, shoes, blankets, items of clothing, first aid supplies, radio batteries and spent ammunition, with the enclosed structures acting as artefact traps (many of these being Argentine but not exclusively so). Artefact deposition was not, however, limited to the structures and to where fighting took place, inasmuch as large quantities of matériel were left behind (most of it by the Argentines) as the British advance on Stanley progressed. These objects ranged from substantial elements of military equipment, such as aircraft wrecks, mobile cookers, vehicles, heavy weaponry (recoilless rifles, mortars, heavy machine guns etc.) to what, on the face of it, are artefacts of a less military aspect, and it is perhaps these that have some of the most interesting stories to tell.

However, before turning to artefacts that remain on the battlefields, it is vital to consider the processes of attrition that lie behind the removal of large amounts of battlefield debris. Battlefield clearance by the British military began in the immediate post-war period and included the collection and detonation of unexploded ordnance (a process that has left tell tale rectangular craters in the ground) and the commandeering of intact hardware such as vehicles. Much of the unwanted material from Mount Longdon was removed and dumped at a location on the southern flank of the mountain, and over the years this has itself been picked over by souvenir hunters.

Souvenir hunting by tourists and members of the current British garrison may be more limited in intensity than those early clear-up operations but it is constantly gnawing away at what could be termed the archaeological resource. There is for instance enough material in the Mount Pleasant military base to furnish a museum (and indeed there has been some thought given to creating one there).

Several local guides provide tours of the battlefields, but when these involve veterans of the conflict, they are more akin to pilgrimages, as the commemoration of fallen comrades provides a strong motivation (Lloyd 1998; Seear 2012), as shown by the growing number of personal memorials on the mountains and elsewhere. These guides, one of whom was on the islands as an adolescent during the war, play a vital role as gatekeepers for the war's heritage and actively use the physical remains of the war, including the scattered artefacts, as vivid *aide-memoires* in the narratives they construct, essentially through the medium of storytelling, while guiding clients through these landscapes of conflict. As the loaded term *pilgrimage* implies, these tours can construe acts of memorialization, encompassing not only the memorials (in their various guises), which provide thought-provoking way stations, but also the structures and artefacts scattered across the landscape. The landscape is more than a stage on which the war was acted out, but there can be no denying that its dramatic character, encapsulated by the castellated rock formations on the mountains, can only serve to further focus the imagination of the casual visitor and stimulate the memory of the returning veteran. These tours transform the battlefields from landscapes of memorialization, populated by monuments, into landscapes of memory, populated by the residues of the events and the people memorialized.

Unfortunately, the removal of artefacts has been so severe in places that the guides have been forced to hide or cache artefacts close to where they lay in order to ensure they are not removed, bringing them out to show their clients as and when required. These include a morphine injector stashed in a rock crevice at the location of the dressing station at the western end of Mount Longdon and a heavily rusted magazine from an FN rifle close to the other end, its poor condition a reminder that decay also plays its part in attrition. Although it is now illegal to remove battlefield relics from the islands, this does not stop Argentine veterans who are fond of digging up their old positions to recover objects associated with their time on the islands in 1982; in theory, any finds could obviously be their own possessions. While some veteran collectors leave the recovered objects on the islands (the present author was shown a helmet dug up from a foxhole at Goose Green by an Argentine veteran), others will post artefacts back to Argentina in order to outmanoeuvre airport customs officials.

From the reconnaissance made by the author in 2012, it was clear that artefact preservation was better on the lesser known sites than on the iconic battlefields, with Tumbledown and Longdon comparing unfavourably with the other mountain battlefields. This might suggest that artefacts from these iconic battles are considered to have a greater 'relic' value' (Novotny 2012, 196–7) than those from other sites or simply be a reflection of the greater number of visitors to these sites.

Souvenir hunting can take different forms, and for the purposes of this discussion several examples observed during site visits in 2012 will suffice to demonstrate this. While being shown the wreckage of an Argentine Dagger aircraft on Pebble Island, which had been shot down on 24 May 1982 by a British Sea Harrier, local guide Allan White noticed that a plate stamped with a maker's mark had been removed from the ejector seat, probably just a few days earlier. As the veritable scrapyard of material on the flank of Mount Longdon attests, not all collected material is moved away from the battlefields. This was also the case with two Argentine 7.62 millimetre cartridge cases and a shell fragment, which a visitor had placed on the gatepost of the memorial to the men of 2nd Parachute Regiment killed in the Battle of Goose Green. It is tempting to regard these as a form of offering, perhaps tying the monument to the actual fighting. However, an unfortunate side effect of this action is that the spatial information provided by the bullets while *in situ* on the battlefield is lost; the cartridge cases would mark the location where a weapon was fired.

Elsewhere in Goose Green there is an example of artefact deposition that has a strong association with memorialization. Lieutenant Nicholas Taylor was buried by the Argentines after his Harrier jet was shot down by ground fire, and today his is the only British military grave on the islands remaining outside the Blue Beach cemetery. When visited by the author, his grave was covered with numerous wreaths and poppy crosses, but what really caught the eye was the small accumulation of aircraft debris sitting immediately behind the gravestone (Figure 10.3).

These pieces of Taylor's aircraft are known to have been deposited by his widow while visiting the grave in 2002 (Ramsey 2009, 153). It might be reading too much into their deposition to regard them as a material allusion to flight and a journey, but there is certainly a well established tradition of the use of metaphor on burial monuments (Pollard, 1999, 36; Tarlow 1999, 47). The accumulation of material is certainly an example of how memorials can take on additional meaning as time passes.

The study of object materiality, cross-cutting between archaeology and anthropology, is now well established, particularly in the sphere of conflict

Figure 10.3 Grave of Lieutenant Taylor. Note aircraft wreck parts behind gravestone (Copyright Tony Pollard).

studies. At the forefront have been explorations of the various meanings to be drawn from trench art created during the First World War, which include objects such as a bullet removed from a wound and then fashioned into a crucifix (Saunders and Cornish 2009, 4). Key here are object biographies, which as in the case of the bullet crucifix can be complex, surprising and long, with changes in functionality and transformations of symbolic meaning holding the capacity to tell us much about the nuances of human behaviour in extreme circumstances. There has however been less interest in the materiality of more mundane objects, which may not at first viewing connote conflict or warfare. One example was encountered in the mouth of a rock shelter with an attached wooden lean-to on the southern side of Mount William. Here a tin lid, possibly from a tin of paint, had been used to stand in for a frying pan or skillet after the addition of a handle formed from bent fencing wire. It would appear that these items have survived the predatory activities of souvenir hunters because of their 'mundane' and in some cases enigmatic character (Pollard 2014, 44). In order to understand the relevance of this object, one must have some knowledge of the straits to which Argentine soldiers were put when it came to surviving extreme conditions with inadequate equipment and food supply. Some men reverted to theft from military stores in Stanley, while others killed sheep and traded them for commodities such as cigarettes (Middlebrook 1989, 221).

Battlefield guides can play a key role in contextualizing these low-key objects within their narratives, as they are unlikely to feature in traditional works of history or even memoirs. Contrary to this, some of the uses to which this recycled material was put are unlikely to leave an archaeological trace and so make us entirely reliant upon the testimony of veterans. For instance, posts and wire from a fence were used to drag two mobile cooker units to the eastern end of Mount Tumbledown (Middlebrook 1989, 222).

It is clear from the foregoing that the *in situ* survival of material culture related to the 1982 war plays into the acts of remembrance associated with battlefield tourism and veteran pilgrimage. These are not sterile deposits but a resource actively drawn upon in order to weave narrative threads, which themselves serve to tie artefacts, memorials and topographic features together in this landscape of memory.

CONCLUSION

Sir Lawrence Freedman, in the closing sentence of the *Official History of the Falklands War*, wrote: 'So while in many respects this conflict still stands out as the last war of a past imperial era, in others it can now be recognized as the first war of the post cold-war era' (Freedman 2005, 747). This chapter has explored other ways in which the war can be remembered. By considering alongside battlefield remains alongside memorials, which thanks to the islands' isolated position are relatively well preserved (cf. Walls 2010, 385), it has been argued that the islands represent a landscape of memory, in which the mediation of battlefield guides and returning veterans creates historical narratives as part of a process of remembrance.

It is the author's aspiration not only to map these remains before too much further attrition can take place but to involve veterans of the war in this process. The material remains of twentieth-century conflicts also survive well on other islands, as can be seen within this volume. The Falkland Islands are unique, however, in that their war took place in 1982 and not 1942–5 (as in the case of Pacific War in the Second World War). When considering any approach to the cultural heritage of war, this brings with it the important premium of veterans who are still relatively young and therefore still capable of adding to our sum of knowledge, especially if, as proposed here, they are engaged in the mapping of the remains. Much could be learned from such an exercise, with the outcomes not only advancing our understanding of the combatant's experience in earlier wars, including the Second World War, with which there are many parallels, but also in helping us to understand the processes of attrition that have impacted conflict landscapes in less isolated regions, such as mainland Europe. On this basis, it seems highly likely that the physical remains of the Falklands War have the potential to advance the archaeological study of twentieth-century conflicts in ways that have only briefly been touched upon here. Far from being islands of no return, as the title of this chapter has suggested, the Falklands are in fact islands of many returns.

ACKNOWLEDGEMENTS

I would like to thank the people of the Falkland Islands for their warm hospitality in 2012, with special gratitude owed to my guides Tony Smith and Allan White. Jon Cooksey, Kevin Harris and Nick Rose have shared their knowledge and experience generously. Any misrepresentations or errors are the fault of the author alone.

NOTES

1. There is a blurring between the terms *monument* and *memorial*, with the former suggesting a structure of some stature and permanence, which can of course function as a memorial, while it would be stretching things to suggest that small and perhaps temporary items, such as the wooden poppy crosses, are anything other than memorials.
2. Personal communication between the author and battlefield guide Tony Smith while on tour of battlefield in December 2012.

BIBLIOGRAPHY

Anon (1982) 'British families protest Falkland burial for troops', *The Michigan Daily*, 2 June.

Carr, G. (2012) 'Examining the memorialscape of occupation and liberation: a case study from the Channel Islands', *International Journal of Heritage Studies*, 18 (2), 174–93.

Dyer, G. (1994) *The Missing of the Somme*, London: Hamish Hamilton.
Freedman, L. (2005) *The Official History of the Falklands Campaign.* Vol II: *War and Diplomacy*, Abingdon, Oxon: Routledge.
Gilbert, J. (2012a) 'Argentina's foreign minister refuses to apologise for Falkland Islands Olympics video', *The Telegraph*, 7 May. Online. Available HTTP: www.telegraph.co.uk/news/worldnews/southamerica/falklandislands/9249953/Argentinas-foreign-minister-refuses-to-apologise-for-Falkland-Islands-Olympics-video.html (accessed 17 September 2014).
Gilbert, J. (2012b) 'Argentine war cemetery in Falkland Islands vandalised', *The Telegraph*, 31 July. Online. Available HTTP: www.telegraph.co.uk/news/worldnews/southamerica/falklandislands/9441941/Argentine-war-cemetery-in-Falkland-Islands-vandalised.html (accessed 17 September 2014).
Holtorf, C. J. (1997) 'Megaliths, monumentality and memory', *Archaeological Review from Cambridge*, 14 (2), 45–66.
Jarman, N. (2001) 'Commemorating 1916, celebrating difference: parading and painting in Belfast', in A. Forty and S. Küchler (eds), *The Art of Forgetting*, Oxford: Berg, 171–95.
Lloyd, D. W. (1998) *Battlefield Tourism*, Oxford: Berg.
Middlebrook, M. (1989) *The Fight for the Malvinas: The Argentine Forces in the Falklands War*, London: Penguin Books.
Novotny, J. (2012) 'Sedition at the supper table: the material culture of the Jacobite Wars, 1688–1760', unpublished PhD thesis, Glasgow, University of Glasgow.
Pollard, T. (1999) 'The drowned and the saved: archaeological perspectives on the sea as grave', in J. Downes and T. Pollard (eds), *The Loved Body's Corruption: Archaeological Contributions to the Study of Human Mortality*, Glasgow: Cruithne Press, 30–51.
Pollard, T. (2002) 'The mountain is their monument: an archaeological approach to the Anglo-Zulu war of 1879', in P. Doyle and M. R. Bennett (eds), *Fields of Battle: Terrain in Military History*, New York: Springer, 117–35.
Pollard, T. (2003) 'The value of enmity: remaking and revisiting historic battlefields in Britain and the United States', *Landscapes*, 4 (2), Bollington, Cheshire: Windgather Press, 25–34.
Pollard, T. (2007) 'Burying the hatchet?—the post-combat appropriation of battlefield spaces', in L. Purbrick, J. Aulich and G. Dawson (eds), *Contested Spaces: Sites, Representation and Histories of Conflict*, Basingstoke, Hampshire: Palgrave MacMillan, 121–45.
Pollard, T. (2014) *Mapping the Cultural Heritage of the Falklands War: Report on Scoping Visit.* Unpublished report, Glascow: University of Glasgow.
Ramsey, G. (2009) *The Falklands War: Then and Now*, London: Battle of Britain International.
Saunders, N. J. (2002) 'Excavating memories: archaeology and the Great War, 1914–2001', *Antiquity*, 76 (291), 101–8.
Saunders, N. J. (2003) *Trench Art: Materialities and Memories of War*, Oxford: Berg.
Saunders, N. J. and Cornish, P. (eds). (2009) *Contested Objects: Material Memories of the Great War*, London: Routledge.
Seear, M. (2012) *Return to Tumbledown: the Falklands-Malvinas revisited*, Nottingham: Critical, Cultural and Communications Press.
Sledge, M. (2005) *Soldier Dead: How We Recover, Identify, Bury and Honor Our Military Fallen,* New York: Columbia University Press.
Tarlow, S. (1999) *Bereavement and Commemoration: An Archaeology of Mortality*, London: Blackwell.
Trigg, J. (2007) 'Memory and memorial: a study of official and military commemoration of the dead and family and community memory in Essex and East London', *Journal of Conflict Archaeology*, 3 (1), 295–315.

Walls, S. (2010) *The Materiality of Remembrance: Twentieth Century War Memorials in Devon*, unpublished PhD thesis, Exeter, University of Exeter.
Walls, S. and Williams, H. (2010) 'Death and memory on the home front: Second World War commemoration in the South Hams, Devon', *Cambridge Archaeological Journal*, 20 (1), 49–66.
Whittingham, C. (2008) 'Mnemonics for war: trench art and the reconciliation of public and private memory', *Past Imperfect*, 14, 86–119.
Winter, J. (1999) 'Kinship and remembrance in the aftermath of the Great War', in J. Winter and E. Sivan (eds), *War and Remembrance in the Twentieth Century*, Cambridge: Cambridge University Press, 40–60.

11 The Coastwatcher Mythos

The Politics and Poetics of Solomon Islands War Memory

Geoffrey White

In 2008 one of the national newspapers in Solomon Islands ran a story with the headline 'Vandalism to Japanese War Monument,' followed the next day by a story titled 'Vandalism Blow to Tourism in Solomons: Bureau General Manager.' These articles reported the ill-fated escapade of would-be thieves who attempted to carry off a rather massive bronze sculpture that adorned the Japanese Peace Memorial perched near the top of Mt Austen overlooking the Solomon Islands capital of Honiara.[1] With the intention of selling the bronze for scrap metal, the thieves removed the statue from its base and managed to carry it about 200 metres down the mountain road towards a nearby settlement area before abandoning their project and leaving the sculpture by the side of the road.[2]

I recall this incident here to introduce the question of local significance of war memorials and monuments that mark the battlefields of Guadalcanal surrounding the capital city of Honiara (see Figure 11.1). Despite the fact that the very existence of Honiara derives from the war,[3] Solomon Islanders today are more likely to think of the Pacific War as 'not our war' and know little of its military history (Zoleveke 1988). Whereas an act of vandalism such as the statue theft may in itself have little meaning, the fact that the general manager of the Solomon Islands Visitor Bureau felt the need to remind his fellow citizens that 'these monuments represent historical events which visitors to Solomon Islands would want to see' suggests that the Japanese memorial, like others installed by United States, Australian, British or New Zealand interests, may have more to do with visitors' interests than those of Solomon Islanders. Although the general manager did note that 'the monument is a landmark in the country's [Solomon Islands'] history', he emphasized that it is there to 'honour those who have lost their loved ones during the Second World War', noting that every year Japanese attend ceremonies there to honour their war dead.[4]

It is certainly the case that war memorials in Guadalcanal have largely been the initiative of foreign powers. In part this is related to the resources and strong commemorative cultures of the militaries that fought the war, including veterans groups (in the United States, Australia and New Zealand) and bereaved families organizations (in Japan). In the 1980s and especially

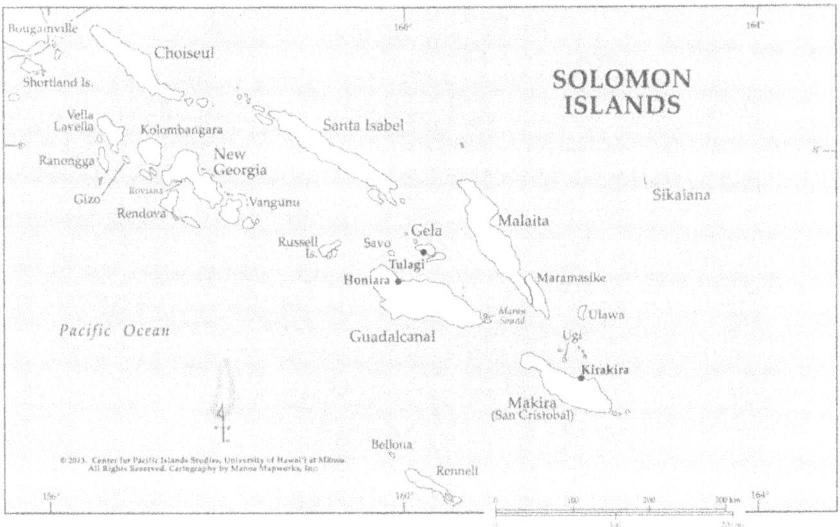

Figure 11.1 Solomon Islands (Copyright Center for Pacific Islands Studies, University of Hawai'i Mānoa. Cartography by Mānoa Mapworks).

the 1990s during the period of WWII fiftieth anniversaries, war veterans supported by national militaries, veterans associations, and government agencies were active in building, dedicating and visiting memorial sites throughout the Pacific, participating in joint (transnational) commemorative events held at island battle sites. But the dominance of foreign initiatives is also the result of a certain lack of interest among Solomon Islanders to commemorate the war in terms of Allied and Japanese military narratives.

In an earlier paper about the commemoration of the fiftieth anniversary of the Guadalcanal battle in 1992 (White 1995), I argued that national war remembrance at the time was driven largely by outside interests, focused especially on the dedication of a monumental new American memorial on Skyline Ridge overlooking the battle sites (partly in response to the Japanese 'peace memorial' that had been erected on a mountain top overlooking the capital in 1980). The fiftieth anniversary activities included the dedication of a statue honouring the most famous Solomon Islander in the war, Sir Jacob Vouza—a native policeman who survived capture and interrogation by the Japanese and went on to fight with the US Marines (Tregaskis 1943). On the surface, the dedication of the Vouza statue would appear to be just the kind of initiative that turns the spotlight onto the native dimension of Solomons' war history—a project that might help build a national public culture, fashioned from elements of shared history. Yet it, too, was largely seen as the project of Australian and American supporters and was criticized as such even before the statue was installed.[5] Editorializing in the main national newspaper, former Prime Minister Sir Peter Kenilorea wrote as follows:

> The Second World War was not our war and Sir Jacob Vouza's proposed statue is a form of 'grease' by Americans to allow the Solomon Islands Government to accommodate the memorial. . . .
>
> What possible benefits do we, as a country, get out of the War Memorial?
>
> This simply reinforces local peoples' sense of inferiority.
>
> The idea to build the monument, its design, the money and the technology all belong to foreigners.[6]
>
> (White 1995, 538)

As the case of the American memorial on Skyline Ridge illustrates, expressions of national war memory in Solomon Islands are extensively shaped by the military histories and commemorative practices of the warring powers. These histories, if they mention Islanders at all, tend to cast them as 'loyal natives'. Indeed, it was the recognition that indigenous histories and the oral archive of stories, songs and dances that record them are largely missing from the global historical record that motivated earlier collaborative research on war memory in Solomon Islands in the 1980s and 1990s (Laracy and White 1988; White, Akin, Gegeo and Watson-Gegeo 1988; White 1989a). A primary goal of that project was to bring more recognition to Islander contributions (scouting, labour support, soldiering) through conferences, publications and media programs. A more fundamental insight to emerge from that project, however, was recognition that indigenous histories and military narratives of victory in the Battle of Guadalcanal are to some degree incommensurable, constructed in different registers for different purposes. This insight also leads to the realization that in national and international contexts, the well known military histories tend to overwrite the more local indigenous histories (even in our own project!) (Riseman 2012).

The politics of remembering that render more localized, indigenous histories invisible to (inter)national publics are typical of colonized societies where war memory is entangled with longer social histories that involve internal conflicts, struggles for political autonomy and religious or regional movements (Fujitani, White and Yoneyama 2001; Cappelletto 2005). To understand the significance of the war for local, indigenous populations is to be able to relate wartime experience to these contexts. In Solomon Islands, it is because of the longer (and more local) context of colonial history that the war is often recalled as a moment of empowerment, the beginning of the end of a rigidly hierarchical colonial system in which European bosses were 'master' and Melanesian natives 'boy'. The war in this context was a different kind of 'turning point', in the march towards independence rather than in the course of military history. Colonial histories often do not articulate well with the military narratives of victory in the Pacific that inform commemorative events and tourism practices. Yet because military histories drive international interest in the war, war tourism (oriented to international audiences) will inevitably reproduce Euro-centred histories with less relevance for local lives, in this case the lives of Solomon Islanders (who mostly experienced the war outside combat situations).

In this chapter it is not possible to adequately address the social and cultural bases for Solomon Islanders' collective memory of the war. However, I take up this question by drawing on previous research as well as considering a recent Coastwatcher Memorial project which dedicated a new monument in 2011 and is working to advance public education on Solomons' war history. This latest project offers an opportunity to examine points of connection and disconnection between indigenous histories and the dominant narratives that underwrite war memory projects initiated both inside and outside of Solomon Islands since the fiftieth anniversary in 1992. In discussing the Coastwatcher Memorial project, I argue that it, too, reproduces much of the conventional Allied narrative evident in WWII tourism but, with the involvement of Solomon Islanders, opens up new possibilities for local voices to speak from indigenous perspectives.

WAR TOURISM IN SOLOMON ISLANDS

Viewed from the perspective of those closest to plans for national development in Solomon Islands, there are many reasons that one would wish to make Guadalcanal a destination for WWII tourism. As a battlefield, the significance of the landscape on Guadalcanal ranks among the most iconic of WWII conflicts. Those aware of the strategic significance of the war in the Solomons have long recognized the value of war sites and objects as cultural and economic resources. The national airline magazine, for example, regularly runs articles that profile battle sites and present stories of travellers rediscovering these places and interacting with local people who gather and sometimes display war artefacts on their property. The puzzle of war tourism in Solomon Islands, however, is that the visibility of WWII in tourism promotion is matched by the lack of development of 'war heritage' in national culture, at least in regard to the war as a military conflict between Japan and the United States and its Allies. Why the apparent gap?

The Battle of Guadalcanal is prominently mentioned in advertising and promotional materials produced by the Solomon Islands Visitor Bureau (SIVB)—a small, underfunded agency tasked with marketing the Solomons as a tourism destination. In its 2012–13 visitor guide, for example, the Visitors Bureau lists visiting WWII sites as one of the 'Solomons' top 10 must dos'. Similarly, WWII appears on a menu of ten 'Tours and activities' on its 'Visit Solomons' website. The guide provides short segments on things to do in each of the Provinces, including Honiara. For Honiara, the thematic importance of WWII is impressive, with more than one page of the three pages devoted to a list of 'World War II tours' (which list numerous war sites, presumably for self-guided tours). Similarly, the website maintained by the provincial government of Guadalcanal is replete with photos and links to WWII sites.[7] Fully half of the introductory overview of the island province is devoted to 'the importance of Guadalcanal in World War II'.

In contrast to the visibility of WWII in tourism marketing, however, there has been almost no government effort to develop the infrastructure of war tourism, such as informative materials, signage and interpretative programmes.[8] The most prominent examples of war history in and around the national capital are the war memorials, such as the Japanese Peace Memorial, put in place by foreign powers. Even though WWII history has been part of government visions for tourism in Solomon Islands from the first years of planning for independence, there have been very few projects or programmes that turn battlefields into historic sites, making them accessible for visitors, either domestic or international. Only diving-related tourism, for which the reefs and lagoons of the Solomons are justly famous, has a strong presence in both marketing and guided visits focused on WWII shipwrecks.[9] In short, only a trickle of hardy travellers manage to find their way through the landscapes and seascapes of war in the Solomons.

To walk into the small office of the Solomon Islands Visitor Bureau in the centre of Honiara (near the entrance to one of the main hotels and across the street from the National Museum) is to discover the distinctly low-key (and low-volume) nature of tourism in Solomon Islands today. One can pick up a copy of the Visitors Guide, ponder a number of posters tacked on a bulletin board and ask questions of the one or two staff who may be present. It is apparent that there are not large volumes of people looking for direction or assistance in discovering war sites of Guadalcanal or beyond.

In the absence of government initiatives or resources devoted to developing war history as a national resource, enterprising individuals have stepped in. The first entrepreneurial spirit on Guadalcanal in the 1980s–90s was Fred Kona, who created the largely open-air Vilu War Museum, gathering plane wrecks and other war remains onto a property outside Honiara (White 1996). His museum also became a popular spot for commemorative ceremonies conducted by both Japanese and Allied veterans (prior to completion of the American memorial in 1992). While there are no regularly scheduled tours to the most famous battle sites on Guadalcanal, two or three individuals do provide guided tours, although they may be difficult to find. In an informal survey of publicity at the Visitors Bureau and in Honiara hotels, I spotted a flyer for 'Historical battlefield tours' in one hotel lobby. The flyer, put out by Sightsee Solomon Islands Tours (operated by a former manager of the Solomon Islands Visitor Bureau), described four different half-day guided tours to different geographic sectors of the Guadalcanal war zones. Also listing various snorkelling and picnic activities, the company offers each tour at a cost of about US$123.

These possibilities aside, most of the limited publicity today regarding guided tours to the battlegrounds of Guadalcanal refers to one person, John Innes, an Australian and part-time resident in Honiara who began his own exploration of the battlefield sites decades ago, cultivating connections with Solomon Islands landowners, as well as war veterans and others interested in Pacific War history. Innes's long-term work walking the battle sites, talking with veterans and researching archives recently resulted in publication of a *Guide to the Guadalcanal Battlefields*, published in ten chapters as separate booklets (Innes 2012). The booklets are accompanied by a DVD video showing Innes at the various sites telling stories and relating histories with a tour group in tow.

Having gained recognition for his specialized knowledge and contacts with all parties interested in WWII history in the Solomons (especially veterans, descendants and relevant military organizations), Innes has played a pivotal role in mediating the return of veterans (until most became too old to travel), speaking at commemorative events and guiding military tours. The American military tour company Valor Tours, for example, lists him as their primary guide for a Guadalcanal battlefield tour that they operate annually, bringing Americans to Solomon Islands in August each year.[10] More locally, the industry website for visiting Solomon Islands lists Innes as the contact for anyone interested in guided tours (although Innes is only a part-time resident in the Solomons), describing him as 'a local WWII history enthusiast who lives in Honiara for most of the year' and for a fee offers 'an informative day tour of the major battle sites around the capital' with the tours running on Saturdays 'when a group comes together.'[11]

In point of fact, John Innes, in his seventies, is spending less time in the Solomons and has begun to think about ways to pass on his knowledge (hence the new series of booklets and DVD).[12] Seeing that local guides lack knowledge of the details of the events that occurred at the various sites and assuming that that reflects a cultural orientation that is not likely to change ('they don't read'), he is now pondering the production of an audio guide that local operators could use in conducting their own tours. John Innes's lament about the absence of informed local interpretation reveals the disconnection between local interests and the military narratives desired by international travellers.

Indeed, the disconnection evident in John Innes's dilemma of continuing the kind of informed interpretation he developed over decades points to a larger cultural divergence between the 'culture' of international war tourism and the culture of local or indigenous histories of the war. The Battle of Guadalcanal is widely represented in articles, videos and websites that create global military histories of WWII. These sorts of media, however, do not reach rural villagers in the Solomon Islands (85 per cent of the population) who generally have low rates of English literacy, have little access to electricity or the Internet and hence do not possess televisions or computers. The gap between local experiences in the war and the military narratives presented in the context of guided tours is one reason for the native view that 'the war was not our war', even though the war had profound effects throughout the Solomons and is widely remembered as a transformative time for personal, local and national histories (Bennett 2009; White and Lindstrom 1989).

Insofar as tourism must always market itself to international audiences, it is inevitable that WWII tourism in the Solomon Islands will focus on the military narratives of invasion and counter-invasion in the epic conflict between the militaries of Japan and the United States and its Allies. These are the stories that interest (English-speaking) foreign travellers. This orientation has long been evident in promotional materials that focus on combat narratives and the bravery and sacrifice of Allied forces, especially the US Marines. The WWII entry in the Visitor Bureau's 'top 10 things to do in Honiara' list is a good example: 'Catch a glimpse of the fierceness of World War II battles and be touched at the bravery of soldiers that made the ultimate sacrifice by visiting war sites in and outside Honiara, on land and underwater' (Solomon Islands Visitor Bureau 2012, 5).

Given tourism's orientation towards the combat narratives of foreign militaries, success in the tourism sector can only pose a predicament for efforts to develop an indigenous approach to war history in public education or nation building. As the postcolonial taste for tourism begins to drive national representations of native culture and history to fit foreign tastes, how will local historical practices survive in relation to new, globalized commodifications of war memory? If in fact indigenous history and Allied/Japanese military histories run on different premises, how will 'the war' be represented in national memory and in those public spaces where representations of collective history seek to fashion an emergent national imagination? One of the risks of developing national war memory in concert with the military histories of the Allies or Japan (as is the case in international tourism) is that Solomon Islanders will be represented objects (and subjects) of other people's narratives. Consider next a case study of an ongoing project aiming to enhance memory of the war in the national imagination.

THE COASTWATCHER MONUMENT PROJECT

One response to the disconnection between local and global representations of the war in the Solomons is to bring more attention to the roles played by Islanders in the Allied war narrative. Indeed, just as that was one of the goals of our collaborative project in the 1980s, so it was a prime motivation for the Coastwatcher Memorial project launched in 2010 with the goal of honouring the Solomon Islanders and coastwatchers who played a key role in the Allied victory. The history of the coastwatchers who, in US Admiral Halsey's words, 'saved Guadalcanal', which in turn 'saved the Pacific War', is well documented in a sizable literature (Feldt 1967; Horton 1970; Lord 1977; Clemens 1998; Feuer 2006). But that literature focuses primarily on the European officers (the 'coastwatchers') rather than on their native 'scouts'. Hence, the idea to recognize the native contributions emerged. Project stationery bears the letterhead, 'Guadalcanal War Memorial. To honour the service and bravery of Solomon Islands Wartime Scouts & Coast Watchers'. The first issue of the project's newsletter announced:

> [O]ur aim at SSCT is to have the Solomon Scouts and Coastwatchers honoured for their bravery as they should be. Their involvement in WWII is better known outside of the Solomon Islands than it is within and so SSCT is working on several strategies to ensure this does not remain the case.
>
> (Anonymous 2010, 2)

Note that the goal of increasing awareness makes reference to the disparity in recognition/knowledge between that which circulates outside Solomon Islands and that inside. This imbalance is frequently referenced by those involved with the project, as they are familiar with the literature on the Battle of Guadalcanal and recognize that most of that, in print literature and audiovisual media, is inaccessible to rural Solomon Islanders.

The force behind this project is an Australian expatriate businessman and long-time resident of Solomon Islands, Bruce Saunders (now Sir Bruce Saunders after receiving a knighthood in 2012).[13] Seeing an opportunity to create a civic project that would build public awareness and foster national unity, Saunders used his business skills, resources and network to create a non-profit entity with an advisory board and fundraising trust (the Solomon Scouts and Coastwatchers Trust, SSCT) to support the project.[14] Within its first two years, the project managed to dedicate a new monument (7 August 2011), obtain a scholarship for a Solomon Islander to pursue graduate study on war history in Australia, and begin the development of new WWII curricular material for use in Solomons schools.

Another stated purpose of the memorial project is to contribute to nation building. The organizers, well aware of the 'problem' of national identity for a country emerging from a period of ethnic violence and state collapse, see WWII as a source of shared history that can inspire a sense of common identity. As stated on the project's webpage, 'The Solomon Scouts and Coastwatchers Memorial Trust was formed in 2010 with the aim of uniting Solomon Islanders in learning about their World War II history—and instilling a sense of pride in the sacrifices made by those who helped the Allied forces change the course of the war in the Pacific.'[15] With these ideals in mind, the organizers named the monument 'Pride of Our Nation' and used that label to headline a series of announcements and media reports, including an electronic newsletter sent to supporters and a special insert in a national newspaper (*Solomon Star*, 12 May 2012). The monument's name, Pride of Our Nation, is now emblazoned across its pedestal.

The use of the pronoun 'their' in the project's statement of purpose ('learning about *their* World War II history') is indicative of the non-native identity of at least some of the sponsor(s). With this in mind, it is useful to ask what histories are represented in the monument and for what audiences? In what ways does the commemorative culture emerging around the monument project speak to the multiplicity of histories of the war at play in Solomon Islands: indigenous and European, local and (inter)national? What types of history are represented, for what purpose and for which audiences among the various constituencies expressing interest in WWII? The main focus, of course, is the military narrative of Allied victory in the Pacific, embodied in the bravery and heroism of the men who served in the war. This is consistent with the function of war memorials generally and with the tradition of WWII commemoration in Solomon Islands, focused on American and Allied (or Japanese) combatants, with some mention of coastwatchers and scouts, most notably the late Sir Jacob Vouza, lionized for his assistance to the US Marines and his role with the Guadalcanal coastwatchers.

One way to look at the current memorial project is that it elevates the native element of war commemoration, which had appeared more as a footnote in the fiftieth anniversary activities of 1992, to centre stage (White 1995). Where the 1992 commemoration had included dedication of the Vouza statue (a project of Allied veterans and designed by an Australian sculptor) and an ill attended medal ceremony for Solomon Islands veterans, the Coastwatcher Memorial project has featured a small number of aging Solomon Islands veterans in its ceremonies[16] and dedicated two plaques listing the names of the members of the British Solomon Islands Defence Force (BSIDF) (insofar as available in official enlistment records, 722 native names and 32 European).[17] Since the BSIDF was a hastily formed unit that disbanded after the war and the Solomon Islands has never had a standing military (only national police), there has never been any organization of Solomon Islands veterans of the Defence Force, or of the Labour Corps or of any body associated with WWII service (although a small association of veterans did group up in the 1990s, with the aim of lobbying for some form of compensation for their service). Despite this absence, the project has fostered the creation of a public discourse of veterans by convincing government administration to proclaim the Guadalcanal landing anniversary Veterans Day and to call the 2011 class of police recruits the 'Solomon Scouts and Coastwatchers' (Anonymous 2011, 2).

Given the importance of such organizations for fundraising and the organization of commemorative activities, the monument project found it easier to coordinate its activities with military and veterans' organizations of the Allied powers than to find representation of Solomon Islanders who served in some formal capacity in the war. Each year's commemorative ceremony, held on 7 August—the day of the American landings on Guadalcanal—has brought extensive diplomatic and military representation from the United States and Australia. The unveiling in 2011 was marked by speeches from the United States ambassador and two military commanders, as did the seventieth anniversary ceremony in 2012, which also included a communication from US President Barack Obama. The 2013 ceremony, which again brought a sizable US contingent (ambassador, commanders, marching band), saw the unveiling of a third segment of the memorial honouring the Royal Australian Navy (formerly in command of the coastwatcher network), represented by a WWII anchor and plaque.

While the war memorial project is, in part, a vehicle for celebrating these international connections, project organizers were also aware of the need for native leadership and included prominent Solomon Islanders in the board of advisors and other important roles. It is important to note that Sir Peter Kenilorea, the former prime minister who was so critical of the American memorial and Vouza statue projects prior to the fiftieth anniversary activities in 1992, was recruited as a member of the Memorial Trust Board and is quoted prominently on the project webpage, referring to the lack of awareness about the war among current generations of Solomon Islanders:

> The name Guadalcanal is still etched into the minds of many American, Australian and New Zealand citizens, and indeed Japan as well. But here in Solomon Islands, many of our young people are not aware of the role their grandparents and great-grandparents played in the building of what is today this wonderful nation Solomon Islands.
>
> (BJS Group n.d.)

Kenilorea's wording here seems significant. He refers to the role of ancestors not narrowly in coastwatching or scouting but in 'building of what is today this wonderful nation Solomon Islands.' Such phrasing can include a wide array of wartime experiences, including work in the Solomon Islands Labour Corps or in sustaining communities facing the hardships of war. Such phrasing could even include the role of ancestors in the anticolonial (British) political movement known as Maasina Rule, directly inspired by wartime experiences but suppressed with mass arrests in 1947 (Fifi'i 1988; Keesing 1978). Such histories reside in a universe of historical storytelling far removed from WWII commemorations as currently conducted in Solomons' public culture.

The focus on coastwatchers in the Memorial project provides a highly symbolic linkage between a more Solomons-centred history and the epic Allied military history, insofar as the coastwatching units had a formal role in the military command structure (with European officers in charge) while reaching into the grassroots of Solomons' society through the native scouts who worked with them.[18] The term *coastwatcher* is itself ambiguous in a way that facilitates this linkage. The primary reference for the term is to the European commanders. At the same time, however, it can be used to refer to the entire coastwatching unit, including native scouts as well as

European commanders. It is in this larger sense that the recent project seized upon the term as the key to connecting native involvement in the war and a well documented record of heroic service. The phrase 'Solomons Scouts and Coastwatchers' used to name the project is generally understood to refer to '[Native] Scouts and [European] Coastwatchers'.[19]

Using the coastwatcher 'mythos' to build native identification with national history is, however, a risky strategy insofar as the coastwatcher literature itself is heavily Euro-centric, concerned largely with the personalities and actions of the European officers in charge, with little information about the Solomon Islanders who served with them, much less the vast network of villagers who assisted the scouts in countless ways. The hierarchy of command, underwritten by the larger and longer hierarchy of colonial administration, is inscribed in histories that write about nameless native scouts and in group photographs with only one name in the caption, that of the European commander (Figure 11.2).

The sometimes hidden dynamics of colonial relations that reside mostly in the background of the monument project are more visible, literally, in the design of the monument itself (Figure 11.3, page 205). The monument,

Figure 11.2 Martin Clemens' coastwatching group, Guadalcanal, October 1942. Standing, left to right: Daniel Pule, Martin Clemens, Andrew Langabaea. Seated, left to right: Olorere, Gumu, Jack Chaparuka, Jack Chaku (Courtesy USMC photo, National Archives [127-N-50505]).

Figure 11.3 Figures in coastwatcher monument (Pride of Our Nation), Pt Cruz, Honiara (Copyright Geoffrey White).

designed and rendered in steel mesh and plaster by Solomon Islands sculptor Frank Haiku, shows four figures tightly clustered together with one European standing tall in the middle, gazing through binoculars as three natives crouch around him, one in front talking on a radio, one at his side pointing in the direction of the European's gaze, and one on the other side in a crouch with a rifle. Only the European wears a hat, and two of the three natives wear a plantation-style waistcloth. By placing the European as the highest, central figure in the motif, the sculpture reproduces a spatial image of the relations of power that were characteristic of the relationship between European commanders and native scouts—relations that for many had developed in pre-war police training but also had roots in plantation society between European bosses ('master') and native labour ('boys').

Viewed through the lens of Allied military history, these visual features are easily interpreted as indicative of native 'loyalty'—a value inscribed on the monument plaque in the words, 'Pride of Our Nation—Honouring the bravery, loyalty and courage of Solomon Scouts & Coastwatchers'. Seen in terms of colonial history, however, the monument's aesthetic may acquire a different significance. In the final section of this chapter, I explore these other interpretations by considering ways in which local stories about native experiences in the war (oral histories) can dramatically reframe events long taken as prime exemplars of the coastwatcher epic.

WAR(RING) NARRATIVES

As already noted, the history of the coastwatchers is well represented in the literature on the Pacific War. In these texts, the coastwatchers justifiably loom large as heroic figures in the Allied military victory. While native actors play an often nameless support role, their presence *is* important in the rhetorical construction of Allied histories of the war. They embody the loyalty of the local population, validating the theme of liberation from Japanese occupation. (Jacob Vouza is iconic in this regard.) War history told from the 'native point of view', however (to borrow a phrase from Clifford Geertz 1974), may sound very different. The most obvious difference, of course, is that most Solomon Islanders did not have any formal military involvement but rather recall experiences of survival in civilian communities or, for nearly 4,000 men, work in the wartime Labour Corps (White, Akin, Gegeo and Watson-Gegeo 1988, 129).[20] But even for those who did serve in military roles, such as the scouts in coastwatching units, wartime experience was deeply embedded in European–native relations and the British colonial context.

The case I want briefly to consider is that of one of the most prominent scouts, Bill Bennett (Bennett 1988), who worked with a coastwatcher named Donald Kennedy.[21] Kennedy was an outsized and controversial figure in the British colonial service who had a reputation for violence, abuse of women and alcoholism. In the war, he quickly gained a reputation for aggressively defending the territory around his base in the western Solomon Islands—so much so that Walter Lord (1977, 271) referred to his group as 'Kennedy's army' (see also Boutilier 1989). In a well known incident called the 'Battle of Marovo Lagoon', Kennedy and some of his 'army', travelling on their armed motor launch, attacked a Japanese patrol, killing all of them (about twenty soldiers), with Kennedy himself suffering a gunshot wound in the thigh. In the context of the war and in the decades that followed, the Battle of Marovo Lagoon came to represent both the bravery of the coastwatchers and the loyal service of the islanders who supported them.

In 1987, however, in a conference on local Melanesian recollections of the war held in Honiara, Solomon Islands, and involving a dozen or so Solomon Islands veterans, Bennett revealed that it was *he*, not the Japanese, who had shot Kennedy under the cover of battle—a result of long simmering resentment for abusive treatment. He first made his 'confession' in the context of a drinking session one evening after the conference but later expanded on it in an interview with the author.[22] This plot twist not only disrupts standard narratives of coastwatching; it draws attention to the need to consider the politics and pragmatics of popular war histories.

Whether a literary war history such as that crafted by Walter Lord or the oral storytelling of Bill Bennett, historical narrative always entails some kind of commentary, more or less explicit, about social identities and relations. Outside the context of war, Kennedy's toughness proved problematic.

He was investigated on at least two occasions for violence towards Islanders and, according to at least one oral history, caused the death of one of his own men. Although decorated for his war service, by the end of the war he was deemed mentally unfit to continue in his position. Yet in histories of wartime coastwatching, Kennedy's violent actions towards his men have been recast as a military asset. Here is Walter Lord's description of Kennedy's relationship with his men:

> Discipline was always tough. In this kind of warfare the lives of all could depend on one man's performance, and Kennedy demanded unswerving obedience . . . [His] harsh punishments did spawn a certain amount of grumbling, but it was not so much resentment as the griping of tough troops who will put up with almost anything for a commander they believe in.
>
> (Lord 1977, 171–2)

It is instructive to read this alongside interviews with Bennett and at least one other of the scouts who served with him. Bennett's relationship with Kennedy was fraught with tension, contradiction and ultimately deep resentment. So much so that at one point, after he had the humiliating experience of being asked to procure a native girl for Kennedy and then publicly punished for his failure, he contemplated killing Kennedy by lobbing a grenade in his room at night (tape-recorded interview with the author, 27 October 1987). Once one has listened to stories such as Bennett's recollections or those of other scouts, one can only look critically at the way in which earlier writers such as Lord easily minimized his brutality or even reframed it as a necessary factor for victory in war.

One reason why the Battle of Marovo Lagoon had always made for such good storytelling (and resisted much critical scrutiny) is that it can be told as a 'heroic history' built around larger-than-life figures such as Kennedy and Bennett who are already familiar as examples of the well known types of European hero and loyal native—prototypic actors who may be found in a large corpus of published war histories and memoirs. Whether one accepts Bennett's revelations at face value or not,[23] the very fact that he chose to tell his story for an audience that included not only visiting academics but also other Solomon Islanders unsettles the loyalty narrative and provides stark evidence of the social and emotional conflicts that could enter into colonial and wartime 'service'. It is also a reminder of the contingent quality of historical representation, dependent on available information and the social and moral frames we use to interpret it.

In vivid contrast to the stories of coastwatcher heroism is the story of wartime *dis*loyalty attributed to a prominent islander (from Santa Isabel Island) named George Bogese (one of just two cases brought to trial). Although not as well known as the war's heroes and certainly not part of the public memory of the war represented on anniversary occasions, the trial and conviction of Bogese for aiding the Japanese figure importantly in the Solomons' war memory as an example of wartime disloyalty and the ensuing punishment in the colonial justice system. Bogese, who had been one of the most prominent Solomon Islanders of his generation and the first to receive training in Fiji as a native medical practitioner, would spend nearly seven years in confinement during and after the war, including three years of a four-year sentence for disloyalty. Bogese was tried for aiding the Japanese attempt to find Donald Kennedy—an act that Bogese denied but, more convincingly, argued that it was he, not Kennedy, who had been mistreated prior to the war because of frictions with Kennedy. Historian Hugh Laracy would later reinterpret Bogese's case by examining the colonial context that framed the accusations against him and subsequent dispensation of justice (Laracy 1991).

The case of George Bogese provides a deeply ironic, even tragic example of the complications that surround the idealized categories of European hero and loyal [and disloyal] native. With Bennett's 'confession', the triangulation of Kennedy, Bennett and Bogese takes on the moral complexity of Shakespearian drama. By admitting that he had shot Donald Kennedy, Bennett not only complicated his own image of loyal service but, potentially, unsettles the very category of loyal native. But how are we to interpret Bennett's account? Is the shooting an aberration, related mainly to his personal problems with Kennedy? Is it something to please European academics keen to discover acts of resistance (as Kennedy's biographer suggests: Butcher 2012)? Or does Bennett's narrative signal something more significant for collective memory, giving voice to indigenous perspectives silenced by dominant Allied narratives built around themes of native cooperation and liberation?

CONCLUSION

The recent Coastwatcher Memorial project offers an opportunity to consider some of the ways in which these multiple histories appear in the public spaces and activities intended to represent Solomon Islands national memory of the war. Just as the project has many parts (construct and dedicate a monument, organize commemorative ceremonies, create educational curricula and support research and study by native Solomon Islanders), so it speaks in multiple voices. On the one hand, the new monument is a focus for international commemorative activities that reproduce the most well known Allied histories highlighting the heroic actions of coastwatchers and their native scouts—a stronger and more elaborated version of the gesture in this direction by the Vouza statue twenty years earlier. On the other hand, the educational and research aspects of the project are creating openings for indigenous voices that often speak in a different register. In this context, the lament first voiced by the veterans in the 1987 conference that 'the war was not our war' (Zoleveke 1988) is finding a place. And even Bill Bennett's 'surprising story' has received mention, even if marginal to the more public and well publicized commemorative activities.

It is noticeable that none of the few Solomon Islands veterans have been asked to speak at the commemorative ceremonies. Even though three individuals have been present and prominently recognized in the ceremonies and picture taking, they did not speak formally (as Sir Gideon Zoleveke had at the unveiling of the Vouza statue nearly twenty years earlier). While age and infirmity might have been a factor, another issue would have been the gap in language and culture. Where former Prime Minister Sir Peter Kenilorea (member of the Memorial Trust Board) spoke at the unveiling ceremony, a granddaughter of a well known coastwatcher Geoffrey Kuper spoke at the seventieth anniversary ceremony in 2012 (when two plaques with names of formally enlisted members of the Solomon Islands Defence Force were added to the monument).

Even two generations away from wartime experience, the presence of Rita Koroi, a Solomon Islands woman speaking at the anniversary, injected an indigenous component to the ceremony that allowed for modest departures from the dominant commemorative culture. In her speech, she tapped into a vein of humour and irony well known in indigenous storytelling to relate incidents in which gaps in knowledge and technology would have made the scouts' work especially difficult, even implying that confusions and miscommunications—in those days, would have come down heavily on the scouts, who were not only lower in rank but in colonial status as well:

> There were many stories of such mishaps and misunderstandings that occurred which even resulted in these men being penalised very heavily. However they still choose to remain loyal to the cause. Most of these brave Solomon Islanders had no idea what the war was about or why it eventuated in their backyard, and though they did not want to be part of the war they had no option but to take up their duties as scouts and coastwatchers.
>
> (Koroi 2012, 2)

In these remarks, Rita Koroi echoes themes evident in the recollections of many of the scouts interviewed in the 1980s–90s. While not entering into the public ceremonies, even Bill Bennett's surprising story surfaced in publicity generated by the monument project. At the time of the unveiling of the monument in 2011, an Australian website devoted to Pacific Islands issues produced several articles about the memorial dedication in 2011, including a section on 'hidden stories', referencing Bennett's attempt to kill Donald Kennedy (Sharp 2011). The blog quoted the Solomon Islands scholar supported by the project, Ms Annie Kwai, to the effect that published histories by European coastwatchers have overshadowed the scouts' experiences, evident primarily in oral traditions. Her reflection, expanded in a just completed MA thesis at Australian National University (Kwai 2013), suggests that the coastwatcher memorial project, in the course of expanding the Allied narrative of victory in the Pacific, is also creating opportunities for some surprising developments in Solomon Islands war memory:

> But how much is really known about the position of Solomon Islanders enveloped by a calamity not of their own making? And how much is known about those islanders who worked closely with the coastwatchers gathering and sharing intelligence for the Allied cause?
>
> From the written memoirs of European coastwatchers . . . a picture emerged of locals—seen by the Allies and the Japanese—as 'civilians' whose loyalty was there to be won. Once trusted, their local knowledge could be used in pursuit of military victory. [However,] this perspective does not generate a comprehensive picture . . . Rather it promotes the story of coastwatchers, while suppressing that of Scouts.
>
> (Kwai, quoted in Sharp 2011)

This study of the making of public memory of WWII in Solomon Islands may be read as a parable for the predicament of local histories and war memory worldwide. Islands, like small communities everywhere, tend to be the focus for intensely produced localized identities, at the same time as they are always entangled in wider forces of globalization. In the resulting tensions between local and global modes of remembering, it is in the arenas of *national* memory making, such as the monument project discussed here, where conflicts or contradictions are most evident. The case of the national commemoration of WWII in Solomon Islands illustrates the way in which local modalities are often subordinated to dominant (inter)national narratives where the moral clarities of combat, with its alignments of allies and enemies, submerge longer histories of colonialism with all their moral complexities. A structural approach to explaining the politics of national memory making (such as deployed by Marshall Sahlins [1985] to interpret early contact histories in Polynesia) might well look to the different sorts of logic that drive indigenous histories (local, oral, socially embedded) as opposed to those of nation states (universalizing, iconic, abstract). Such an approach is useful in directing our attention to the ease with which state-centred modes of remembrance articulate with the narratives of other nation states (Australia, Britain, United States), thus amplifying the visibility and prestige of state-centred histories through transnational collaboration and tourism development.

Having said this, it is important to point out that this chapter's focus on the arenas of national memory making runs the risk of minimizing the importance of local memory practices, adding to the hegemony of literate state-centred histories. The war in Solomon Islands is recalled in multiple histories that are themselves organized, socially and spatially. Where national commemorations of the war occupy state centres in the form of monuments and tourism development planning, indigenous histories are continually reproduced in local storytelling, singing and performing. But these multiple histories are not bounded or insulated one from the other. Just as national practices may subsume and recontextualize the local, so indigenous storytelling, singing and performance are equally effective in localizing and 'socializing' WWII by embedding its events in local histories (White 1989b). In the case of national memory in Solomon Islands, where war history meets colonial history, the war did not arrive in an archipelago of isolated islands. The 'conjunctures' set in motion by wartime encounters were already foreshadowed in histories of colonialism and local historiographies that had long accommodated contact events such as the arrival of Christian missions or colonial government. This chapter's discussion of recent developments in national war remembrance suggests that indigenous narratives have begun to emerge in the public spaces of postcolonial Solomon Islands, with uncertain possibilities for the future. It seems that islands—even metaphorical islands—are as useful for thinking about circulation and connection as difference and dominance.

ACKNOWLEDGEMENTS

A version of this chapter was presented at the Islands of War, Islands of Memory conference at the McDonald Institute for Archaeological Research, University of Cambridge, UK, 6–7 April 2013. I am grateful to the organizers for the opportunity to present this work and to participants for comments on an earlier draft. I also want to thank David Akin and host communities in Solomon Islands, including organizers and participants in the Coastwatcher Memorial project, especially Sir Bruce Saunders and Ms Annie Kwai, for information and conversations that have contributed to this chapter.

NOTES

1. The bronze statue in question, titled 'Sound of the Tide,' is of a muscular male standing in a classical pose. It was sculpted in the 1930s by Eikichi Takahashi, a Japanese artist who died as a soldier on Guadalcanal. The statue was donated by his hometown, Ishinomaki City, where a small museum is devoted to his work.
2. See www.solomontimes.com/news/vandalism-to-japanese-war-monument/1255.
3. The city of Honiara today owes much of its infrastructure—roads, bridges, wharves, airfield—to the construction of wartime facilities in the area surrounding a strategically placed airstrip started by the Japanese in 1942. The airfield, now the country's only international airport, was the focus for American landings on 7 August 1942—the first Allied offensive in the Pacific War. The battles that ensued involved six months of deadly naval warfare, aerial

bombardment and grueling jungle fighting before the Japanese were defeated, leaving over 20,000 dead, more than the total native population at the time. The US Marines and Army, who left some 5,000 dead of their own, look at the battle as one of the epic conflicts in their own military histories.

4. See www.solomontimes.com/news/vandalism-blow-to-tourism-in-solomons-bureau-general-manager/1260. The creation of the Japanese memorial on the Mt Austen ridge top in 1980, just two years after Solomon Islands gained its independence from Great Britain, is a case in point. Consistent with the genre of 'peace memorials' constructed by Japanese citizens in the post-war period, the Mt Austen memorial includes plaques written in Japanese and English and honours 'all the spirits of those who sacrificed their lives in World War II'. The plaque affixed to the central pillar of the memorial reads as follows:

   SOLOMON PEACE MEMORIAL PARK
   REST IN PEACE
   At this place repose all the spirits of those
   who sacrificed their lives in World War II
   at Guadalcanal and the entire Pacific area.
   This represents a requiem
   for their souls and serves to remind us
   [of] all their patriotism
   which they dedicated to their mother countries.
   We pledge here to establish eternal peace on earth.
   25 October 1980, erected
   28 October 2011, renovated
   Pacific War Memorial Association

5. If further evidence were needed of the weak local reception of the Vouza statue dedicated in 1992, a new project has emerged in the wake of the Coastwatcher Memorial project, to erect a new statue to Jacob Vouza that will portray him in the uniform he was well known for (rather than in the dress of a plantation labourer as the Australian sculptor opted for) and will be located in his home region of Guadalcanal, not outside the police station on the other side of Honiara (Frank Haiku [sculptor], interview with the author, 15 July 2013).
6. Sir Peter went on to criticize the American initiative as rekindling the conflict between America and Japan, only this time over memory. Indeed, foreign war memorials give former combatants a means of reinscribing their own narratives in the soil where they once fought. In the 1992 Guadalcanal anniversary, American and Japanese interests organized parallel commemorative activities, leading Kenilorea to conclude his editorial, 'I think we have already had enough of USA vs Japan during the last war' (*Solomon Star*, 28 April 1989, 7).
7. See www.guadalcanal.com.
8. Although difficult to estimate from available statistics, fewer than 5,000 people visit the Solomon Islands in a given year as leisure travellers, mostly in the niche market of ecotourism, diving and adventure travel. Vanuatu, an island Melanesian country with half the population of Solomon Islands, has ten times that number. In 2012 (the seventieth anniversary of the landings on Guadalcanal), tourist arrivals for the second quarter were 1,547, with more than half from Australia (Statistical Bulletin: No: 11/2012, *Visitor Statistics* [Second Quarter 2012] Solomon Islands National Statistics Office. Honiara: Ministry of Finance, 5).

   In addition to these gaps between local and (inter)national memory of the war, the most obvious reason for the lack of development of WWII tourism in Solomon Islands is the lack of development generally. The problem with poor infrastructure and amenities is summed up in a guidebook to WWII sites in the Pacific: 'Few Pacific War sites carry as much historical significance as Guadalcanal, but the area suffers as a whole due to Honiara, the least attractive—might as well say ugly—destination city covered in this book' (Thompson 2002, 111). Efforts to develop a tourism industry were drastically curtailed by political violence that engulfed the capital Honiara and much of Guadalcanal in the late 1990s, leading to a coup in 2000 followed by the intervention of a multinational peacekeeping force in 2003.

9. In addition to the various companies specializing in diving-related tourism listed on the Visit Solomons website, the website of Solomon Airlines includes a top-level menu item labeled 'wwii history' that links to an interactive map titled 'Flysolomons Wreck Finder.' Available HTTP: www.flysolomons.com/index.php?option=com_content&view=section&id=14&Itemid=135,
10. Guadalcanal Province website lists Valor Tours as the recommended contact for anyone interested in a guided tour of the battlefields. Valor Tours is an American tour company specializing in military tourism that offers annual guided visits to battle sites around the world, primarily where American troops have fought (see www.valortours.com/tours.html). For Guadalcanal, the company offers a package tour of about ten days at a cost that includes travel from Los Angeles in the United States (2014 tour price for a single person: US$4,200). Guadalcanal Province, it seems, does not expect inquiries from more casual travelers with a somewhat less consuming interest in WWII history.
11. See www.welkamsolomons.com/honiara/things/ww2/.
12. John Innes, interview with the author, 16 July 2013[0].
13. Sir Bruce Saunders is self-described as 'an expat of Australian descent' for whom 'the Solomons are now his home' (Fuss, E. 'SA expat knighted in the Solomon Islands,' ABC North and West South Australia, 19 July 2012).
14. The project website (www.bjs.com.sb/ssc.html) is hosted as a menu item on the website of Sir Bruce Saunders' company, the BJS Group of Companies, which offers an array of services targeting especially employees and projects connected with Australian assistance to Solomon Islands (AusAid and RAMSI).
15. See www.bjs.com.sb/ssc.html.
16. Hence the importance of the three (!) Solomon Islander veteran scouts who were able to attend the dedication ceremony in 2011, with their aged and stooped bodies acting as authentication of the monument's purpose. Realization of the discursive importance of their presence explains the inclusion of the names of those three individuals on the dedication plaque affixed to the front of the monument.
17. The tenuous way in which the memorial project learned of relevant work done in the 1980s WWII project (much of it based in the same city, Honiara) reveals the spectral quality of indigenous war memory in Solomon Islands. Despite a major international conference on Islander recollections of WWII convened in Honiara in 1987; the publication of a book of WWII oral histories, *The Big Death: Solomon Islanders Remember World War II*, a University of the South Pacific book in English and Solomon Islands Pijin English (White *et al.* 1988); and a special issue of a local journal, *'O'o: Journal of Solomon Islands Studies* (1988) devoted to WWII, the Coastwatchers Memorial project learned of the project only when one of the aging veterans who had participated in the 1987 conference showed project organizers a conference photo in 2010. That was followed by an email to the author with a request for assistance with records of the names of Solomon Islanders who served in the British Solomon Islands Defence Force.
18. One of the episodes that added to the coastwatchers reputation in Pacific War history was the rescue of young Lieutenant John F. Kennedy (and subsequent

US president), whose patrol boat, *PT 109*, was sunk by a Japanese cruiser in the Western Solomons in 1943 (Donovan 1961; Donovan and Warner Bros. Pictures (1923–1967) 1963; Ballard and Morgan 2002).

19. This neat racial dichotomy is blurred by at least one individual of mixed ancestry, Geoffrey Kuper, who was considered native in the racial ideology of the times but who occupied a command position in the coastwatching network. 'Geoffrey Kuper was the only Coastwatcher in charge of a station who was born in the Islands' (Lord 1977, 155). His granddaughter, Rita Koroi, has played an important role in the coordination of the coastwatching project, delivering the keynote speech at the monument dedication ceremony in 2011.
20. At some point in its evolution, the monument project expanded its stated goals to include the contributions of ordinary people in their daily lives, including, minimally, 'not betraying the coastwatchers' hiding places'. In the March 2013 newsletter, one of three goals is listed as: 'Recognise the important contribution of the many people who helped the Scouts and Coastwatchers with information, food, shelter and rescues, and by not betraying their hiding places to the Japanese' (*The Monument: The Newsletter for the Solomon Scouts & Coastwatchers Trust* [SSCT]. 6, March 2013, 2).
21. Donald Kennedy was not related to US President John F. Kennedy.
22. I first interviewed Bill Bennett at his home in Kakambona (Honiara) in 1984 with Solomon Islander political scientist David Gegeo. At that time, Bennett related the Marovo story much like the one published by Walter Lord. However, in a 1987 conference on Melanesian recollections of WWII (in Honiara) and again in a follow-up interview with me two months later, he told the story of how it was he who had shot Kennedy during the battle. In doing so, he made it clear that it was difficult for him to talk about something that had been hidden so long.
23. Kennedy's biographer, Mike Butcher, does not. In his book (2012), he speculates that Bennett made up his confession in the context of a conference of academics eager to hear stories of postcolonial resistance. Since I recorded the interviews with Bennett in the course of a larger project on indigenous histories of the war (during three field trips in 1984, 1987 and 1988), it is possible to interpret his narrative not as a sole performance but as a text in relation to a wider range of stories elicited in about seventy interviews (approximately 100 hours) conducted over the course of several years with men who had been involved in scouting, labour corps or other wartime activities.

BIBLIOGRAPHY

Anonymous. (2010) *The Monument: The Newsletter for the Solomon Scouts & Coastwatchers Trust (SSCT)*, 1 (July).

Anonymous. (2011) *The Monument: The Newsletter for the Solomon Scouts & Coastwatchers Trust (SSCT)*, 3 (March).

Ballard, R. D. and Morgan, M. H. (2002) *Collision with History: The Search for John F. Kennedy's PT 109,* Washington, DC: National Geographic Society.

Bennett, J.A. (2009) *Natives and Exotics: World War II and Environment in the Southern Pacific*, Honolulu: University of Hawai'i Press.

Bennett, W. (1988) 'Behind Japanese lines in the Western Solomons', in G. White, D. Akin, D. Gegeo and K. Watson-Gegeo (eds), *The Big Death: Solomon Islanders Remember World War II*, Suva, Fiji: University of the South Pacific.

BJS Group. (n.d.) 'Solomon Scouts and Coastwatchers' Online. Available HTTP: www.bjs.com.sb/other.html (accessed 2014).

Boutilier, J. (1989) 'Kennedy's "Army": Pacific Islanders at war, 1942–1943', in G. White and L. Lindstrom (eds), *The Pacific Theater: Island Representations of World War II*, Honolulu: University of Hawai'i Press.

Butcher, M. (2012) *'. . . when the long trick's over': Donald Kennedy in the Pacific*, Kennington, Victoria: Holland House.

Cappelletto, F. (2005) *Memory and World War II: An Ethnographic Approach*, Oxford and New York: Berg.

Clemens, M. (1998) *Alone on Guadalcanal: A Coastwatcher's Story*, Annapolis, MD: Naval Institute Press.

Donovan, R. J. (1961) *PT 109: John F. Kennedy in World War II*, New York: McGraw-Hill.

Donovan, R. J. and Warner Bros. Pictures (1963) *PT 109*, motion picture, n.p.: Warner Bros. Pictures.

Feldt, E. A. (1967) *The Coast Watchers*, Sydney and London: Angus and Robertson.

Feuer, A. B. (ed). (2006) *Coast Watching in World War II: Operations Against the Japanese on the Solomon Islands, 1941–43*, Mechanicsburg, PA: Stackpole Books.

Fifi'i, J. (1988) 'World War II and the origins of Maasina Rule', in G. M. White, D. Gegeo, D. Akin and K. Watson-Gegeo (eds), *The Big Death: Solomon Islanders Remember World War II*, Suva, Fiji: Institute of Pacific Studies.

Fujitani, T., White, G. and Yoneyama, L. (eds). (2001) *Perilous Memories: The Asia Pacific War(s)*, Durham, NC: Duke University Press.

Geertz, C. (1974) '"From the Native's point of view": on the nature of anthropological understanding,' in Clifford Geertz, *Local Knowledge: Further Essays in Interpretive Anthropology*, New York: Basic Books, 55–70.

Horton, D. C. (1970) *Fire over the Islands: The Coast Watchers of the Solomons*, Sydney: A. H. and A. W. Reed.

Innes, J. (2012) *Guide to the Guadalcanal Battlefields*, John Innes: available via jinn1064@bigpond.net.au.

Keesing, R. M. (1978) 'Politico-religious movements and anticolonialism on Malaita: Maasina Rule in historical perspective,' *Oceania*, 48, 241–61.

Koroi, R. (2012) *The Monument: The Newsletter for the Solomon Scouts & Coastwatchers Trust (SSCT)*, 5 (September).

Kwai, A. A. (2013) *Islanders in the Second World War: A Solomon Islander Perspective*, MA subthesis, history specialization, Canberra: Australian National University.

Laracy, H. (1991) 'George Bogese: "Just a bloody traitor"?', in G. White (ed), *Remembering the Pacific War*, Honolulu: University of Hawai'i Center for Pacific Islands Studies.

Laracy, H. and White, G. (eds). (1988) *Taem Blong Faet: World War II in Melanesia. Special Issue of 'O'O: A Journal of Solomon Islands Studies*, Honiara: University of the South Pacific Centre.

Lord, W. (1977) *Lonely Vigil: Coastwatchers of the Solomons*, New York: Viking Press.

Riseman, Noah J. (2012) *Defending Whose Country? Indigenous Soldiers in the Pacific War*. Lincoln: University of Nebraska Press.

Sahlins, M. (1985) *Islands of History*, Chicago: University of Chicago Press.

Sharp, S. (2011) 'Recognition comes late for Scouts "rounded up and thrown into the fight"'. Online. Available HTTP: www.telingamedia.com/index.php?option=com_content&view=article&id=76:recognition-comes-late-for-scouts-rounded-up-and-thrown-into-the-fight&catid=32:wwii-history&Itemid=28 (accessed 1 March 2013).

Solomon Islands Visitor Bureau. (2012) *Solomon Islands: Visitor Guide 2012–2013 Edition*, Honiara: Solomon Islands Visitor Bureau.

Thompson, C. (2002) *The 25 Best World War II Sites, Pacific Theater*, San Francisco: Greenline Publications.
Tregaskis, R. (1943) *Guadalcanal Diary*, New York: Random House.
White, G. M. (1989a) 'The politics of remembering: notes on a Pacific conference', *Cultural Anthropology*, 4, 194–203.
White, G. M. (1989b) 'Histories of contact, narratives of self: wartime encounters in Santa Isabel,' in G. M. White and L. Lindstrom (eds), *The Pacific Theater: Island Representations of World War II*, Honolulu: University of Hawai'i Press, 43–71.
White, G. M. (1995) 'Remembering Guadalcanal: national identity and transnational memory-making', *Public Culture*, 7, 529–55.
White, G. M. (1996) 'War remains: the culture of preservation in the Southwest Pacific', *Cultural Resource Management Special Issue*, 19, 52–6.
White, G. M., Akin, D., Gegeo, D. and Watson-Gegeo, K. A. (eds). (1988) *The Big Death: Solomon Islanders Remember World War II*, Suva, Fiji: University of the South Pacific.
White, G. M. and Lindstrom, L. (eds). (1989) *The Pacific Theater: Island Representations of World War II*, Honolulu: University of Hawai'i Press.
Zoleveke, G. (1988) 'The war was not our war. Taem Blong Faet: World War II in Melanesia', *Special Issue of 'O'O: A Journal of Solomon Islands Studies*, 4, 75–8.

Section III

Islands of War, Islands of Dark and Difficult Heritage

12 The Sacred and the Profane

Souvenir and Collecting Behaviours on the WWII Battlefields of Peleliu Island, Palau, Micronesia

Neil Price, Rick Knecht and Gavin Lindsay

INTRODUCTION

The Republic of Palau is a proudly independent nation of some 21,000 people that lies to the south-east of the Philippines and north of New Guinea (Figure 12.1). Its chain of islands, largely enclosed by a sheltering coral reef, makes up the westernmost region of Micronesia. Situated near the equator, Palau's climate is generally stable all year round, mixing tropical sunshine with a little daily rain, enough to maintain the shimmering greenery and freshen the air. Unsurprisingly, the islands have become something of a tourist paradise, attractive partly for their almost stereotypical Pacific beauty and partly for the unparalleled diving offshore.

Some people also come for other reasons, principally focused on one of the southernmost islands: the tiny outcrop of Peleliu. Only five square miles in size, today Peleliu resembles many of its neighbours as a dot of limestone and coral, covered with jungle and mangrove, fringed with white beaches along a turquoise sea. However, some seventy years ago in 1944, Peleliu looked very different—a smoking wasteland of soiled white rock, its vegetation burned away by more than two months of desperate fighting, its indigenous people dispersed, its sacred cultural landscapes obliterated by one of the worst battles in the entire Pacific Theatre of WWII. The strange legacies of this conflict and the island's unexpected status as the best preserved battlefield from the Western Allies' struggle against the Empire of Japan have brought visitors there in a small but steady stream ever since.

This chapter concerns the reasons for their presence and the behaviour of a few of them at the intersection of memory and material culture, in their encounter with the wartime debris that still covers the island. With a beginning during the actual fighting—often only seconds after the death of enemy troops—the collecting (and definition) of 'souvenirs' is a little studied aspect of contemporary warfare that deserves closer attention, and it played a major role in the conflict narrative of Peleliu. The resulting material culture is reinvented as honoured heirlooms in veteran families and

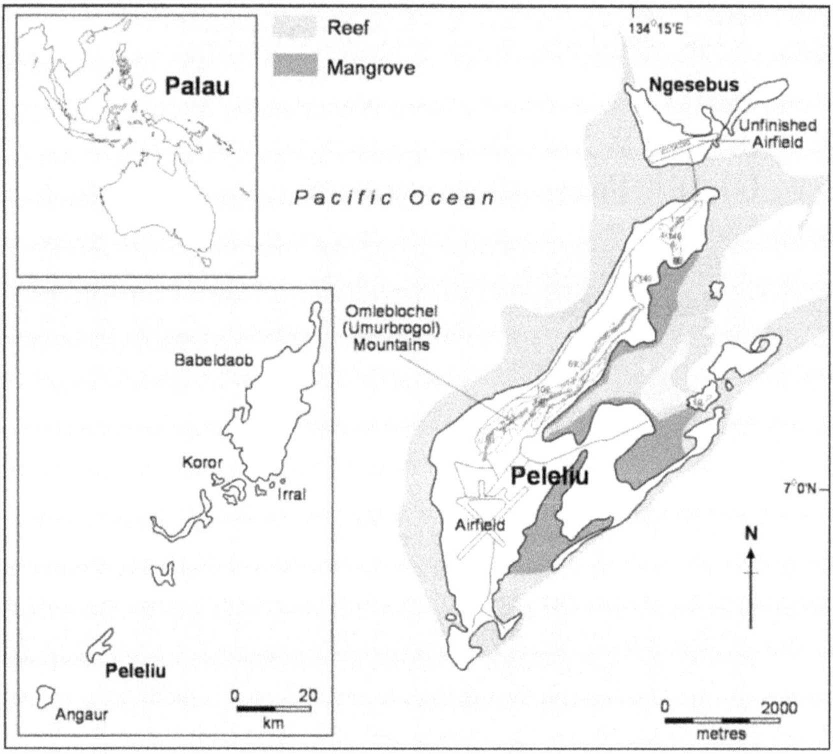

Figure 12.1 Peleliu Island in the Republic of Palau, western Micronesia (Map by Jenny Johnston. Image used courtesy of the University of Aberdeen).

sometimes as valuable commodities for sale, partially decontextualized but nonetheless still a component of the battle's archaeological heritage. Similar activities continue today, with a complex and contested relationship to the (in)appropriate and (dis)respectful recovery of human remains. Against a contemporary backdrop of geopolitical concerns that still resonate from the war, conflicting memories of the Battle of Peleliu are thus reified as both the sacred and the profane. We attempt to illuminate the phenomenon here, using the archaeological, social and cognitive landscapes of the island as our case study, starting with some brief but necessary background to the battle itself.

15 SEPTEMBER 1944: ANOTHER D-DAY IN THE PACIFIC

When WWII began, the Palau islands had already been under Japanese occupation for more than thirty years and served as the capital of the

Nan'yō, the Micronesian mandate of the Imperial throne (Peattie 1988). In the aftermath of Pearl Harbor, the islands were heavily garrisoned against the potential and then actual Allied advance by air and sea. By 1944 and the fall of the Marianas that put US bombers in range of the Japanese homeland, Palau found itself on the very front line (for the larger picture of the Pacific Theatre, see Costello 1981; Hastings 2007).

Peleliu was among the smaller islands of the Palau group, but it held the most developed airfield and was garrisoned accordingly. The Japanese made best use of the available ground, concentrating their defence on more than 600 caves and tunnels dug into the Omleblochel mountains, a confusing maze of limestone peaks, ravines and gullies that formed the spine of the island. The American command believed that Peleliu's capture was key to MacArthur's planned invasion of the Philippines and that the island could be quickly overrun in no more than four days. They were wrong on both counts: the Battle of Peleliu would prove tragically irrelevant to the wider strategic objectives, but its defenders fought to the very end, necessitating over two months of combat that the official history of the Marine Corps declares to have been their toughest fight of the war. (We present here only the briefest background to the battle; for a more extensive survey, see Moran and Rottman 2002; Price and Knecht 2012, 2013 include full references to the Peleliu literature.)

From the landings on 15 September 1944, over the seventy-four days of fighting before the island was declared secure (though some Japanese holdouts remained until as late as 1954), the full savagery of the battle can be appreciated through its grim statistics. The American forces consisted of 9,000 Marines in combat roles with 9,000 more in support, later reinforced by 11,000 men from the 81st Infantry; of these, one Marine battalion lost 71 per cent of its men, many regiments were reduced by half, and overall the American forces took almost 30 per cent casualties. According to conventional military doctrine, this represents the outer limit of what any unit can sustain and continue as a viable force. The Peleliu casualties were proportionately among America's worst of the war; for comparison, the US fatalities on the tiny island exceeded the Allied death toll on all five Normandy D-Day beaches combined. Almost half of the American survivors were rotated home with what would today be diagnosed as post-traumatic stress syndrome.

Their opponents fared far worse. Of the 11,000-strong Imperial Japanese Army and Navy garrison, only 19 men survived; a further 34 emerged from the jungle years later, the last formal surrenders of WWII. Approximately 3,000 forced labourers also died, the majority from Korea but also including people from Okinawa, nearby Micronesian islands and as far away as Java.

The fighting was exceptionally vicious, involving clearing each cave and bunker individually using explosives, flamethrowers, burning liquid fuel and hand-delivered napalm; near the end, the remaining defended caves were sealed, entombing their occupants alive. US records show an expenditure on Peleliu of 118,000 grenades, 150,000 mortar bombs, 123,000 artillery shells, and a staggering 15.5 million rounds of 0.30, 0.45, and 0.50 calibre bullets by the Marines and infantry alone. To this may be added the 5,849 tons of ordnance fired by the Navy and the 806 tons of bombs dropped from aircraft, while the Japanese expenditure was probably equal to all the rest combined. As a result, archaeological investigation has thus far been limited to pedestrian surface surveys, in addition to exploration of caves in the company of a professional demining crew.

The indigenous cultural and sacred landscape of the island—formed over millennia of settlement and tradition—was erased, with far-reaching social and psychological consequences for the native Palauans. However, beyond Palau itself, the tragic combination of appalling losses incurred largely in vain meant that Peleliu—the 'forgotten corner of Hell'—was long overlooked in comparison with other Pacific War battles such as Guadalcanal and Iwo Jima, which were less ambiguous fights with clearer objectives and outcomes and with readily communicable narratives. Against this background, our purpose here is to explore the different ways in which the former combatants, the indigenous population, bystanders and external stakeholders have chosen to remember the fighting through their relationships with its material remains.

THE ARCHAEOLOGY OF THE BATTLE

The Peleliu battlefields have been the subject of two comprehensive archaeological surveys, the first in 1981 by D. Colt Denfeld (1988) and the second in 2010 by the authors (Knecht, Price and Lindsay 2012; Price and Knecht 2012, 2013), with more fieldwork completed in late 2014. The nine-day survey in 2010 documented 285 WWII sites over a relatively small area of the 'central combat zone' in the island interior. Many of these sites were structural remains such as standing buildings, foundations, defensive installations, caves, wrecked vehicles and aircraft, as well as large assemblages of military equipment or hardware or concentrations of human remains.

The Battle of Peleliu began on the landing beaches on the south-western coast of the island, where military debris still protrudes from the sand and the rusting carcasses of landing craft remain along the treeline of dense secondary-growth jungle. Pillboxes and other prepared Japanese gun positions, hewn out of the coral bedrock, overlook the beaches which were the scene of heavy American casualties. Episodic tropical storms such as the recent (2012) Typhoon Bhopa continually shift marine sands and expose buried features such as Japanese landmines, as well as quantities of live hand grenades probably dropped into the surf by Marines assaulting the most heavily defended positions.

Beyond the beaches lies a relatively flat and low-lying expanse of ground, now largely reclaimed by jungle but originally the location for a substantial

Japanese airfield complex, a primary objective for the American forces. Amidst the tropical undergrowth are the shell-damaged remains of the concrete technical, administrative and domestic Japanese building complexes, as well as the post-battle American occupation and expansion of the airfield. Aircraft graveyards stretch for hundreds of metres around the perimeter of the field, representing a host of Japanese and US planes.

Throughout this already complex and layered landscape is the more ephemeral but omnipresent evidence of the battle itself. Hastily constructed firing positions, mortar pits, aid stations and rifle pits are especially abundant within the central combat zone where fighting continued for months after the landing. The surfaces of the dirt roads, constructed by the Marine engineers in the aftermath of the battle, are still studded with thousands of crushed cartridge cases, bullets, shrapnel fragments and pieces of miscellaneous military hardware.

Along the interior spine of the island, the central combat zone is marked by the steep, unforgiving limestone ridges, valleys and sinkholes typical of tropical karst topography. These features were brilliantly utilized and augmented by the Japanese Imperial Army and Navy, who forced labourers to construct a dense network of subterranean defences within the mountain. By exploiting natural cave features, creating new ones and furnishing them with multiple entrances, firing positions and interconnecting tunnels, the Japanese succeeded in turning the upland core of Peleliu into a heavily fortified labyrinth of caves and connecting passages between ridge and valley systems.

Here the fighting was fiercest and afterwards was beyond the reach of post-battle clean-up or development. The density and quality of artefact concentrations are dramatic evidence of the ferocity of the months of close-quarters combat that took place here. Large piles of empty ammunition storage cases, discarded mess kits and water bottles perforated by shrapnel, the rubber soles of worn-out boots, bullet-punctured helmets and the ruptured remains of napalm canisters can be seen among the battered artillery pieces and the more prosaic remains of Japanese garrison life such as sewing machines and bicycles.

For the Japanese defenders, each cave served as a home as well as a fighting position, and this is clear in the material assemblages that survive inside. One enters a cave by alternately crawling on knees and belly, down slopes and past metal barrel barricades peppered with bullet holes, twisted heavy machine guns and dismembered field artillery pieces. Past the entrances, the narrow passageways are strewn with artefacts ranging from cookware, glass *sake* bottles and medical supplies to ammunition stockpiles, gas masks and personal combat equipment. Many of the caves follow a set of standard layouts, depending on their function and affiliation with the Japanese Imperial Army or Naval forces. The most elaborate command caves feature rooms and niches off the main passageways, some of which were originally fitted out with wall cladding and furniture.

The caves were nearly impervious to bombs and shellfire, with fighting in the narrow limestone valleys resembling urban combat. Stark evidence of the nature of the cave assaults can be seen in the form of discharged blackened residues on the walls left by flamethrowers and petrol bombs, charred human remains and partially melted personal effects of the Japanese defenders. The dust in the caves clings to one's clothing and still smells of burned hair and bone. Confronting the wartime past in Peleliu is an experience that both residents and visitors must come to terms with. More than seventy years after the event, the archaeological record on Peleliu retains its extraordinary power to convey the horrific nature of WWII island combat (Price and Knecht 2012, 2013).

MATERIAL MEMORIES: COLLECTING EXPERIENCE

This profusion of wartime remains may at first appear to represent a relatively pristine record, an apparently 'untouched' battlefield as the combatants left it. In fact, the archaeological integrity of this and other similar insular fields of combat has a special definition, in that the surface artefacts, however well preserved they may be, are in reality rarely *in situ* but have instead remained in episodic motion from the time of the battle until the present day.

This process of site disturbance—though, of course, not always conceived as such—began as part and parcel of the fighting itself, when material culture and features were repeatedly damaged, moved and transformed by shelling, bombing and weapons fire. The configuration of battlefield remains was also altered by American heavy equipment used to remove defensive and natural obstacles for troops and vehicles as the battle progressed. By way of drastic example, the damaged airfield was being repaired by Navy Seabees (CBs, or construction battalions) even as Japanese fire continued to rain down upon it. Furthermore, renovation of the garrison areas and roads continued nonstop until at least two years after the war had ended, resulting in a WWII archaeological record that is stratified in complex ways.

The nature of the terrain here naturally tended to favour highly portable war material culture, and it was the smaller pieces of Japanese militaria that were subject to souvenir hunting by the US forces that took place even under heavy fire. Fallen Japanese soldiers were virtually field-stripped for souvenirs seconds after being killed or in many cases while they still lay wounded and struggling.

One of the best first-person combatant accounts of the Peleliu campaign comes from Eugene Sledge, who fought with the 5th Marines and described his experiences in what is often been claimed to be one of the finest WWII memoirs ever written. In *With the Old Breed* (1981), he depicted the souvenir collecting on the battlefront:

> During this lull the men stripped the packs and pockets of the enemy dead for souvenirs. This was a gruesome business, but Marines executed it in a most methodical manner. Helmet headbands were checked for

> flags, packs and pockets were emptied, and gold teeth were extracted. Sabers, pistols, and *hari-kari* knives were highly prized . . . Rifles and other larger weapons usually were rendered useless and thrown aside. They were too heavy to carry in addition to our own equipment. They would be picked up later as fine souvenirs by the rear-echelon troops. . . . The men gloated over, compared, and often swapped their prizes . . . It wasn't simply souvenir hunting or looting the enemy dead; it was more like Indian warriors taking scalps.
>
> (Sledge 1981, 118–9)

Another category of material that was sought after by American troops was that of photographs, personal snapshots taken from the bodies of enemy dead. We have treated this subject more extensively elsewhere (Price and Knecht 2013, 210), but here we see a curious transformation of items intended for private reflection, changing meaning within different contexts. For the Japanese, they were memories of home and family, as well as souvenirs of their military service. For the Americans, they were trophies of fallen enemies, often literally picturing their faces and also a symbol of what they hated: many of the photos showed the Peleliu defenders in their previous station of Manchuria, associated with war crimes in the eyes of their opponents. Some American troops in the Pacific Theatre went further and took more literal trophies of the dead (Harrison 2006).

Strangely, souvenir hunting actually took a significant toll on those who practised it, as casualties were incurred while looking for objects. This is a particular feature of island battles in that the sheer scalar compression of the fighting meant that those seeking artefacts often exposed themselves (and others) to extreme danger. It can seem bizarre to contemporary eyes that men were willing to unnecessarily hazard their lives in the middle of a fight, simply to gain choice trophies. Peleliu preserves a sobering example in the overturned wreck of a Sherman tank on Hill 210, which was destroyed by a Japanese improvised explosive device while attempting to rescue Americans who had crossed enemy lines in search of souvenirs (Price and Knecht 2012, 31). The tank crew were all killed; similar risks were run by medical personnel who came to the aid of wounded looters in the field and by military police who attempted to stop the search for mementoes.

Even after the main battle was over, American souvenir hunters died when they entered caves still occupied by scattered Japanese survivors. In addition, intelligence gathering was often hampered when potentially valuable documents and weaponry were taken before they could be analyzed. The following comes from a report by Charles C. Byrd, of the 1st Medical Battalion:

> There was a road that passed the blockhouse and hit the road from the airport. That was a pick up point for the wounded, before we put up our tent hospital. Later on they put MPs there to stop souvenir hunters,

> but they would slip around the guards and get up in the cave area and get killed or wounded. We took turns being litter bearers and then [the Japanese] would get some more shots in. I dreaded it. One day we got two [souvenir hunters] that was shot up real bad and we fixed them up. I heard one of them tell the other one he was sorry he talked him in to going. The other reached his hand down for a shake and said he wouldn't have missed it for nothing.
>
> (Thomas n.d.)[1]

There are also numerous diaries and personal recollections that make clear the contempt of the fighting veterans for those in support roles who sought to gain souvenirs of combat in which they had not participated (Sledge 1981 contains several examples).

Post-battle souvenir hunting continued among personnel from the American garrison that remained on the island until 1947, although the Japanese cave interiors seemed to have been less impacted—probably because they were foul places for many years, littered with rotting food and the remains of their defenders.

PELELIU TODAY

Clearly, the island of Peleliu is now a very different place from the defoliated, war-scorched landscape that was carved out by the intense fighting of 1944: the vegetation has regenerated, homes have been remade and livelihoods reforged. Despite these recoveries, however, the unrelenting legacy of war continues to impact upon present-day Peleliu.

The destruction of boundary markers and distinctive natural features once used to indicate land divisions has been one of the most crippling impacts of WWII on the post-war recovery of Peleliu and continues to affect reconstruction and development today (Murray 2006, 248). The majority of the *chad ra Beliliou* (people of Peleliu) live in the small village of Kloulkubed located in the north of the island, to which the evacuated population were returned in 1946. This village, originally consisting entirely of prefabricated corrugated steel Quonset huts (similar to Nissen huts) until they were levelled by Typhoon Luis in 1964, was constructed by the US military as a replacement for the five native villages of Peleliu that were destroyed during WWII. Despite their relocation, the *chad ra Beliliou* continue to maintain a connection to the traditional villages through membership by matrilineal decent. The enduring concentration of the population in Kloulkubed can be seen as a direct consequence of the war, with the resulting land ownership issues and the scale of resources that would be required to rebuild what has been lost.

Peleliu's pre-war reliance on the fertile garden soils and abundant fishing grounds has also been irreversibly changed as neither supports the population as it once did. The environmental impact on the soils and water from thousands of white phosphorus shells, high explosives and other pollutants such as bullet lead, petroleum, DDT and other insecticides has taken its toll on fish numbers and soil fertility with only a few small garden plots and taro patches having been put back to crop since the end of WWII (Murray 2006, 214–8). There is now a high dependency on imports to sustain the population of Peleliu, and islanders have begun to seek alternative means to support themselves and their families, one of which has been through exploitation of the very material that has effected such a dramatic change on their island home in the first place.

As the American garrison departed, some buildings and equipment were dismantled and salvaged, but much was left behind and even in recent years has represented an important source of supply for the Palauan population. It is probably fair to say that most households on Peleliu today still make use of some form of WWII material culture, ranging from building supplies to personal items. Japanese and American corrugated metal is used on many walls and rooftops, heavy aluminium military cookware is still abundant and many Peleliu school children used Japanese mess kits as lunch boxes until relatively recently. A Japanese staff car was driven on the island until the 1970s. The Palauan approach to collecting WWII remains tends towards the pragmatic, in stark contrast to American and Japanese collecting behaviours that focus on memorializing the battle.

Over and above these responses, the presence of wartime debris represents a disquieting intrusion into the spiritual landscape of the islanders, an alien and unwanted overlay onto ground that was and remains a wellspring of identity and rootedness. Coupled with this, the overwhelming reality of ordnance in enormous quantities is seen neither as collectible militaria nor items of hallowed memory but rather as a particularly lethal form of litter (Price and Knecht 2013).

International groups and individuals with an amateur interest in military history and WWII in the Pacific have formed the backbone of battlefield tourism on Peleliu since the 1990s. With the release of the American HBO miniseries *The Pacific* in 2010, which featured the Battle of Peleliu in three central episodes, the island has witnessed a significant increase in battlefield tourism (Ambrose 2010; see also Price and Knecht 2012, 2013). The steady stream of visitors has brought additional income to the small island economy, providing a welcome support for local businesses. The interest in wartime material has also encouraged the creation of dedicated tourism attractions, such as guided tours and a battlefield museum which the islanders opened with support from the Peleliu State Government in September 2004. The financial benefits and development potential of such a well preserved conflict landscape has therefore not been lost on the inhabitants of Peleliu, who have lived in and amongst the ruins of WWII for the past seventy years; the desire to capitalize further on their wartime heritage is still keenly felt (Murray 2006, 2). This intersection of commercial interests and local heritage is also clearly a factor in the archaeological investigation of the island and the resulting inventories of its remains.

The steep topography of the island's interior, where the battle was at its most sustained, was and remains inaccessible to vehicles. The battlefield detritus in this area is the least disturbed, especially within the more remote Japanese defensive caves. Even here, however, for the more adventurous groups and individuals who choose to stray further from the roads and trails, the sense of isolation and the illusion of discovering sites and artefacts for the first time has sadly led to a higher degree of interaction with the material remains and a propensity towards modern-day souvenir collection. The archaeological record has thus been impacted by collecting behaviours which reflect multiple and culturally based views of the Battle of Peleliu.

Visits from aging American and the few remaining Japanese veterans have become less frequent in recent years, and most visitors today have at best only a second-hand knowledge of the war. Nevertheless, because of the remarkable preservation, today's visitors often share an emotionally powerful and visceral experience as they physically encounter the past on the battlefield. Both American and Japanese visitors can occasionally feel a sense of entitlement to taking a souvenir home with them as kind of private memorial. For the Americans, there seems to be a vicarious sense of kinship to the victors on Peleliu and a proprietorial attitude to a share in the spoils. This may explain why most Americans tend to focus on collecting *Japanese* militaria, despite the abundant material remaining from both sides. Other visitors collect for purely acquisitive and even commercial interest. Aircraft graveyards, with their rare parts, are particularly vulnerable to looters and have a potentially substantial commercial value.

While private and public American memorials exist on Peleliu (see discussion in Price and Knecht 2013), no actual graves remain of American combatants. In 1947, more than 1,000 American dead were repatriated from the military cemetery on Orange Beach to collective Pacific War burial grounds in Honolulu and Manila.

In many ways, Peleliu, with its wartime heritage, remains a contested place between the two once warring powers, as each of the governments and their associated agencies maintain different attitudes and stakes in the archaeological remains. As a brief illustration of these divergent interests over Peleliu, in 2000 the Japanese national aid body, Japan International Cooperation Agency (JICA), conducted an economic development study on Peleliu which made no reference at all to the war, the remains on the island or their potential as a tourism resource (Murray 2006, 346). In contrast, the United States National Park Service has been pursuing the potential of establishing a large battlefield historic park there, a notion that continues to be explored today in cooperation with the Peleliu War Historical Society, an American non-profit organization (and the grant holder for our surveys on the island).

Since the 1950s, Japan's principal government-endorsed interaction with war material on Peleliu has been through the intermittent collection of human remains and associated personal effects, a process closely monitored by archaeologists from the Republic of Palau's Bureau of Arts and Culture (Price and Knecht 2013, 235). Repatriation of human remains from Peleliu and Palau was begun in the 1960s by privately funded Japanese groups and continues to the present day. Despite this slow but steady effort, human remains are still abundant in the caves and on the jungle floor of the rugged interior. In recent years, Japanese repatriation teams have been funded by the Japanese government but are still composed largely of amateur volunteers accompanied by a physical anthropologist. Material culture on the battlefield and caves is literally pushed aside during these searches, and anecdotal evidence suggests that small pieces are often collected by individual participants but are considered almost as relics rather than collectible commodities.

For many Japanese, Peleliu remains a graveyard and as such is a sacred place. The hopeless odds and fatalistic defence of Peleliu almost literally to the last man are also respected as representing key values of patriotism and self-sacrifice. War memorials to the fallen defenders span the whole political spectrum, from personal shrines to large Shinto complexes, in addition to others sponsored by nationalist organizations; inside the caves, some veterans and military groups have deposited small Buddhist figurines in the deep dark (this material is reviewed at length in Price and Knecht 2013).

Much of this commemoration also involves the movement of artefacts from items of equipment picked up and left as offerings on the shrines to material from the beaches (especially sea shells) brought inland and deposited as gifts around the standing buildings of the former airfield. Similarly, soil—and especially beach sand—is retrieved by relatives of the dead and taken back to Japan in lieu of a body. Known as 'Peleliu dust', this forms an important part of memorial ritual in the home islands (again, discussed in Price and Knecht 2013).

Finally, a further category of material remains slowly circulates around the island, as visitors pick up interesting objects and then later redeposit them elsewhere on Peleliu. Informal interviews with visitors suggest that momentary curiosity and the well intentioned desire to bring home a memento of costly sacrifice give way to a guilty feeling that such action is inappropriate, leading to the artefacts' decontextualized redeposition.

Other visitors deliberately try to be helpful in placing items of unexploded ordnance in prominent places (usually in bombed-out buildings or on wrecked vehicles) so that people will not step on them. Apart from being hazardous in its own right, this further serves to disturb the integrity of the battlefield.

ARCHAEOLOGICAL CHALLENGES AND POTENTIAL

How do these behaviours affect the surviving material record, its documentation and interpretation? The archaeological reconstruction of events on the Peleliu battlefield presents both challenges and opportunities. Aside from the close atmosphere and stifling heat, the sheer mass of material and lack of soil make it difficult to differentiate between sequences of events at a site or to apply traditional methodologies to their interpretation. Many of the caves changed hands frequently, as both sides advanced and retired through the protracted days of intense fighting in the Omleblochel. As shortly as minutes after a cave was taken, the artefact assemblages would have witnessed disturbance from souvenir hunters, and in many of the more accessible caves a narrow path of cleared floor can be found snaking its way back into the gloom, indicating where more recent visitors have rummaged through the floor deposits.

Despite such a history of site disturbance, where the *in situ* nature of artefacts cannot be fully relied upon, a great deal of fresh insights into the combatant experience on Peleliu can nevertheless be gathered (Knecht, Price and Lindsay 2012, 137). Comparative analysis of historical and present-day photographs of cave interiors demonstrate that their contents can shift dramatically, but the remaining inventory is still distinctly legible in archaeological terms (Figures 12.2 and 12.3). Some categories of objects were always more prized than others and are thus largely absent today. These particularly include the distinctive Type 14 Nambu pistols, *katana* swords, other rare weapons, flags and anything with Japanese writing. However, more broadly issued uniform and equipment items remain, including webbing, marching boots, pouches, service respirators, mess kits and water bottles. Unfired Japanese and US small arms ammunition is also present in very large quantities, as it represents a low-value item for looters and is also not deemed sufficiently hazardous to be removed by UXO clearance teams.

Figure 12.2 A passage leading into the interior of the Japanese Navy command cave in October 1944 (United States National Archives and Records Administration, photo RG127-MW-729/107863, USMC photographer Fitzgerald; in the public domain).

Figure 12.3 The same cave passage in 2010; note the movement of artefacts (Photo by Rick Knecht, image reproduced courtesy of the University of Aberdeen).

Although disturbed, the movement and continual sorting of materials on the battlefield in themselves illuminate the multiple ways in which the wartime past has been processed. Artefacts have clearly been cast to the edges of passageways or heaped in recesses but will rarely have been moved more than a couple of metres from their place of origin. This is in part a matter of practicality and safety, as souvenir hunters in such close environments will have worked systematically and carefully so as not to cast items onto live ordnance. Taphonomic processes notwithstanding, the research potential remains great on Peleliu in that valid interpretations of what activities took place in particular sections of cave passageways, rooms or niches are still possible from the artefact types present. In particular, the small arms ammunition offers great promise as a means of identifying certain classes of combatant that may have been operating from a cave during the battle. The standard issue, low-looting-value items such as mess tins also contribute to our broader understanding of the combatant experience, serving as quantifiable and potent reminders of the number of men who would have left their non-essential equipment behind at a cave before embarking on an offensive operation, never to return.

What becomes evident from the current situation is that the ubiquitous material legacy of recent conflict on Peleliu resonates with a wide range of participant groups and individuals, each with their own set of priorities, perceptions and principles. For some, Peleliu forms a stage where the political tensions that still exist between the dominant powers in the Pacific can be played out through the memorial landscape. For others, the remains are a resource with revenue potential or a connection with the past, a place of lost souls or an arena of reflection. The material that survives in such quantity and quality of preservation therefore continues to perform a powerful role in Peleliu's present and future narratives, holding a multitude of meanings for a range of stakeholders engaging with it in a variety of different ways.

ACKNOWLEDGEMENTS

Our work on Peleliu was undertaken for the US National Park Service American Battlefield Protection Program, on a grant awarded to the Peleliu War Historical Society; we wish to thank Steve Cypra and D. Colt Denfeld for their generous support. The work was subcontracted through the Palauan government's Bureau of Arts and Culture (BAC), and we are honoured by the trust placed in us by our Palauan hosts. We work both with and for them and extend our grateful thanks to former Governor Kangichi Uchau, present Governor Temmy Shmull, and the State Legislature of Peleliu, together with former Speaker Desengai Matsutaro. The hereditary chiefs of Peleliu honoured us with permission to work on their lands. From the BAC, we thank Director Sunny Ochob Ngirmang, Calvin Emesiochel, Errolflynn Kloulechad and Gener 'Eigner' Moon Sangalang; on Peleliu, we would also like to thank Tangie Hesus, Jolie Liston, Reiko's Guesthouse and Alex at the Store in Kloulklubed village. As ever, we would most particularly like to thank Steve and Cassandra Ballinger and their UXO disposal operation at Cleared Ground Demining–Hiob Keptot, Morgan Matsuoka, Yosko Ngiraked and James Ongklungel; they kept us safe in the field and are improving the lives of Peleliu's citizens at the conscious risk of their own. David McQuillen looked after our health and safety with great good humour and generously shared his trove of archive information on the battle. At the University of Aberdeen, thanks to Jenny Johnston of our cartographic department for creating the map and to all our archaeological colleagues there for advice on the project. Professor Sir Ian Diamond, principal of the university, has been very supportive as our work on Peleliu developed.

Disclaimer: this material is in part based on work assisted by a grant from the United States Department of the Interior, National Park Service. Any opinions, findings and conclusions or recommendations expressed in this material are those of the authors and do not necessarily reflect the views of the Department of the Interior.

NOTE

1. Oral report collected on a Peleliu veterans tribute website, with specific URL: www.thomas5.com/tribute/Vets13.html.

BIBLIOGRAPHY

Ambrose, H. (2010) *The Pacific*, Edinburgh: Canongate.

Costello, J. (1981) *The Pacific War 1941–1945*, London: Macmillan.

Denfeld, D. C. (1988) *Peleliu Revisited: An Historical and Archaeological Survey of World War II Sites on Peleliu Island*, Micronesian Archaeological Survey 24. Saipan: Division of Historic Preservation.

Harrison, S. (2006) 'Skull trophies of the Pacific War: trangressive objects of remembrance', *Journal of the Royal Anthropological Institute*, 12 (4), 817–36.

Hastings, M. (2007) *Nemesis: The Battle for Japan 1944–45*, London: Harper.

Knecht, R., Price, N. and Lindsay, G. (2012) *WWII Battlefield Survey of Peleliu Island, Peleliu State, Republic of Palau*, Archive report lodged with the Bureau of Arts and Culture, Koror, Palau, and the US National Park Service, Guam. Online. Available HTTP: www.peleliuhistorical.org/Peleliu_REPORT.pdf (accessed 7 October 2014).

Moran, J. and Rottman, G. L. (2002) *Peleliu 1944: The Forgotten Corner of Hell*, Oxford: Osprey.

Murray, S. C. (2006) 'War and remembrance on Peleliu: islander, Japanese, and American memories of a battle in the Pacific War', unpublished PhD thesis in anthropology, Santa Barbara, University of California.

Peattie, M. R. (1988) *Nan'yō: The Rise and Fall of the Japanese in Micronesia, 1885–1945*, Honolulu: University of Hawai'i Press.

Price, N. and Knecht, R. (2012) 'Peleliu 1944: the archaeology of a South Pacific D-Day', *Journal of Conflict Archaeology*, 7 (1), 5–48.

Price, N. and Knecht, R. (2013) 'After the typhoon: multicultural archaeologies of World War II on Peleliu, Palau, Micronesia', *Journal of Conflict Archaeology* 8 (3), 193–248.

Sledge, E. B. (1981) *With the Old Breed: At Peleliu and Okinawa*, Novato: Presidio Press.

Thomas, G. M. (n.d.) 'A Tribute to Michael A. Lazaro and All Other Peleliu Veterans'. Online. Available HTTP: www.thomas5.com/index.html (accessed 12 September 2014).

13 War Remnants of the Greek Archipelago

Persistent Memories or Fragile Heritage?

Nota Pantzou

Heritage sites such as Auschwitz-Birkenau, the German Nazi concentration and extermination camp in Poland, Robben Island in South Africa and the Topography of Terror in Berlin compete in visitor figures with acclaimed tourist attractions such as Stonehenge, the Athenian Acropolis and Petra. Visitor numbers speak for themselves. In 2011, Auschwitz-Birkenau received 1,405,000 people; 'a record figure' in the history of the site.[1] Similarly, the Topography of Terror, advertised as 'one of the most frequently visited places of remembrance in Berlin', attracted 900,000 visitors last year.[2] The list of sites of atrocities and war that have become transformed into heritage destinations grows ever longer as traumatic memories fade, governments' policies shift and tourist fascination increases, with the landmarks of the Sarajevo siege[3] and the Spanish Civil War ghost town of Belchite being among the most recent entries.

The growth of tourist demand goes hand in hand with a steady increase of academic publications and studies. Although visits to places related to death and violence dates back to antiquity (Stone 2006, 147), dark tourism as an independent field of inquiry emerged in the early 1990s. Since then, site typologies have been suggested and issues of demand and supply have been addressed in an effort to map the phenomenon and lay the theoretical and practical framework (e.g. Bigley, Lee, Chon and Yoon 2010; Biran, Poria and Oren 2011; Kang, Scott, Jeonglyeol and Ballantyne 2012; Stone 2006; Stone and Sharpley 2008; Strange and Kempa 2003). Yet this legacy of death, distress and human suffering equally concerns academics and professionals from the disciplines of archaeology and heritage management in recognition of its immense historical and educational value, as well as power of reconciliation. The abundance of terms one encounters in this body of literature is striking. *Difficult* (Logan and Reeves 2009), *dark* (Biran, Poria and Oren 2011), *painful* (Jasinksi 2013), *dissonant* (Ashworth and Hartmann 2005), *negative* (Meskell 2002), *traumatic* (Pantzou 2011) and *hurtful* (Dolff-Bonekamper 2002) are some of the adjectives employed to define the ways in which these tangible and intangible cultural resources are perceived and approached. All the proposed terms mirror the profusion of ideas and the striving of scholars to lay the ground for academic and public discussion, to raise awareness and to shed light on a past (recent or remote) that divides people as much as it unites them. No matter what term is used, there is considerable agreement among scholars that this heritage is extremely susceptible to the politics of obliteration and strategies of forgetting when compared with sites and monuments that celebrate life and human achievements (Ashworth and Hartmann 2005; Carr 2007; Pantzou 2011). However, there are numerous proofs that steps are taken to help these material remains and stories resume their rightful place in history and people's memory, namely the attainment of the World Heritage status (Logan and Reeves 2009), the abundance of projects both in university or non-academic contexts[4] and the widespread media coverage that this theme involves.

The numerous examples from the literature illustrate the omnipresence of this heritage in diverse cultures, societies and political contexts. However, under the scope of this edited volume and this chapter, attention is first focused on islands. For King (1993, 14) 'an island is a most enticing form of land'. Enticing to tourists, without a doubt, but islands are also appealing to governments and authorities of all types, who have been exploiting their boundedness to set up penal colonies, slave centers, quarantine stations, military bases and concentration camps, to name a few uses (Strange and Kempa 2003; Tunbridge 2005). The second point of focus here is the theme of war. With these two ingredients in mind, I selected Greece as an ideal backdrop for investigating the impact of war and politics on small island communities.

Greece is a country of 6,000 islands, islets and skerries, whose modern history is marked by transnational wars and civil conflicts. Considering the country's long island tradition and the prominence of 'islandscapes' in tourist imagery and national narrative, this exploratory research proposes to discuss national and local approaches to island conflict heritage and the prospects of dark tourism. Using a combination of sources and findings, attention is drawn to the Aegean exile islands. More precisely, this chapter is divided into three sections. The first part presents the Aegean region's centrality in Greek tourism and its contemporary exile legacy. Then, in an effort to examine national attitudes towards this past, measures taken towards its preservation are delineated in brief. The next step is to introduce and analyze a survey conducted by the author, the aim of which was to obtain a preliminary estimate of public's views on this traumatic heritage. To gain insight into the local perspective, an empirical study of municipal websites was performed and is presented in the last section of the chapter. The intention is to record any signs of grassroots activities planned and executed towards this heritage's promotion and protection. This initial investigation of the heritage of political internment will hopefully reveal the degree to which insularity can affect its sustainability and the intensity of conflict memories in state policies, in public imagination and in these remote and peripheral contexts.

STORIES OF EXCLUSION AND HUMAN SUFFERING FROM AN EDENIC LANDSCAPE

In the Renaissance, the Aegean Sea was perceived as 'the prince of seas' and was known as Archipelago (Ceccarelli 2012, 26). Gradually, the word *archipelago* acquired its present meaning as 'any sea, or sheet of water, in which there are numerous islands'.[5] Nowadays, the term *Aegean Sea* is used to describe the space that contains the island groups of Cyclades, Dodecanese, Sporades, the North Eastern Aegean islands and Argosaronic islands.[6]

What is the first thing that comes to people's mind when they think of the Aegean islands? How is the Aegean Sea imagined? Whitewashed houses with blue doors and shutters, traditional villages hanging from cliff tops, blue sky and the crystal blue waters, warm sunny weather, picturesque beaches, unspoiled landscape and renowned archaeological sites seem like the obvious answer (see also Berg and Edelheim 2012; Buhalis 2001). Indeed, these images dominate the Greek National Tourism Organisation's (GNTO) campaign. On the GNTO's website, wallpapers, banners and brochures of the region, these stereotypical images of easy living, scenic traditional villages and the tangible evidence of country's golden era are perpetuated in an effort to disseminate fixed polished images of the national self and promote one of the country's major industries: tourism. As stated on the website of the Association of Greek Tourism Enterprises (SETE), the tourism sector alone employs one out of five Greeks and contributes 16.4 per cent of the gross domestic product.[7] Konstantinos Andriotis (2004, 119), professor of tourism, pinpoints that insular regions have benefited remarkably from tourism activities, increased numbers of tourist arrivals, state subsidies and the GNTO's favourable treatment. It goes without saying that islands are one of the strongest selling points of the country. Baloglu and Mangaloglu's study (2001, 6) about tourism destination images of Greece and other Mediterranean countries held by US travel intermediaries is revealing to this end. According to their responses, islands are considered as one of the most typical characteristics of Greece, together with 'sunshine, blue skies, beaches', 'relaxing' atmosphere and 'ancient ruins' (Baloglu and Mangaloglu 2001, 6).

Cristoforo Buondelmonti is credited as one of the first 'tourists' who visited the Aegean Sea, entranced by Greek antiquities and ancient geography. Many travellers followed. Yet it is only from the mid-1970s onwards that Greece, as well as the Aegean islands, was established as a holiday destination and brand (Buhalis 1999). Between the 1920s and 1970s, the region was far from the tourist imagery and idealistic views of sea, sun and sand. Aegean islands were places of confinement and exclusion. They functioned as 'exories' (*Εξορίες*), exile locations.

The institution of exile is as old as Greek democracy, but it reached 'pandemic' proportions during the twentieth century. Then exile as a procedure for oppressing and punishing political opponents and supporters of the left was institutionalized and became a common practice among successive governments (Panourgia 2009; Voglis 2002). More concretely, during this period, thousands of civilians—men, women and children—were detained for political reasons in islandscapes all around the Aegean. These places and their history embody a time when the Greek nation was tormented by a civil strife, two world wars and a borderland conflict with Turkey (1919–22).

In the biography of exile and imprisonment in the contemporary Greek context, five distinct phases can be defined that will facilitate this study. Before 1920, only individual cases of banishment are recorded (first phase). The political banishment of Benarrogias and Gionas to Naxos in 1914 is one such example (Bournazos 2009, 37).[8] The second phase lasted from the 1920s to the early 1940s. This was the time when the first wave of large numbers of exiles occurred in arid and remote islands, among them Ai Stratis and Anafi (Kenna 2001). The third phase of exile covered the period 1946–9, a time when the country was torn by a civil war. Then the first internment camps opened in the Aegean islands of Giaros, Makronisos, Chios (Konstantopoulou 1976), Trikeri and Ai Stratis, hosting thousands of people of all ages and backgrounds (Bournazos 2009; Voglis 2002). In the fourth phase (beween 1950 and 1963), only a rather small community of dissidents remained in exile, whilst the camps in Ai Stratis, Giaros and Makronisos closed one after the other (Margaritis 2009, 20). The fifth and last phase of the phenomenon coincided with the Military Junta (1967–74). This was a period when the Giaros camp and prison were reused and two military buildings on the island of Leros were requisitioned to meet the pressing need for detention spaces. Aegean islands were an unlimited source of places and spaces for incarceration and banishment. It was an inexpensive solution since this peripheral, scarcely populated and difficult to access region could accommodate large numbers of enemies of the state whenever the mainland facilities were deemed inadequate. On those occasions, the detainees either rented spaces from the locals or lived in makeshift camps or prisons.

Scholars estimate that about one hundred locations served as prisons or exile places during the turbulent twentieth century (Bournazos 2009, 39); thirty-five out of those loci are Aegean islands (Figure 13.1).[9] The type and duration of use varied from island to island and from time to time.[10] Yet a classification is possible and will be extremely useful here, providing a basis for this research. In the Aegean region, for almost fifty-five years, there existed:

- Indoctrination camps in deserted islands, such as Giaros and Makronisos,
- Concentration camps in the scarcely populated islands of Ai Stratis and Trikeri,
- Camps in small or medium-sized islands like Chios, Ikaria and Leros (Partheni and Laki),
- Small communities of exiles or individual banishment cases under strict supervision by local police in Andros, Alonissos, Amorgos, Anafi, Antikythera, Antiparos, Gavdos, Santorini, Ikaria, Ios, Kimolos, Kythera,

Kythnos, Lemnos, Mykonos, Milos, Naxos, Paros, Samothraki, Serifos, Sifnos, Sikinos, Skyros, Syros, Skiathos, Tinos, Folegandros and Fourni,
- Prison compounds in Aegina and Giaros.

As this list shows, Greece's most famous tourist attractions were popular exile destinations because of their island qualities. At the same time that people remained bounded in these Aegean islands, the Greek political leadership invested in tourism projects for the promotion of Greece as a travel destination, driven by an urge for openness and extraversion. At the outset of the Greek Civil War, the GNTO launched one of its more ambitious projects: the creation of a network of public hotels, the Xenia hotels. The construction between 1953 and 1966 of these forty hotel compounds of unique architectural

Figure 13.1 Map of the Aegean Sea: the circled islands served as places of exile during the twentieth century (Creative Commons, Future Perfect at Sunrise, additions by author).

design was aimed at putting new destinations on the tourist map, boosting the economy and developing adequate tourism infrastructure. Tourist arrival numbers show that tourism-oriented state projects were successful. From 33,333 tourist arrivals in 1950, Greece received 1,131,730 visitors in 1966 (Buhalis 2001, 445). Not surprisingly, numbers slightly dropped in 1967, when the Junta seized power (996,473). By 1973, a year before the restoration of democracy, numbers rose again to the phenomenal figure of 3,177,682. Today Greece holds the seventeenth position worldwide in international arrivals, with domestic tourism functioning as a complementary but considerable force. So is there space in the tourist product offered to visitors (local and foreigners) for alternative stories that are not tied to the Edenic projections of Greece, forty years after the country's exit from the dark era of exile? Before even attempting to provide an answer to this question, I will present state efforts towards the preservation of these material manifestations. Emphasis is placed especially on legal protection measures on the grounds that laws transform remnants of the past into heritage assets by assigning them public value (Carman 1995, 20).

STATE VALUATIONS OF EXILE MONUMENTS

To unfold the story of the thirty-five exile islands in a few paragraphs is impossible and not desirable in the context of this preliminary effort. However, this account of state policies towards the exile heritage will be enhanced by a brief mention of the stories of Makronisos, Giaros and Ai Stratis. This selection is justified by these islands' role as iconic markers of exile in public imagination and the fact that they are among the few examples where concrete actions were taken towards safeguarding them.

Makronisos, a strip of land opposite the Attica port of Lavrio, is often referred to as a purgatory since it is considered one of the darkest and most violent places in Greek contemporary history. In 1989, almost thirty years after the camps' fall into disuse, the island of Makronisos in its entirety was designated a historic site by the Greek Ministry of Culture (MoC):

> because it has marked the history of contemporary Greece and represents a place of memory not only for those who experienced the horrors of the 1946–1953 period, but for all the Greeks and mainly for the younger generations since it construes a symbol of disapproval of the civil war, of violence and all kinds of oppression, a monument to freedom of speech and ideas.
>
> (My translation, MoC *et al.* 1994, 7)

In the last twenty-five years, development studies were conducted, conferences were organized, a museum was established in Athens by former exiles, a film and documentaries were produced, but the material remains of the camp have been left to decay (see MoC *et al.* 1994) (Figure 13.2). Thus, the burgeoning academic and public attention has not yielded results from a preservation and presentation point of view. Moreover, despite this historic site being protected by law, there are hitherto few signs that it has become 'archaeologically important' for the Greek state (Carman 1995, 19).

Figure 13.2 A gate of Makronisos indoctrination camp today (Copyright Pantzou).

The island of Giaros shares a similar destiny. This inhospitable and unpopulated island, north of the capital of the Cyclades, Syros, is known as the Dachau of the Mediterranean. Its notoriety derives from the inmates' harsh and cruel conditions of detainment during both the Greek Civil War and the Junta periods and from the British-designed prison compound, which was built by internees, that dominates the barren landscape to the present day. After the camp's official closure in 1975, the island served as a naval fort for training activities until its declaration as a historic site in 2002 (in its entirety) (Stavridis 2006). Since then, Giaros has been in the news for the wrong reasons. In 2011, the Central Council of Modern Monuments approved the decrease of the limits of the protection zone for the designated site to many people's surprise and disappointment. A year later, stories about the island's transformation into a wind farm surfaced, causing a sensation and stirring a debate. The MoC, principal stakeholder in the management of Greek cultural heritage, confirmed the story by stating in a letter addressed to the Greek parliament that the Ephorate of Contemporary and Modern Monuments is in favour of renewable energy sources. The alteration of site boundaries and subsequent downgrading of this property were hence made in the name of progress.

Hundreds of miles to the north is located Ai Stratis. Used for detainment from the 1920s to 1960s, it represents a microcosm of political banishment (see Ministry of the Aegean 2000; Pantzou 2011). This north Aegean island, whose name has become synonymous with exile, can enlighten further the ambiguous position that this difficult heritage holds in national imagination and state politics. Ai Stratis hosts the Museum of Democracy,[11] a modern history public museum which was inaugurated in 2007 and is housed in the community's old infirmary/school.[12] This scheduled building is the place where thirty-three political exiles died from starvation during WWII and stands as an undisputed landmark to exile legacy. Notwithstanding this, the monument's dark symbolism is totally omitted in the justification for inscription as a historic monument, taking instead into account its importance for the local community. This is also the case for the Aigina prison (1880–1985) and the monastic compound on Trikeri island, where exiled women were ostracized from 1949 until 1953. For both sites, political imprisonment is one phase in their long biography, and MoC, as a national agency, typically gives priority to values that celebrate the glorious Greek past and its achievements. For this reason, from the Aigina prison complex, only the main building and a church were deemed of national importance on the premise that these two structures constituted representative architectural examples of the early nineteenth century, the time of the establishment of the Greek state.[13] In addition, the monastery of Trikeri acquired legal protection solely due to its value as a medieval relic. A further example of MoC's tendency to concentrate its efforts and centre attention only on monuments that fit the national agenda is on its official portal: Makronisos, Giaros and the Museum of Democracy are absent from the cited list of monuments and museums, as well as from the interactive culture map.[14] In contrast, the GNTO endorses a different line of approach as far as its website is concerned. It shares with its online visitors[15] various facets of Greece's modern political struggles, from the early days of exile to the seven-year dictatorship. Stories about political dissidents are mixed with Greek mythology and appealing descriptions of the natural environment, promoting a more fresh and vivid image of modern Greece to suit GNTO's needs for projecting the country's alluring qualities.[16]

The aforementioned examples underline the wavering of the Greek state between negation and acceptance of these contested heritage resources. Irrespective of the listing, the proposals for these monuments' rehabilitation and the formation of working groups, little has been done so far towards the maintenance of this past's remains as a whole. As a result, this heritage is at risk, and immediate action is required. It is encouraging to know that a good number of buildings and structures, both scheduled and not, still survive. In Ai Stratis, Trikeri, Ikaria, Gavdos, Anafi and Leros, visitors can still discern ruins of the camps and other structures, visit carceral spaces or see village houses standing unscathed that once hosted inmates. Nonetheless, this is not the case for several mainland sites. In the perfecture of Attica, for example, a great number of monuments have been demolished and deleted from the map. The case of Averof prison is enlightening in that respect. The notorious prison facilities were taken down in early 1970, and, quite ironically, in their place the Supreme Court of Greece was erected. This abrupt transition beyond question deprived the site of its tangible features, although the memories, stories and artefacts preserved by former inmates retain and keep the aura of the site alive. The Vourla prison in Piraeus, from which twenty-seven political prisoners escaped in 1955, suffered the same fate, and today a newly built block of flats stands on this spot.

While these sites of distress were erased from the landscape, several dark monuments were 'reborn' and rehabilitated in Attica. Their negative associations were washed away, and, as empty spaces, they reentered social life and social memory anew. The MoC, for instance, occupies such a building. It is the former Security Police headquarters in Bouboulinas Street, Athens, where many people were detained, tortured and interrogated during the Junta period. The list of examples is relatively long: a prison turned into a conservatory,[17] the Gestapo interrogation offices transformed into a cosmetic store and a former army base turned into an art gallery and theatre. Similarly, several buildings of political oppression in various Greek mainland locations have been demolished or reused, such as the notorious prison of Acronafplia in Nafplion, which was taken down in order to allow space for the Xenia Hotel, and the Yedicule prison compound in Thessaloniki, which today hosts the ninth Byzantine Archaeological Ephorate. As far as the evidence shows, islands sites appear to have better survival rates than continental ones. Does this mean that the insularity of islandscapes can be considered a significant factor in the sustainability of this heritage? In the periphery of the Greek state, the Aegean islands experience serious disparities in health access, all kind of infrastructures, services and ship or air connections to the mainland (Konsolas, Papadaskalopoulos and Plaskovitis 2002). Decision-making processes are relatively slower, and projects are often more costly. Additionally, islands often enjoy a slower pace of life and change. Perhaps all of these parameters together create the conditions for this conflictual past's 'well-being'. Yet a further exploration of this idea is required in order to unravel the complexity of the politics of memorialization and oblivion. Even if this discussion is introductory, it has demonstrated that, by and large, the management strategies for these cultural properties are still at an early stage, judging by the limited state interest and fragmented interventions. Optimistically, the findings of the survey presented next will illustrate whether the timing is right to boost the importance of the legacy of political banishment.

A PRELIMINARY SURVEY OF PUBLIC ATTITUDES

The previous section examined national attitudes through the frame of the existing legal framework for heritage protection. As public opinion is of equal interest in the context of this research, a pilot survey was conducted on the public understanding of exile heritage. To narrow down the target audience, attention was focused on undergraduate and postgraduate students in the fields of heritage management and tourism management. They are the next generation of custodians and decision/policy makers; therefore they should have a stake in the protection and promotion of these cultural remains. Their views will shed light on the following lines of inquiry:

- How familiar are people with exile sites and the practice of political banishment?
- Are the tangible and intangible aspects of exile seen as integral parts of Greek cultural heritage?
- To what extent do they agree or disagree with measures taken in favour of its protection and promotion or the dissemination of knowledge about this difficult period of Greek history?

For the purposes of this survey,[18] a questionnaire was designed and administered to ninety-six students from January to September 2013.[19] The collected sample consists of fifty-seven final-year students from the Department of Heritage Management and New Technologies (University of Patras), twenty-seven students from the Master's Programme in Heritage Management (Athens University of Economics and Business/University of Kent) and twelve MBA Tourism Management students (University of Piraeus).[20] The gender distribution was 77 per cent female and 23 per cent male. The respondents' ages ranged from 21 to 47, the vast majority of whom were between 21 and 24 years old.

The questionnaire opened by asking whether the respondents had ever heard of Makronisos (seven undergraduate students responded that they did not know Makronisos at all) and was followed by a subquestion that attempted to measure the respondents' knowledge level about the island's history (intended only for those who answered yes in question 1). The reasons for selecting Makronisos lie in its strong political connotations and the fact that until recently it was the only exile site mentioned in Greek history textbooks. Out of eighty-nine respondents, in question 1b, fifteen undergraduates and five postgraduate were not aware of Makronisos's exile past.[21]

Participants then were asked whether they had heard about political banishment and, if so, what were their main sources of information about the topic? The aim was to understand their awareness of the avenues for obtaining information about the practice of banishment. Only four students (three undergraduate and one undergraduate) had never heard about exile. Most students have learnt about it in various ways. Family and school were mentioned as two of the major sources of information, whereas one-third of the participants said that they had also learned about the topic in university. Fewer were those who learnt about exile through media and friends.

In question 3 of the questionnaire, a list of the thirty-five exile islands was given, and participants had to circle the ones they were aware of. The purpose was to identify their degree of familiarization with these cultural properties. Only one student answered that all thirty-five sites are places of exile. About half of the students (49 per cent) recognized between one and four sites. Close to 30 per cent of them knew more than five sites, and 21 per cent did not recognize a single site (three postgraduates and seventeen undergraduates). Nearly 80 per cent of the undergraduate and postgraduate students were familiar with Makronisos and Giaros. The results confirm that these two islands evoke, like no other place, the history of political oppressions. Other exile islands popularly recognized by the students were Ai Stratis, Leros and Alonissos.

As the findings from questions 1, 2 and 3 show, undergraduates exhibit the least awareness of the topic. This might be linked to how this traumatic era of modern Greek history is portrayed in the school curriculum. Greek pupils learn about contemporary historical events in the sixth grade of elementary school, third grade of secondary school and third grade of high school.[22] In 1995, in the history textbook *Ιστορία Νεότερη και Σύγχρονη* [*Modern and Contemporary History*] (Pedagogical Institute 1995, 11th edition) used in the last grade of high school, eight pages were dedicated to the Greek Civil War; whereas in the 2007 edition (*Ιστορία του Νεότερου και του Σύγχρονου Κόσμου* [*History of the Modern and Contemporary World*]), the coverage of the events was limited to a paragraph (Pedagogical Institute 2007a). The section about the civil conflict in the 2007 history textbook *Στα Νεότερα Χρόνια* [*In the Modern Times*] (Pedagogical Institute 2007b, vol. 3) for the sixth grade of primary elementary school was also of a similar length. Suffice it to say that in all these editions, references to exile places are limited or absent. The story of exile obviously does not conform to the official version of Greek national history or to the resistance narrative, which stresses the perceived uniqueness of 'the unity and fighting spirit of the Greek people when they confront threats' (Nakou and Apostolidou 2010, 116; see also Hamilakis 2003). Fortunately, in the 2012 edition of the history textbook *Ιστορία του Νεότερου και Σύγχρονου Κόσμου* [*History of the Modern and Contemporary World*] (sixth grade of primary school), six pages touch upon the topic of the 1946–9 civil conflict (Pedagogical Institute 2012a). Furthermore, several references to exile can be found not only in the history textbook *Νεότερη και Σύγχρονη Ιστορία* [*Modern and Contemporary History*] (Pedagogical Institute 2012b) of the third grade of secondary school but also in the textbook *Κείμενα Νεοελληνικής Λογοτεχνίας* [*Modern Greek Literature*] (third grade of high school) (Pedagogical Institute 2012c). These suggest that in the new educational policies, there is space for 'useful kinds of histories' (Nakou and Apostolidou 2010, 126), and attempts are made to overcome national amnesia about dark moments of Greek history through the incorporation and dissemination of this conflictual heritage.

In the final question, students had to indicate their level of agreement with five statements. The objective was to assess the value that students place on exile as a cultural resource and their stance on preservation and promotion issues. A symmetric five-point agree–disagree scale was used to score responses on the following statements:

The tangible and intangible remains of exile are integral parts of Greek cultural heritage.
Exile places have to be protected.
These sites should be restored and become accessible to the public.
This chapter of modern Greek history should be taught in schools.
These locations could become tourist attractions.
This is a dark period of modern Greek history and it should be forgotten.

The majority of students saw with a positive eye the protection and opening of such sites to visitors. Most of them also concurred with the idea that these heritage properties deserve a place in Greek national policies and identity narratives. Just three students (one postgraduate student and two undergraduates, about 3 per cent) did not perceive exile as a component of Greek heritage. These students are also not in favour of the introduction of these sites into the heritage realm. Finally, it is noteworthy that 86 per cent of the twenty respondents who did not know Makronisos or any other exile place did not object to the prospect of implementing restoration plans and preparing the sites for tourist audiences.

Greek society seems to be little by little coming to terms with these contested cultural assets. In the last decade, signs of renegotiation and reconciliation with the history and materiality of political prosecutions have frequently emerged in the Greek media, on the Internet and in academia. In fact, the latest signs exhibit an unprecedented drive to reflect upon and record the phenomenon. Archives are being formed, books are being published and exile memories are being revived. In this spirit, the first mass cultural event was hosted in Makronisos in the summer of 2003. A concert dedicated to the memory of exiles was performed by Mikis Theodorakis' orchestra. Theodorakis, a former exile himself and emblematic figure of Greek politics, has inspired a couple of generations, and his songs have accompanied numerous demonstrations and public meetings. This successful concert was repeated ten years later. In the midst of the Greek crisis, on 14 September 2013; 6,000 people attended the event which, according to Theodorakis, constituted 'revenge for all those people who have spent time in the island during those periods' (Yannaka 2013).

GRASSROOTS APPROACHES TO EXILE HERITAGE

Admission to the most recent of Theodorakis' concerts in Makronisos in 2013 was free for the public, thanks to generous municipal financial support and voluntary work provided by the Hellenic Red Cross, Médecins Sans Frontières Greece and scouts. Grassroots initiatives like this one exemplify how this heritage is remembered and approached at the local level. Exile memories linger over local communities and can have an impact upon their everyday lives in a concrete manner. In other words, municipalities are and should be seen as principal stakeholders in the negotiation of this past. Therefore, in this last section of the study, official municipal portals are scrutinized. These websites provide a window into local conceptualizations of the past and careful representations of local identity. The objective is to collect information that manifests any kind of interest pertaining to the projection or preservation of the historical or tangible aspects of exile.

Current findings suggest that minor interest exists. Following a thorough search of these thirty-five exile islands' municipal portals, references to exile were made solely by the municipalities of Kea, Lemnos, Leros, Ai Stratis and Syros. The information found on the Internet concerns six sites, respectively: Makronisos, Lemnos, Leros, Ai Stratis, Giaros and Syros. Most municipal websites' history sections contain information that extends from prehistoric times to the 1821 War for Independence. Interestingly, at the same time that municipalities are not foreseeing a place for modern exile in their agenda, they include on their websites historical references to practices of exile from the Roman or Byzantine periods. The Aegean region can, in fact, boast of being the area with the longest tradition of exile. For example, according to a legend, Santorini, one of the top Greek tourist destinations and heritage attractions, is named after Saint Eirini, who died on the island while exiled in the fourth century BCE. Maybe such omissions occur because the era of political oppressions in Greece came to an end just forty years ago, and, as

Uzzell and Ballantyne (1998, 508) note, '[P]hysical and psychological distance from events can accentuate or moderate (in our case) one's emotional involvement, knowledge, concern and of course action'. Nevertheless, exile seems to be a consistent theme and significant element of these islands' history, both in the remote and/or recent past.

The municipality of Ai Stratis is among the few that envisions a concrete role for this heritage resource. It is a small island in the northern Aegean Sea with just 270 inhabitants. The island's authority does not negate the exile past, instead openly embracing it in an effort to boost the local economy and to establish Ai Stratis as a travel destination.[23] On the official municipal website, the island is presented as an 'off the beaten path' gateway, where prominent members of the Greek intelligentsia, such as the acclaimed poet Giannis Ritsos and the distinguished actor Manos Katrakis, were banished together with other dissidents, leftists or not.[24] Moving forward and embracing new technologies as a means of achieving its goals, reaching out to the public and breaking the distance barriers, the community of Ai Stratis joined forces with the team of Greece Virtual (greecevirtual.gr) and designed a virtual tour of the island's sights. To attract visitors, beautiful panoramic views are accompanied with text that advertises the island as a place 'of special scientific interest, due to its natural habitat', with 'superb beaches', 'spectacular sunsets', where 'the first Greek public museum dealing with a period of recent Greek history of crucial political and social importance' is located. By thus propagating stereotypes of Greece's tourist promotion, the local mayor and island's community attempt to blend island qualities with local history, creating an attractive product for tourists seeking alternative experiences.[25]

But are there any visitor groups systematically visiting these islands out of an interest in exile heritage? Annually, Ai Stratis, Giaros, Trikeri and Makronisos are visited by local committees of leftist political parties, history clubs and societies comprising former exiles, their relatives and individuals who all together act as volunteer custodians of these memories and physical remnants (Figure 13.3).[26] These short and 'unofficial' visits take the form of pilgrimages and include events and activities that aim to honour the memory of former exiles. Sporadic visits by other interest groups, such as students or researchers, are also not uncommon. Due to these people's enthusiasm and dedication, memorials have been erected, commemorative plaques placed *in situ* and makeshift museums established. These grassroots endeavours have complemented state actions in what is, so far, a unique fashion. It is not an overstatement to say here, that, on several occasions the task to preserve these cultural resources for posterity has even fallen entirely on them.

In September 2013, the travel section of the Sunday edition of the daily Greek newspaper *Kathimerini* advertised a day tour to Makronisos. This is the first package guided tour to an exile site, advertised and sold by a tourist agency that I am aware of. This 12 October 2013 tour to Makronisos, according to the travel agency, is a repeat due to high demand.[27] The first

Figure 13.3 A former exile sharing his experiences during a visit to Markonisos (Copyright Pantzou).

trip took place in June 2013, and, to their knowledge, no other travel agent offers such a product. Does this new tourist product signal the emergence of a dark tourism trend in Greece? Could present conditions favour the development of a new type of travel destinations? Could dark tourists get enlisted among other special interest tourists that the Greek islands attract? It is unquestionably too early to say, and further research is required. Up to the present moment, to my knowledge, no systematic scholarly work has been carried out from a Greek perspective. Only scant and general mentions are made of the phenomenon of dark tourism in the Greek media. Taking into account that in Greece there are a great number of dark sites, ranging from ancient battlefields to WWII concentration camps and Civil War mass graves, it is only a matter of time before Greek case studies will be introduced in theory and practice in the area of dark tourism.

CONCLUSION

The Aegean islands bear evidence of a traumatic legacy that has left its mark upon the landscape and its people. Contrary to popular belief, exile sites stand as proof that island life in the Aegean region is not always 'ideal'. Stories about adversities and dark monuments lurk behind emblematic places of

Greek culture like Santorini, Paros, Naxos or Mykonos. Regardless of time, distance and the state's intentional or unintentional failure to care for these monuments in the past, this suppressed heritage and painful history have, though, persisted and entered the public sphere. The fact that the GNTO and some island municipalities regard exile as a chapter worthy of mention of national and local history, respectively, signifies that the healing process has been accelerated. These loci are no longer exclusively in the custody of a specific group of people who have been taking initiatives, motivated by personal memories, academic interest or political affiliation. The foundation of the Museum of Democracy and the existing plans for some sites and buildings' restoration attest further to the acceptance of this difficult past (see also Mamoulaki, Chapter 3 in this volume). But are the islands' contrasting qualities the ones which dictate the sustainability of memories at a national or local level and the burgeoning public and academic interest? It is difficult to say with certainty since a juxtaposition of the presented data, with information on the treatment of sites on the mainland, is needed. Nevertheless, as for the survival of island properties in comparison with mainland ones, according to existing evidence, the former seem to be less vulnerable than the latter, given the insularity of islandscapes and their slow pace of change.

The surviving island sites associated with political and social exclusion represent a valuable information source concerning the development of the camp system and the realities behind the architecture of detainment in Greece. In particular, the intensity of the phenomenon of exile in the Aegean elucidates further the centrality of islandscapes in mechanisms of persecution and imprisonment, providing in this way a platform for interesting analogies to be drawn between these properties and examples from various geographical regions and cultural realms. Finally, the availability of abundant data and the number of standing structures waiting to be explored and investigated indicate how future research in the area could contribute to the fields of conflict archaeology and internment archaeology.

As fast as the interest in exile places from a heritage and tourism perspective spreads, equally rapidly the memories of exile fade as the keepers of the history—the former exiles and political prisoners—pass away. One cannot overlook the fact that we are now reaching a critical point, where mechanisms have to be activated and strategies designed towards preserving and documenting exile heritage in its entirety and raising awareness before it is too late. Overall, it is encouraging that young people who are working or who are going to be employed in the field of heritage and tourism management are favourably disposed towards the preservation and promotion of these neglected cultural resources. They can step in and make up for lost ground. It is also good news that the history of exile has been introduced in school textbooks, laying the foundation for the future. It is rather unfortunate that the healing process is accelerated at a time of grave recession for the Greek state. Optimistically thinking, the current circumstances might help us move away from a stance that all initiatives and interventions must be state funded and state instigated and inspire people of all ages to open up to new collaborations and prospects.

The next step is to involve former exiles in this research. It is essential that former detainees voice their opinion about the management of this past. It would also be extremely interesting to conduct a follow-up survey focusing on students from fields not related to heritage management and tourism and, of course, to do fieldwork at a local level, ideally by concentrating on selected islands. This way we will gain a better insight into Greece's position in the discussion about the processes of negotiation and remembrance of war heritage in island contexts.

ACKNOWLEDGEMENTS

Many thanks to Katerina Stefatou, Neni Panourgia, Dimitris Papadopoulos, Georgia Zouni and Myrsini Pichou for their support and advice. Particular thanks are due to all survey participants. Finally I sincerely thank Gilly Carr and Keir Reeves for putting together this conference and edited volume, as well as for the stimulating, interesting discussions.

NOTES

1. This data comes from the official website of the Auschwitz-Birkenau memorial and museum. Online. Available HTTP: http://en.auschwitz.org/m/index.php?option=com_content&task=view&id=953&Itemid=8 (accessed 7 October 2013).
2. This number appears on the main page of Topography of Terror website. Online. Available HTTP: www.topographie.de/en/ (accessed 7 October 2013).
3. For more on dark tourism in Sarajevo, see Kamber, Karafotias and Tsitoura (2013).
4. Under the research area of Critical Heritage studies, the University of Gothenburg has initiated projects on Heritage in Conflicts and Crises. Online. Available HTTP: www.criticalheritagestudies.gu.se/clusters+and+heritage+academy/globalizing-heritage/heritage+in+conflicts+and+crises/ (accessed 7 October 2013).
5. This is a definition of archipelago provided by *Oxford English Dictionary* (*OED*) (accessed 7 October 2013).
6. Evia and Crete are part of the Aegean region too. However, they are not included in this study because of their size. An example of Crete's political banishment heritage is the Itzedin prison at Chania.
7. Online. Available HTTP: http://sete.gr/EN/Home/ (accessed 7 October 2013).
8. E. Venizelos' archive has in its collections a telegraph by these two political exiles. While detained in Naxos, they wrote to Eleftherios Venizelos, then prime minister, to request their release. Online. Available HTTP: www.venizelosarchives.gr/rec.asp?id=994 (accessed 7 October 2013).
9. Exiles places and prisons are scattered all around Greece. Yet there is not a clear image of the exact number of locations. As for the Aegean region, for the purposes of this research, I used the list provided by the Foundation of the Hellenic World. Online. Available HTTP: http://www2.egeonet.gr/aigaio/forms/fLemmaBodyExtended.aspx?lemmaID=6984 (accessed 7 October 2013).

10. For instance, Ai Stratis served as a detention place for individual banishment cases, small communities of dissidents and thousands of political exiles during the Greek Civil War.
11. The Museum of Democracy has a very informative website. Online. Available HTTP: www.mouseiodimokratias.gr/english/exhibition2.asp (accessed 7 October 2013).
12. A Museum of Political Exiles of Ai Stratis also exists in Athens. The museum was founded by former exiles and is housed in space provided by MoC.
13. Aigina is the first capital of Greece. The former Aigina prisons were housed in an orphanage built by Ioannis Kapodistrias, Greece's first governor. In recent years, proposals have been drafted and projects designed for the rehabilitation of the buildings and the setting up of a diachronic museum *in situ*. Online. Available HTTP: http://aeginahistory.blogspot.gr/p/blog-page3.html (accessed 7 October 2013).
14. Online. Available HTTP: www.oddysseus.culture.gr.
15. The official portal of GTO is www.visitgreece.gr.
16. The Gavdos example is representative of GNTO's approach. Online. Available HTTP: www.visitgreece.gr/en/greek_islands/gavdos (accessed 7 October 2013).
17. The infamous prisons of Oropos (operated during the Junta period) host today the municipal conservatory. Online. Available HTTP: www.oropos.gov.gr/tourismos/istorikoi-horoi/amalieio/item/amalieio.html (accessed 7 October 2013).
18. This questionnaire was inspired by previous research done about dark tourism visitors (Bigley, Lee, Chon and Yoon 2010; Biran, Poria and Oren 2011; Kang, Scott, Jeonglyeol and Ballantyne 2011). Yet this preliminary survey moved beyond issues of supply and demand and endeavoured to acquire an understanding of public knowledge and interest in exile heritage.
19. A four-question (personal information excluded), close-ended and self-administered questionnaire was distributed to the students.
20. The response rate was 100 per cent.
21. Questionnaires were distributed to students (MA in heritage management and BA in cultural heritage management and new technologies) prior to my lecture on traumatic heritage and exile sites in order to ensure that I will not influence their responses. The questionnaire was administered to MBA students in tourism management by a colleague.
22. In the Greek educational system, students attend elementary school (*dimotiko*) for six years, secondary school (*gimnasio*) for three years and high school (*likeio*) for three years.
23. The official website of the municipality is www.agios efstratios.gov.gr/default.asp?typos=about_agios_efstratios (accessed 7 October 2013).
24. Online. Available HTTP: www.agios-efstratios.gov.gr/default.asp?typos=about_agios_efstratios.
25. Schoolchildren from Ai Stratis have set up their own blog. They have dedicated a section to Ai Stratis as a place of exile. Online. Available HTTP: http://gym-ag-efstr.les.sch.gr/IstoriaNisiou.htm (accessed 7 October 2013).
26. Group visits of that character have been realized in the past in other islands too, such as Chios and Ikaria.
27. Personal communication/phone call to the tourist agency, September 2013.

BIBLIOGRAPHY

Andriotis, K. (2004) 'Problems of island tourism development: the Greek insular regions', in B. Bramwell (ed), Coastal Mass Tourism: Diversification and Sustainable Development in Southern Europe, Sydney: Channel View Publications, 114–32.

Ashworth, G. J. and Hartmann, R. (eds), (2005) Horror and Human Tragedy Revisited: The Management of Sites of Atrocities for Tourism, New York: Cognizant.
Baloglu, S. and Mangaloglu, M. (2001) 'Tourism destination images of Turkey, Egypt, Greece, and Italy as perceived by US-based tour operators and travel agents', *Tourism Management,* 22, 1–9.
Berg, I. and Edelheim, J. (2012) 'The attraction of Islands: travellers and tourists in the Cyclades (Greece) in the twentieth and twenty-first centuries', *Journal of Tourism and Cultural Change*, 10 (1), 84–98.
Bigley, J. D., Lee, C-K., Chon, J. and Yoon, Y. (2010) 'Motivations for war-related tourism: a case of DMZ visitors in Korea', *Tourism Geographies*, 12 (3), 371–94.
Biran, A., Poria, Y. and Oren, G. (2011) 'Sought experiences at (dark) heritage sites', *Annals of Tourism Research*, 38, 820–41.
Bournazos, S. (2009) 'Μισός αιώνας πολιτικής εξορίας στην Ελλάδα: Σκέψεις για την ιστορική μελέτη του φαινομένου' '[Half a century of exile in Greece: Thoughts about the historical study of the phenomenon]', in P. Panagiotou and L. Dimakopoulos (eds), Τόποι Εξορίας: Ένα Σημερινό Βλέμμα *Topoi Exorias: Ena Simerino Vlema* [*Places of Exile: A Modern Approach*], Athens: Ekdoseis Alexandrei, 35–47.
Buhalis, D. (1999) 'Tourism on the Greek islands: issues of peripherality, competitiveness and development', *International Journal of Tourism Research*, 1, 341–58.
Buhalis, D. (2001) 'Tourism in Greece: strategic analysis and challenges', *Current Issues in Tourism*, 4 (5), 440–80.
Carman, J. (1995) 'The importance of things: archaeology and the law', in M. A. Cooper, A. Firth, J. Carman and D. Wheatley (eds), Managing Archaeology, London: Routledge, 17–30.
Carr, G. (2007) 'The politics of forgetting on the island of Alderney', Archaeological Review from Cambridge, 22 (2), 89–112.
Ceccarelli, P. (2012) 'Naming the Aegean Sea', *Mediterranean Historical Review*, 27 (1), 25–49.
Dolff-Bonekamper, G. (2002) 'Sites of hurtful memory', Getty Conservation Institute Newsletter, 17 (2), 11–6.
Hamilakis, Y. (2003) '"Learn history!" antiquity, national narrative, and history in Greek educational textbooks', in K. S. Brown and Y. Hamilakis (eds), *The Usable Past: Greek Metahistories*, Lanham: Lexington Books, 39–67.
Jasinski, M. E. (2013) 'Reinforced concrete, steel, and slaves. Archaeological studies of prisoners of World War II in Norway—the case of Romsdal Peninsula', in H. Mytum and G. Carr (eds), Prisoners of War: Archaeology, Memory and Heritage of 19th and 20th Century Mass Internment, New York: Springer, 145–65.
Kamber, M., Karafotias, T. and Tsitoura T. (2013) 'The Sarajevo siege dark heritage: protection, interpretation, management and tourism issues', unpublished MA dissertation, heritage management, University of Kent/Athens and University of Economics and Business, Athens, Greece.
Kang, E-J., Scott, N., Jeonglyeol, T. and Ballantyne, R. (2012) 'Benefits of visiting a "dark tourism" site: the case of the Jeju April 3rd Peace Park, Korea', *Tourism Management*, 33 (2), 257–65.
Kenna, M. (2001) *Η κοινωνική οργάνωση της εξορίας: Πολιτικοί κρατούμενοι στον Μεσοπόλεμο* [*Social Organization of Exile: Greek Political Detainees in the 1930s*], Athens: Alexandria.
King, R. (1993) 'The geographical fascination of Islands', in D. G. Lockhart, D. Drakakis-Smith and J. Schembri (eds), *The Development Process in Small Island States*, London: Routledge, 13–37.
Konsolas, N., Papadaskalopoulos, A. and Plaskovitis, I. (2002) Greek Regional Development in Greece, New York: Springer.

Konstantopoulou, A. (1976) 'Για το στρατόπεδο γυναικών της Χίου [Regarding the female camp at Chios]', in V. Theodorou (ed), *Στρατόπεδα Γυναικών* [*Female Camps*], Peristeri: Athanassiou, 37–80.

Logan, W. and Reeves, K. (eds), (2009) *Places of Pain and Shame: Dealing with Difficult Heritage,* New York: Routledge.

Margaritis, G. (2009) 'Μνημεία του πρόσφατου χθες [Monuments of contemporary past]', in P. Panagiotou and L. Dimakopoulos (eds), *Τόποι εξορίας: Ένα σημερινό βλέμμα* [*Places of Exile: A Modern Approach*], Athens: Ekdoseis Alexandreia, 17–20.

Meskell, M. (2002) 'Negative heritage and past mastering in archaeology', *Anthropological Quarterly,* 75 (3), 557–74.

Ministry of Culture, Ministry of Public Works, SADAS-PEA, Technical Chamber of Greece, ICOMOS, National Technical University of Athens, and PEKAM. (1994) *Μακρόνησος: Ιστορικός Πολιτιστικός Τόπος* [*Makronisos: A Historic Cultural Site*], Athens.

Ministry of the Aegean. (2000) *Αη Στράτης: Φωτογραφικά ίχνη (1940–1970), Αρχείο Βασίλη Μανικάκη* [*Ai Stratis: Photographic Traces (1940–1970)*, Vasilis Manikakis's Archive], Piraeus: Ministry of the Aegean.

Nakou, I. and Apostolidou, E. (2010) 'Debates in Greece: textbooks as the spinal cord of history education and the passionate maintenance of a traditional historical culture', in I. Nakou and I. Barca (eds), *Contemporary Public Debates over History Education*, International Review of History Education (6), Charlotte, NC: Information Age Publishing.

Panourgia, N. (2009) *Dangerous Citizens: The Greek Left and the Terror of the State*, New York: Fordham University Press.

Pantzou, N. (2011) 'Materialities and traumatic memories of a 20th century Greek exile island', in A. Mayers and G. Moschenska (eds), *Archaeologies of Internment*, London: Springer.

Pedagogical Institute. (1995) Ιστορία Νεότερη και Σύγχρονη [*Modern and Contemporary History*], Athens: Organisation for the Publication of School Books.

Pedagogical Institute. (2007a) *Ιστορία του Νεότερου και του Σύγχρονου Κόσμου* [*History of the Modern and Contemporary World*], Athens: Organisation for the Publication of School Books.

Pedagogical Institute. (2007b) *Στα Νεότερα Χρόνια* [*In the Modern Times*], Vol. 3, Athens: Organisation for the Publication of School Books.

Pedagogical Institute. (2012a) *Ιστορία του Νεότερου και Σύγχρονου Κόσμου* [*History of the Modern and Contemporary World*], Athens: Organisation for the Publication of School Books.

Pedagogical Institute. (2012b) *Νεότερη και Σύγχρονη Ιστορία* [*Modern and Contemporary History*], Athens: Organisation for the Publication of School Books.

Pedagogical Institute. (2012c) *Κείμενα Νεοελληνικής Λογοτεχνίας* [*Modern Greek Literature*], Athens: Organisation for the Publication of School Books.

Stavridis, S. (ed). (2006) *Μνήμη και Εμπειρία του Χώρου* [*Memory and Experience of Place*], Athens: Alexandria.

Stone, P. R. (2006) 'A dark tourism spectrum: towards a typology of death and macabre related tourist sites, attractions and exhibitions', *TOURISM—An Interdisciplinary International Journal*, 54 (2), 145–60.

Stone, P. and Sharpley, R. (2008) 'Consuming dark tourism: a thanatological perspective', *Annals of Tourism Research*, 35 (2), 574–95.

Strange, C. and Kempa, M. (2003) 'Shades of dark tourism. Alcatraz and Robben Island', *Annals of Tourism Research*, 30 (2), 386–405.

Tunbridge, J. E. (2005) 'Penal colonies and tourism with reference to Robeen Island, South Africa: commodifying the heritage of atrocity', in G. J. Ashworth and

R. Hartmann (eds), Horror and Human Tragedy Revisited: The Management of Sites of Atrocities for Tourism, New York: Cognizant.

Uzzell, D. and Ballantyne, R. (1998) 'Heritage that hurts: interpretation in a post-modern world', in D. Uzzell and R. Ballantyne (eds), Contemporary Issues in Heritage and Environmental Interpretation: Problems and Prospects, London: Stationery Office.

Voglis, P. (2002) *Becoming a Subject: Political Prisoners During the Greek Civil War*, New York: Berghahn Books.

Yannaka, S. (2013) 'Οι μνήμες "ζωντάνεψαν" στην Μακρόνησο [The memories were revived in Makronisos]', *To Vima*, 31 August. Online. Available HTTP: www.tovima.gr/relatedarticles/article/?aid=153470 (accessed 7 October 2013).

14 Post-War Legacies in the Island of Kythera

Oblivion Versus Historical Memory

Irene Lagani

INTRODUCTION

The dramatic growth of the neo-Nazi organization Golden Dawn in Greece in recent years has been attributed mostly to the profound financial crisis, to the International Monetary Fund's imposition of severe austerity measures, and to the large number of immigrants who have altered the character of the centre of Athens. While Golden Dawn's growing influence has made an impression even in areas of mainland Greece, such as Distomo and Kalavryta, which witnessed the carnage of victims by Nazi occupation forces, very few people have wondered about the extent to which the ignorance of history has contributed to the organization's growth.

Kythera, an island at the far south-western edge of the Peloponnese known for its strong winds and wild seas has no victims to show for the 'Nazi atrocity' during the Occupation (Figure 14.1). Visitors do not encounter even so much as a monument referring to this era, and those wishing to learn more will need to demonstrate great zeal and perseverance. A realization of the risk inherent in an ignorance of history, particularly in our own day as neo-Nazi propaganda endeavours to conceal the gap in historical knowledge, has led here to an investigation of the Occupation in Kythera. Although the percentages recorded for Golden Dawn during general elections do not appear especially worrying on the island, field research certainly leaves no room for complacency.

The goal of this chapter is not to highlight the uniqueness of Kythera vis-à-vis other islands, since research reveals both differences and similarities with other islands. Rather, my goal is principally to contribute to the restoration of history, to the highlighting of heroes-as-symbols, and to filling in a gap in memory. My purpose is to highlight the reasons for the denial of the past and contribute to an acceptance of both its positive and negative parts. My goal is knowledge of and reconciliation with the past, as well as a rejection of ignorance and oblivion, given that the danger is already in sight. This is to argue for an investigation of the stance of the younger generation, which has inherited the absence of promotion and complete devaluation of those who fought against fascism and Nazism during WWII on the island.

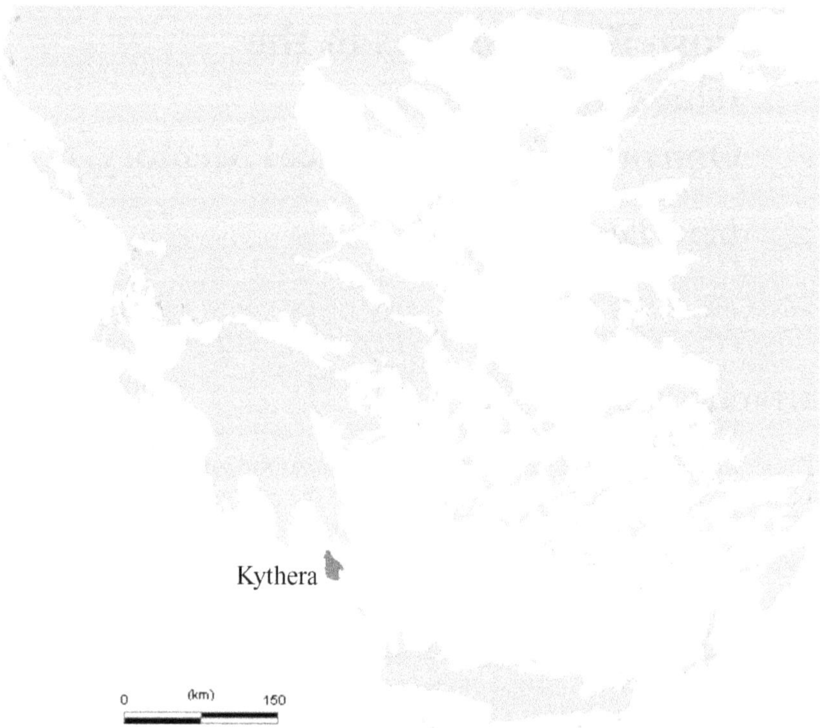

Figure 14.1 Map of Kythera, Greece (Copyright and courtesy visitkythera.com).

For this reason, I will be guided in the present through an investigation of the past.

In contrast to Greek islands such as Crete, Lefkada, Syros and others, where the memory of the Occupation and WWII resistance is cultivated and where local fighters received honours, nothing similar has occurred on Kythera. Despite the fact that civil war on the island during the Occupation was avoided in the end and the deep chasm between inhabitants appears at first sight to have been averted, in actuality a veil of silence has covered this period and continues to do so today.

To what is this silence owed? Is it connected with the island's history, its long Occupation by foreigners, with a feeling of dependency, fear and insecurity cultivated over the course of centuries? What happened on Kythera during the war that would dictate oblivion or an aversion to the WWII past? In this chapter I explore whether this attitude is generally held by Kythera's inhabitants or whether there is a distinction between the north and south of the island (the so-called Inner and Outer Municipality), a distinction that relates not only to the war period but to earlier times as well. I also examine to what extent the post-WWII conditions in Greece contributed to the marginalization of events during the Occupation and resistance on Kythera. Finally, I look at how the period of occupation and resistance is employed today on Kythera by the Kytherians themselves, as well as why the celebration of the island's liberation continues to form a bone of contention even today.

The methodology employed here was developed to explore the potential implications of ignorance among the younger generation about WWII on Kythera and to investigate whether this ignorance makes them more vulnerable to far-right propaganda. Fieldwork was conducted 2011–3. More than thirty people aged between 14 and 92 were interviewed. The research presented here was based on the collection of both written and oral testimony of those who were alive during the Occupation. It also involved questions addressed to the young concerning how they saw the phenomenon of Nazism and the Nazis, the Occupation, and the resistance, primarily at the local and national levels, the causes of the increasing influence of Golden Dawn, and their attitude towards the latter.

HISTORICAL BACKGROUND

Kythera, known from antiquity and the Homeric epics as the birthplace of the goddess Aphrodite (Ησίοδος [Hesiod] 1959, 188–99), was in the West connected with a dream-like, distant place where love and eternal peace reigned supreme (Aroni-Tsichli 2003, 38; Vernant 1965, 328 and passim). This image has inspired Western painters, writers and even musicians.[1] Although Petrarch included Kythera in his poems in the fourteenth century and Goethe inserted the island in his *Faust*,[2] the French painter Antoine Watteau (1684–1721) with his well known paintings *Voyages à Cythères* contributed the most to identifying Kythera as an idyllic, far-off place[3] associated with pleasure and the eternal quest. The symbolic nature of Kythera as an ideal and utopian place has been preserved in the European collective memory down to the present. One of the slogans of the Paris 1968 student uprising was '*Allons tous unis vers Cythères*'.[4]

Kythera is situated among three very different Greek regions: the Peloponnese, Crete and the Ionian Islands. After the end of the Fourth Crusade (1204) and the fall of Constantinople to the Levantines, the island passed into Venetian rule for six centuries. Venetian and, later, English rule (1815–64) had a decisive impact on the island. Kythera was to follow the history of the Ionian Islands, cut off from mainland Greece.

The influence of Kythera's particular historical and cultural past should not be overlooked in the course of events during the Occupation (1941) and liberation (1944) of Greece or in how the Kytherians themselves understood this period. The historical and cultural particularities of Kythera refer on the one hand to social conflicts of the past between nobles and peasants and on the other to the submissive stance of locals towards each of their foreign protectors, who arrived to strengthen both individual as well as collective security in the face of piracy (Kalligeros 2008, 182).

Through Venetian and English rule, an imaginary line divided the island as its nobles came mainly from the south part of Kythera (*Mesa Demos*: Inner Municipality) while its peasants were settled in the north (*Exo Demos*: Outer Municipality). This division was preserved during WWI with the monarchists in the place of the nobles prevailing in the south and the liberals (Venizelists) in the north (Kalligeros 2008, 178).

This imaginary line continued to exist during WWII. During this period, in the northern part of the island, a resistance movement developed against the Germans and Italians, who invaded in May and June 1941, respectively. In the southern part, where the island's capital of Chora was situated, the residents, believing that they ran the risk of retaliation by the Germans, obstructed the growth of a resistance movement in their region. Even before the outbreak of WWII, Kythera had experienced the intense shock of the economic crisis of 1929. One of the island's largest banks failed, and significant capital, chiefly from the savings of Kytherian immigrants in Australia who had later returned to the island, evaporated. Uncertainty about the island's future increased the outflow of immigrants, with the result that on the eve of the war, the island's population had fallen to 8,178 from 14,605 during the interwar period. In contrast, during the war, Kythera was treated as a refuge by Kytherians who had settled in Athens; a number of them decided to take refuge on the island to escape the famine decimating the population of Athens. The island's population thus once more increased, reaching 14,000 in 1944 (Kalligeros 2008, 187).

THE IMPACT OF THE GREEK NARRATIVE ON KYTHERA

For many years in Greece, the period of Occupation and resistance during WWII was a highly controversial issue. The early conversion of Greece into a 'hot' field of confrontation between two opposing ideologies that clashed and foreshadowed the tensions of the Cold War contributed to this period (the Occupation of 1940–4 and the Civil War of 1946–9) becoming an object of political propaganda or, at best, a taboo subject. In the course of WWII, an ideological-political conflict developed between the major resistance organizations of EAM (National Liberation Front) and EDES (National Republican Greek League). Although EAM encompassed a broad political spectrum, the Communist party undoubtedly played a decisive role in the organization. As in other European countries (such as Yugoslavia and France), the resistance organizations that suffered the greatest losses to occupying forces during the war were those of the left and the Communist party, and they looked forward not only to liberation but also to political and social transformation in the post-war period. The same was true of the resistance organization EAM and its military wing ELAS (National People's Liberation Army), which, according to German sources, inflicted the greatest losses on the invaders.[5] However, during the liberation, since the probabilities were high of the Communists (who controlled the greater part of the country) prevailing, intervention by the British led by Churchill played a decisive role in restoring the pre-war status quo with the return of the monarchy. The ensuing Civil War (1946–9), which for decades continued to be called the 'bandits' war', and the international context (i.e. the Cold War) left no room for healing the wounds left by the war, the Occupation, and the Civil War that ensued.

As usual, the winners punished and marginalized the 'losers'. After the fall of the dictatorship (1974) and throughout the post-dictatorship period, the spirit of oblivion, which looked to bridge the gap between 'winners' and 'losers', prevailed. For the same reason, during the post-dictatorship period, the years 1946–9 were renamed the Civil War in place of *symmoritopolemos* (the 'bandits' war') and 'foreign conspiracy against Greece', both terms used officially for more than two decades (Lagani 2013, 23).[6]

Forty years had to pass before recognition by Law 1285/1982, passed by the Greek Parliament in 1982, which acknowledged the contribution of those who had struggled against the invaders during the German occupation (1941–4) and for their resistance to be characterized as 'national'.[7] It may be noted that until then there had remained in effect the law of the dictatorship (1967–74), which recognized as members of the resistance those who betrayed the fighters and collaborated with the Nazi occupation forces during WWII (Voglis 2007, 437–56).[8] According to Prime Minister Andreas Papandreou (founder of the Panhellenian Socialist Movement, PASOK), who took the initiative to recognize the national resistance, 'Greece was the only country in Europe that had not paid its debt to the resistance, despite the fact that the Greek people as a whole resisted throughout the German occupation', adding that 'it is our duty to render honour to those who have been forgotten, those who have been persecuted'.[9] Thus, in reality, this was the recognition of the resistance activity of EAM.[10] As a result, 280,000 fighters of the national resistance were rehabilitated and were awarded pensions, and 45,000 political refugees returned to Greece by the end of the 1980s.

This did not mean that official recognition would reverse the situation prevailing locally and in closed communities like that of Kythera. Moreover, the official policy during the post-dictatorship period favoured oblivion for the sake of national reconciliation.[11] To date, Kythera has followed the path of oblivion in its public history, given that its traumatic memory has not healed.

TESTIMONY OF KYTHERIANS REGARDING THE PERIOD OF WWII

To comprehend the cause of the long silence covering public discourse in this period, one needs to return to Kythera during the years 1941–4. Events during this period have been recorded by Kytherians who experienced them first-hand, and their testimony has been published at various times (e.g. Stathis 1947; Kassimatis 1957; Chlambeas 1983; Kassimatis 2002 and Dapontes 2005). Despite the fact that the authors of these books belonged to different political spheres, they 'conversed' among themselves, allowing the cross-checking of evidence and the filling in of gaps. Interestingly, the older books have long been sold out, with the result that they are not available in the island's bookstores. However, in their pages one discovers how close Kythera came to a generalized civil war, which in the end was averted—in contrast with what occurred in other parts of Greece—where ultimately civil war could not be prevented.

On the basis of these sources and oral testimony from survivors, one may conclude first that the only resistance organization active in Kythera was EAM. Second, regarding the resistance activity of the organization on the island, it has been confirmed that on 25 September 1943, EAM captured 150 Italian soldiers together with their arms and ammunition in the northern part of the island (Potamos). After disarming them, they removed them from the island. The disarming and retreat of the defeated Italians from the island, the work of the rebels (*antartes*), was considered a 'sensational event' and aroused enthusiasm among inhabitants. The region's authorities also participated actively in the disarming of the Italians (Chlambeas 1983, 15; Dapontes 2005, 35; Kasimatis 1957, 228).

Third, despite the fact that nearly all the sources make reference to other acts of resistance on the part of EAM—including a ban on the collection of the 32 per cent tax on olive oil production imposed by the Occupation government on the island, the requisitioning of German ships carrying food, and the capture of German and Italian soldiers—it is clear that the organization did not enjoy popular support throughout the entire island but only in the north (Kasimatis 1957, 221 and 229).

The residents of South Kythera (Inner Municipality) argued that the region was vulnerable to the Germans, and residents feared that in case of a clash between the rebels and Germans, the latter would retaliate against them. It is true that this measure (retaliation) had been implemented by the Germans in other regions of Greece, with many victims. However, in South Kythera, EAM remained weak because, for the most part, its residents were actually opposed to the character of the organization, its ideological orientation and its dominance. The Germans knew this and never engaged in retaliatory measures in the region for the actions of the rebels. Indeed there are incidents reported of the residents' offering help to the Germans when some soldiers who had escaped capture by EAM were looking for the road back to their outpost (Chlambeas 1983, 37).

Furthermore, some Kytherians known for their anti-EAM and anti-Communist sentiments believed they were in danger of future purges by EAM and did not exclude collaboration with the Germans to exterminate the organization. In the end, conflict between rebels (*andartes*) (from the

Peloponnese and Potamos) and the residents of Mesa Demos could not be avoided. On 6 July 1944 the rebels, targeting the German outpost in the south (Agia Elessa), found themselves facing their local opponents, who had set up ambushes. The result of this armed conflict was the dissolution of the organization in the south and the interruption of communications between south and north for a certain period of time. There were no victims, but it was later revealed (on 24 July) that there was a plan of action against EAM. These plans were not revealed by Giorgos Chlambeas, the general secretary of EAM on Kythera, to the organization's senior ranks in order to avoid the punishment of the guilty parties, which could have ignited civil war on Kythera (Chlambeas 1983, 38). He followed, a 'policy of appeasing the passions' as he himself characterized it. Giorgos Chlambeas would be accused by his comrades of 'rightist opportunism'; his contribution to averting civil war would not be recognized by Kytherians who belonged to the opposite ideological camp (Chlambeas 1983, 34). The latter focused their attention on the burning of three houses for intimidation purposes by EAM in the Inner Municipality, which was one of the black marks against EAM's activity on the island. The incident appeared to justify the fears of the island's more conservative inhabitants and offered itself for ideological exploitation. It was described in the following way by a resident of the south:

> The sight of the smoke and flames from the burning houses was terrible and sad for the mild and refined Kytherian customs. It was a day of universal popular mourning, and will remain in popular tradition as one of the island's saddest vicissitudes.
>
> (Kasimatis 1957, 252)

This act of arson, however, had an entirely different significance for the followers of EAM. It was considered to be a symbol of EAM's prevalence throughout the entire island following the clash of EAM members with the Germans in South Kythera (Kontolianika) on 25 July 1944. They also saw it as a punishment against those who undermined the organization, who they suspected of collaborating with the occupiers.

The defeat of the Germans showed that they did not have the strength to expel the rebels even from the south, and the masters of the situation on the island were the rebels (*antartes*) (Kasimatis 1957, 252). With the Germans' defeat, EAM had succeeded in gaining control of the entire island, and civil war had been averted. According to accounts, Giorgos Polymenakos, the commander of ELAS on Kythera, was one person among others who contributed significantly to averting civil war. He provided an amnesty to all those accused of collaborating with the Germans in a plan to undermine EAM on the island, and he did not execute fifty Kytherians as he had been ordered to do by his superiors.

Clearly, if the orders from the mainland Greek Lakonian coast opposite Kythera had been implemented, civil war would not have been avoided.

Chlambeas argued that in any case, these orders were based on 'inadequate information', 'the unsuitability of (certain) individuals' and their 'extremist positions' (Chlambeas 1983, 34). Nor should it be overlooked that those issuing orders from the coast of Lakonia were, with the exception of Polymenakos, ignorant of the place's particular characteristics and the psychological makeup of Kytherians (Kasimatis 1957, 253). The incompatibility of the orders with Kytherian reality and, finally, their non-implementation should also be attributed to the distance separating Kythera from metropolitan Greece, which in the end proved to be even greater than it appears on maps. From 1 to 4 September 1944, there was tough fighting between the rebels and Germans on the island. The Germans withdrew from Kapsali (on 4 September) only after receiving assistance from a warship and the air force and under continual rebel fire. Following the Germans' disorderly retreat from the island, Polymenakos, the ELAS officer and the military commandant of Kythera who had coordinated the four days of battle, raised the Greek flag at Ai-Giorgi's church within the castle (*Kastro*) of Chora (Figure 14.2) (Chlambeas 1983, 41; Kasimatis 1957, 255–63; Dapontes 2005, 53–9). On 15 September, the English fleet moored at Kapsali, and Kythera became the first region of Greece to be liberated (Figure 14.3).

Figure 14.2 The Castle of Chora where the Greek flag was raised by Polymenakos, 4 September 1944 (Copyright I. Lagani).

Figure 14.3 Landing of British troops, Kapsali, Kythera, 15 September 1944 (Photo by M. Sofios, courtesy of M. Dapontes).

SELECTIVE MEMORY IN THE PUBLIC HISTORY OF THE ISLAND

While the legendary pirate Barbarossa and the destruction of the former capital of Palaiochora in 1536 (Hadjidimitriou 2000, 80) form an integral part of the history of Kythera, more recent events with sociopolitical ramifications in our own time are not included in the public narrative of local history. The result is that important events that do not accord well with the 'peaceful' nature of the island that is projected outward continue to remain on the margins of history. Characteristic examples of marginalized history include the slaughter of nobles by villagers at Kastro in 1800 (Kalligeros 2008, 74–6; Leontsinis 1987), as well as the Occupation period during WWII. More specifically, regarding the war, despite the existence of written and oral testimony from different sources including first-hand reports, resistance activity by Kytherian members of EAM against the Italians and German invaders and their contribution to the Allied cause has been doubted or systematically suppressed. The reluctance to grant public recognition at the local level to the EAM resistance tends to devalue its role, leaving those who fought to liberate the island without vindication. Finally, it tends to consign to oblivion an important chapter in the island's modern history which sealed relations among the members of local society, as shown by the reactions it continues to arouse today.

HERITAGE AND MEMORY ON KYTHERA

The Occupation period remains a taboo subject today, one which dictates a certain silence. Moreover, up to now the psychological consequences of the foreign occupation of the island have not been investigated. Had this happened, then it might have assisted in a better understanding of the attitude on the part of the Kytherian community towards the occupation by the Axis powers. The sympathetic stance by one part of this community towards their German occupiers, though not openly expressed in public places, is summarized as testimony or evidence in a recent book by a Kytherian who lived through the Occupation as a child (Kasimatis 2002). This volume projects an embellished image of the German occupants, while simultaneously highlighting the weaknesses of EAM members.

It is characteristic that while as a rule in Greece the recollection of the Nazi Occupation is described in very sombre tones, it appears that this is not the case in Kythera. Residents of the Mesa Demos (Inner Municipality) acknowledge today that 'the Nazi boot must not have been as heavy in Kythera as in other parts of Greece' (Argyropoulos 2002, 66), and they do not hesitate to discern cultivation, even humane feelings in their German occupiers—a description at variance with the generalized image of the Nazis in Greece. The way a young Kytherian schoolboy from Mesa Demos describes his chance meeting with a German captain in his recently published memoirs is characteristic:

> We had some trouble understanding each other, both of us using a hodgepodge of languages. I explained to him that my destination was at exactly the same point as his (our house was only 100 meters away from the German detachment). Walking slowly along, we continued our conversation, still communicating with the help of the same mix of languages and certainly this fine walk with him was rather pleasant. At some point he left me speechless when I heard him reciting Homer '*Ἄνδρα μοι ἔνεπε μοῦσα πολύτροπον . . .!*' Of course I didn't know Homer, for I was still in one of the lower grades, but later when I was taught it, I fully grasped the significance of this incident. Nevertheless, the surprise I felt is indescribable, even more so because I was simultaneously flooded by feelings of pride that a German occupier, indeed one holding the rank of captain, had a rich knowledge of ancient Greek

> authors, and had occupied himself with Ancient Greek culture. When we arrived, we exchanged a warm farewell and parted in friendly fashion.
>
> (Kasimatis 2002, 66–7, 92, 102)

Also characteristic is the scene described by the same writer in the port of Kythera shortly before the liberation, when a German Nazi officer reportedly expressed sympathy and concern for the writer's younger brother, who was to travel by ship with his family under adverse conditions (Kasimatis 2002, 102). On the other hand, the EAM members are accused of having no religious sentiment. Indeed, even here the Germans are presented as being more respectful of the island's religious traditions than the members of EAM. The writer notes:

> Throughout the Occupation, the Monastery of Agia Elessa held services on only one occasion. This happened at Christmas 1942, and after permission had earlier been sought from the Germans. It is also noteworthy that the Germans themselves followed the service with great devotion. It should also be clarified that the Germans showed profound respect for the monastery building.
>
> (Kasimatis 2002, 92)

Within this climate, it goes without saying that any resistance that took place against the Germans is underestimated, and those who betrayed the German Nazis to the members of EAM were despised. The revulsion described by EAM opponents over the execution of German Nazis is characteristic of this stance (Kasimatis 2002, 80, 85).

These 'recollections' reflect to a large extent the feelings for the Germans on the part of residents of the Inner Municipality during the Occupation, reproduced today by the their children, who bring to light moments of their childhood through the eyes of their families. The problem is that reminiscences of the Occupation which today present the German conquerors in a sympathetic light by focusing on their human side have a tendency to disconnect them from Nazi ideology and criminal activity, thus obscuring the causes of WWII. Schools for their part do not help the younger generation to form a clear picture of the Occupation. Local history is not taught in the schools, and consequently the period of Occupation—locally—does not fall within the framework of Greek, even less of European, history. For this reason, the fact that civil war in Kythera was in the end averted—in contrast to mainland Greece—has not been assessed or given enough importance. Despite the chasm between the northern and southern parts of the island, the non-involvement of foreign actors (in this case, the Germans) in the opposition between Kytherians, as well as the absence of victims during conflicts between EAM and its opponents, contributed to the non-proliferation of civil war. The moderation and wisdom shown by the Kytherians at critical moments contributed significantly to the defusing of tensions.

Nonetheless, the obscuring of this period, the frequent adulteration of the truth by word of mouth and the view that reconciliation is built on oblivion and not on memory was reproduced in Kythera—as it was to some extent in the rest of Greece—for the period of the Occupation and war.

However, we can reflect on how much more difficult life would have been in Kythera if civil war had not in the end been averted. Despite the fact that civil strife did not spread to Kythera and there were no victims, the consequences for the losers (the former resistance fighters who had fought the Germans and Italians on Kythera) were identical to those experienced by members of the resistance in the rest of Greece. Not only was their activity against the Germans not recognized as 'national' or 'patriotic', but they were also accused of having unpatriotic motives and of seeking to secure power and to impose a Communist regime in Greece upon liberation.

After the war, many members of the resistance were forced to immigrate to major urban centres or abroad to avoid persecution. More specifically, many Kytherian leftists emigrated to Australia, where their friends and relatives had already settled before the war. The testimony of Kytherian resistance fighters about the hardships they suffered by remaining on the island are no different from those of other Greeks who found themselves on the 'losing' side (Chlambeas 1983, 15). Both Polymenakos and Chlambeas, who contributed to averting civil war in Kythera, belong to this category. The former was executed a few years after liberation, while the latter's family was scattered and went as far as Australia to avoid persecution. Chlambeas himself abandoned Kythera and settled in Athens, where he published his experiences in the wake of the national resistance's official recognition. In these, he notes *inter alia*:

> After December 1944, as a resistance fighter from 1941 to 1944, I too suffered the consequence of persecutions . . . Following my acquittal on all charges by the judicial system, my Kytherian 'friends' began their own persecution against me, which lasted for many years. [This persecution took the form] either of books, slander and print materials generally, or oral threats. As a result, I was constantly being called to the Security Departments, independent of the fact that I had no involvement in politics following the Occupation. I was even displaced for seven months on the basis of a decision by the President of the Community of Potamos in 1947, for completely incredible and unsubstantiated things.
>
> (Chlambeas 1983, 43)

Untold stories are legion. Indeed, in contrast to the continuing promotion and interest in those who played a leading role in WWII, the freedom fighters who opposed Nazism and were later unjustly persecuted for their actions remain on the sidelines of history and rarely attract the spotlight. Thus, the reality experienced by the island's ordinary inhabitants is that the tensions and conflicts between them are marginalized—not because these have been overcome but because they continue to be open wounds, as exemplified by the statement made by one of those interviewed, 87-year-old Giorgos: 'I won't forget as long as I live'. Discussions about this period are confined to the family setting as opposed to being conducted in public. Younger inhabitants normally have only a fuzzy image of the Occupation, either accepting unquestioningly the version conveyed to them by their elders or rejecting it when it does not agree with the prevailing (majority) view.[12] Many of the young are indifferent to something that happened a half-century ago and uncomprehending of the connections between 'then' and 'now'.

Thus, at least on the surface, the tension and disagreement about what happened during the Occupation are of concern only to the elderly, to those who lived through it and resist consigning it to oblivion. The ignorance and indifference of the young towards this period are perceived by many older residents as a 'protective shield' against a controversial past that could bring to light unwanted stories that would 'rock the island's boat'.

Today, thirty years after the official recognition of the national resistance in 1983, the only thing recalling the war period on the island is a wall plaque in Kapsali harbour referring to the landing of the English in September 1944. There is not the slightest mention of the resistance struggle against the Italians and Germans. It is only in the old municipal office of Potamos that a white sheet of paper atop a table, under glass, put there some years ago, recalls the rebels' capture of 150 Italians.

Within the context of the silence that prevailed concerning the Occupation, for many years there was not even any discussion about celebrating the anniversary of the island's liberation. Why have Kytherians not celebrated the liberation of the island all these years? Perhaps we might argue with a local inhabitant that:

> The Venetian occupiers were right to use the famous expression '*Tutto il mondo uno mondo e il Cerigo*[13] *un altro mondo* [All the world is one (world), and Cerigo (Kythera) is another (world)]'.[14]

However, the uniqueness of the island remains to be shown, since to date there has been no research and comparative study of the islands of Greece (the Cyclades, Dodecanese, Ionian Islands, Crete) regarding the similarities and differences in how they experienced the consequences of Occupation and war and how they have healed their wounds—or not.[15] Kythera, although it never knew the excesses of civil war, offers fertile ground for the examination and analysis of the effects of ideological polarization in local society, despite whatever differences it may have with mainland Greece. However, when one studies the incidents of acute conflict recorded in Kythera before and after liberation, one realizes that to a certain extent, the tension, divisive problems and the 'extreme manner' of confronting these, were 'imports' from mainland Greece (the coast opposite Kythera) and that they were unsuited to the temperament of Kytherians, as they themselves in any case might claim (Dapontes 2012, 86; Stathis 1947, 152).

Several years ago, an elderly Kytherian who participated in the national resistance as an 18-year-old boy summarized the situation on the island as follows:

> While the Greek state has officially recognized our National Resistance, most of us on the island insist that we have not recognized the National Resistance, and disagree whether the liberation should be celebrated as a major event of our island, which was the first part of Greek territory to lift the brutal German Occupation and be liberated on 4 September 1944, and the place from which the liberation of the rest of our homeland began. And while every Kytherian should feel a unique sense of pride in this major historic event, not only does that not happen, but we have not agreed whether or not to celebrate this historic and honourable anniversary.[16]

Recently, an effort to establish an annual celebration of the anniversary of the island's liberation met with difficulties in a dispute between the Inner and Outer Municipalities concerning the celebration's date. Representatives of the Outer Municipality insisted on establishing 4 September—the day the Germans withdrew—while those of the Inner Municipality insisted on transferring the celebration to 14 September, which coincides with the Elevation of the True Cross and is a religious holiday *par excellence* as well as the start of the school year. Both sides' insistence on their own opinion is mostly due to the wish of one side to 'restore historical reality' with local recognition of the contribution of those who took part in the national resistance against the Occupation forces and to the wish of the other to avoid rubbing salt in old wounds by recognizing the role of the resistance and honouring EAM resistance fighters. In an open letter posted on the Internet and addressed to the mayor of Kythera, who espouses the second opinion, it is noted that:

> It is distressing today to witness this inadequate attempt to celebrate the liberation on the wrong date . . . It is not comforting to hear vague stories, fainthearted voices and a distorted, historically groundless narrative that would have us forget conditions during the Occupation . . . It is our responsibility towards the generations to come to know, and to believe that on this island, many remained silent out of fear, some

> collaborated with the invaders, and others fought for freedom. There is a need to analyze the conditions that gave rise to ardent fighters. There is a need to demonstrate the courage of those who struggled to defeat fascism on Kythera.
>
> (Orfanakis 2008)

The consequences of ignorance for the younger generation, a generation that had not experienced the war and the struggle against fascism, became all too clear during the recent financial crisis plaguing Greece. Polls indicate that the xenophobic Golden Dawn, one of the most virulent neo-Nazi parties in Europe (Psarras 2012), has grown into Greece's third political party since the beginning of the economic crisis in 2008.[17] Golden Dawn has found supporters among young people on the island, which has been hard hit by unemployment.[18] This trend appeared to be brought to a halt by the criminalization of the activities of Golden Dawn MPs following the death of an anti-fascist rapper.[19] Prolonged silence concerning an era in which both heroes and German collaborators emerged has unfortunately led to what has eloquently been termed as a 'holocaust of memory'.[20] The failure to pay tribute to the heroes of the resistance and to erect monuments to those who averted civil war on the island reflects the consequences of a crisis which is not only financial but also profoundly cultural.

Students on Kythera accept that they do not learn the history of their island in school because 'it is not included in the curriculum'. Nor is the gap left by the schools covered by the public sphere. Furthermore, students' general knowledge about WWII, the Occupation and the resistance remains very limited when based exclusively on what they learn in school (Lagani 2008, 359–60 and 365). The writing and authorization of school textbooks, including history texts, falls within the jurisdiction of the Greek Ministry of Education. It is characteristic that in Greece, the history of the resistance, despite its extent, remained outside the curriculum for about forty years following WWII. This peculiarity was due to political conditions that were formed in the country after the war. The inclusion of the Occupation and resistance period in schoolbooks took place after the official recognition of the resistance as a national one in 1982 (which is when the law of dictatorship effectively finished). Since then, a major tendency and concern for school textbook writers has been to blunt the earlier oppositions and achieve national accord and unity. Events that could divide people or be characterized as controversial were either not mentioned and avoided or cited but without comment and certainly without interpretation. Collaboration with the occupation forces, the contribution of resistance organizations to the Allies, the causes of the civil war during the Occupation and the role of foreigners all remain more or less taboo subjects (Lagani 2008, 359–70). Thus, expanding discussion on these topics is at the teacher's discretion. Some lyceum students in Kythera suspect that 'there may be something deliberate behind the silence . . .' or 'maybe some people want some things like the cooperation of some with the Nazis to be forgotten. Besides, most of us have never heard about this . . .', as an 18–year-old student, Kostas, put it. Another student observed that 'the resistance has been identified with the communists, and this is why it's never discussed. Communism and fascism are identified as one and the same today . . . I've heard that the Germans' retaliation was due to EAM's resistance activity; that is, if there had been no resistance there would have been no victims. . . .' In response to my suggestion that if there been no resistance against the Germans, the outcome of the war would have been different, I was interrupted by the observation of 17-year-old Giorgos, who asked:

> Then why is resistance activity undervalued on Kythera? Since we're ignorant of the history of the resistance on the island, we don't feel proud of our grandfathers' stance . . . It's as if nothing happened here. Everything's nullified. As if all those who fought never existed. Isn't this the greatest defeat over time?

Twenty-six-year-old Maria was sceptical, noting that 'Kytherians are rather introverted. Usually they don't express what they really think. Perhaps we are afraid to look back. We prefer to look ahead . . .'

Despite being a neo-Nazi party, Golden Dawn has been legitimated by its entrance into Parliament. It is seen by the island's young people more as a party of protest against the corruption of the political system and the memoranda imposing austerity. The risk it represents is underestimated and is not considered to be connected with the past and Nazi Germany. It is characteristic that during conversation, my young interlocutors avoided referring to the members of Golden Dawn as 'neo-Nazis', preferring to call them 'nationalists' or even 'super-patriots'. Of course, our discussion had taken place before Golden Dawn's activity was criminalized, when the television channels were still routinely promoting the organization's 'social action', for example the organization of soup kitchens exclusively for destitute Greeks, the provision of home assistance to helpless senior citizens and others. Such promotion doubtless contributed to ensuring the organization's social standing. Since then, the attitude of the young may have changed since the paramilitary nature of the organization is being revealed day by day. This, however, is not certain, since the prosecution and criminalization of the activity of Golden Dawn may have results that are the opposite of what is expected. In the search of the old difference between Inner and Outer Municipalities, my younger collocutors were anything but enlightening, since many of them appeared unaware of it. I resorted to election results, both general and municipal, and found that it is difficult to discern traces of this difference—at least at the voting level—and this in only a very few districts.

In the face of young people's ignorance—an ignorance which saddens others to observe—I wondered whether there was in fact a risk of a revival of old differences. The recommendation by 87-year-old Manolis, 'Don't open old wounds', which is equivalent to oblivion—the tactic faithfully followed till now—is proving to be anything but a painless and sensible solution, as the price being paid for this choice appears dangerously high.

ACKNOWLEDGEMENTS

I would like to thank Manolis Dapontes, an eyewitness and active member of EAM, for the interviews he granted me and for his valuable assistance. Many thanks to Michalis Orfanakis for the interview he granted me in Potamos (Outer Municipality), Kythera, and to all those who stood by me in this effort. I would also like to thank my colleague Aroni-Tsichli Kaiti from Kythera for her article and our talks on collective memory.

Many thanks to the organizers of the conference, Gilly Garr and Keir Reeves, for their invitation and overall support, as well as the participants and attendees for providing the opportunities for discussion.

NOTES

1. Three opera ballets by André Capra (1660–1744), Philippe Ramau (1683–1764) and Christoph Williband von Gluck (1714–87); two comic operas (*Prix de Cythères* and *Cythères assiegées*) by Charles-Simon Favart (Aroni-Tsichli 2003, 44–5).
2. Kythera finds a place in Goethe's work due to Aphrodite; also in that work is the honeymoon spent by Paris and Helen which, according to myth, they spent at Kythera until a fair wind blew and they sailed to Troy (Kalligeros 2008, 211).
3. Watteau's paintings *Embarquement pour Cythères* or *Pélerinage à l'ile de Cythères* are in the Louvre, Berlin, and Dresden, while *L'ile de Cythères* is in Frankfurt (Aroni-Tsichli 2003, 43).
4. According to Aroni-Tsichli, this slogan is also heard as '*Allons tous universitaires à Cythères!* [Let's all us academics go to Kythera!]' (Aroni-Tsichli 2003, 52).
5. According to a German report, EAM was 'the main support of the entire resistance movement against the Axis Powers' and 'represented the greatest danger to the occupation forces'. See 'Report by the German services on the situation in Greece in 1943' (Kremmydas 1985, 322).
6. Both during and after the Civil War, Greek governments maintained that the KKE (Communist Party of Greece), in collaboration with Communist governments in neighboring countries, threatened the territorial integrity of Greece, aiming at the secession of the Greek regions of Macedonia and Thrace in favor of Yugoslavia and Bulgaria.
7. Speech delivered by the Prime Minister Andreas Papandreou to the Greek Parliament, Recognition of the National Resistance, 17 August 1982. Online. Available HTTP: www.youtube.com/watch?v=bemQvjovsFc and www.youtube.com/watch?v=IcCIWLIAIz4 (accessed 4 July 2013).

8. Legislative Decree 179/1969, 'concerning the National Resistance'.
9. Papandreou, Recognition of the National Resistance, 17 August 1982.
10. It was no accident that Papandreou's speech was delivered on the anniversary of the execution of 315 EAM resistance fighters by Nazi troopers in Piraeus (Kokkinia) and the dispatching of another 1,200 to German occupation camps.
11. According to Papandreou, 'the main goal of recognition of the national resistance was to achieve national unity' (Papandreou, Recognition of the National Resistance, 17 August 1982).
12. Interview with Y. D., son of a member of EAM, who rejected his father's view on resistance, Chora, Kythera, August 2013.
13. The island's medieval name.
14. Interview between author and Manolis Dapontes, Kythera, August 2012.
15. In the studies we have in mind (e.g. that by Papastratis 2001), resistance activity on the islands during the Occupation and military operations are the chief topics of examination.
16. Interview between author and Dapontes, Kythera, August 2012.
17. *What's Wrong with the Greeks?* [Swedish documentary]. Online. Available HTTP: https://www.youtube.com/watch?v=HdCEY2uNk88 (accessed 18 January 2014).
18. Interviews taken in Potamos, Livadi and Mylopotamos during 1912–3.
19. Pavlos Fyssas was assassinated after a mass demonstration on 18 September 2013.
20. Stelios Kouloglou (2012).

BIBLIOGRAPHY

Argyropoulos, G. (2002) *Αναμνήσεις από τα Κύθηρα της Κατοχής* [*Memories from Kythera During the Occupation*], Athens: Etaireia Kythiraikon Meleton.

Aroni-Tsichli, K. (2003) 'Η φαντασιακή πρόσληψη των Κυθήρων στην Ευρωπαϊκή Συλλογική Μνήμη [The imaginative conception of Kythera in the European collective memory]', *1st International Congress of Kytherian Studies*, Kythera, 20–4 September, 5, 35–53.

Chlambeas, G. (1983) *Τα Κύθηρα στην τετραετία 1941–44* [*Kythera During the Period 1941–44*], Athens: Plethron.

Dapontes, M. I. (2005) *Η παρεξήγηση* [*The Misunderstanding*], Kythera: Privately published.

Dapontes, M. (2012) *Ένας Αιώνας Κύθηρα* [*One Century of Kythera*], Kythera: Privately published [original edition being reprinted].

Hadjidimitriou, T. (2000) *Kythera/Photographs by Tzeli Hadjidimitriou*, Heraklion: Crete University Press.

Hesiod. (1959) *Θεογονία II* [*Theogony II*], P. Lekatsas (trans), Athens: Kytheraika, 188–99.

Kalligeros, E. (2008) *Συνοπτική Ιστορία των Κυθήρων* [*Concise History of Kythera*], 4th ed, Athens: Kytheraika.

Kasimatis, I. (1957) *Από την Παλαιά και Σύγχρονη Κυθηραϊκή Ζωή* [*Of the Old and Contemporary Life on Kythera*], 2nd ed, Kythera: Privately published.

Kasimatis, G. (2002) *Φύλλα ημερολογίου* [*Pages from a Diary*], Athens: Mavridis.

Kouloglou, S. (2012) *Το Ολοκαύτωμα της Μνήμης* [*The Holocaust of Memory*], Online. Available HTTP: https://vimeo.com/62354095 (accessed 1 December 2013) and http://vimeo.com/61827733 (accessed 1 December 2013).

Kremmydas, V. (1985) *Ιστορία Νεότερη-Σύγχρονη* [*Modern and Contemporary History*], *Γ' Γυμνασίου* [3rd class textbook of Lyceum], Athens: OEDV.
Lagani, I. (2008) 'Η αντίσταση μέσα από τα εγχειρίδια ιστορίας της Ελλάδας και της Γαλλίας [The resistance through the Greek and French textbooks of history]', in R. V. Bouschoten (ed), *Μνήμες και Λήθη του Ελληνικού Εμφυλίου Πολέμου* [*Memories and Oblivion of the Greek Civil War*], Thessalonica: Epikentro, 349–70.
Lagani, I. (2013) '*Nostos* and Cold War political considerations: the Greek Civil War refugees', *Journal of the Hellenic Diaspora*, 39, 7–28.
Leontsinis, G. (1987) *The Islands of Kythera: A Social History 1770–1863*, Athens: Leontsinis.
Orfanakis, M. (2008) *Πορφυρίς* [*Porphyris*], Kythera. Online. Available HTTP: www.porphyris.wordpress.com/tag/ιστορία/ (accessed 18 September 2013).
Papandreou, A. (1985) 'Αναγνώριση της εθνικής αντίστασης [*Recognition of National Resistance*]'. Speech by Andreas Papandreou to the Greek Parliament, 17 August 1982. Online. Available HTTP: www.youtube.com/watch?v=bemQvjovsFc (accessed 4 July 2013).
Papastratis, P. (2001) 'Οι Κυκλάδες και η Σίφνος την Περίοδο της Κατοχής [Cyclades and Sifnos during the Occupation]', in Πρακτικά Α΄Διεθνούς Σιφναϊκού Συμποσίου, Σίφνος [*Minutes Book of the First Sifnos International Symposium*]. 25–8 June 1998, Etaireia Sifnaiikon Meleton: Athens, 2, 385–96.
Psarras, D. (2012) *Η Μαύρη Βίβλος της Χρυσής Αυγής* [*The Black Book of Golden Dawn*], Athens: Polis.
Stathis, K. (1947) *Τα Κύθηρα Σκλαβωμένα 1941–1946* [*Kythera Enslaved 1941–1946*], Kythera: Privately published.
Vernant, J. P. (1965) *Mythe et pensée chez les Grecs: études et psychologie historique,* Paris: F. Maspéro [Greek translation (1975), Athens: Olkos].
Voglis, P. (2007) 'Η δεκαετία του 1940 ως παρελθόν: μνήμη, μαρτυρία, ταυτότητα [The 1940s as past: memory, witness, identity]', *Τα Ιστορικά* [*Annals*], 47, 437–56.

15 Crete
Visual Memories of War

Maria Kagiadaki

INTRODUCTION

Historical narratives about Crete, the Aegean island to the south of mainland Greece, include changing rulers and struggles against enemies (Figure 15.1). War memories are traced both in the island's rich tangible and intangible heritage and strongly influence the Cretan mentality. While the revolts against the Ottomans for unification with the Greek mainland constitute a strong part of the island's heritage and collective memory, this chapter focuses on WWII in general and on the Battle of Crete and the occupation period specifically.

On 20 May 1941, the German Parachute Corps launched an extensive airborne attack against the Allies (British, Australians, New Zealanders and Greeks) that resulted in the island's occupation by the first day of June (Koliopoulos 1978). Notwithstanding the various evaluations of this fierce battle and its impact on both sides, the key aspect commemorated today is the brave participation of the unarmed locals. Resistance started almost immediately at the start of the occupation. For the Cretans, the Battle of Crete and the Resistance constitute integral parts of their local history that seem to testify to their free, persistent, brave and rebellious nature and to shape their collective memory.

This chapter analyses the expression of war memories in Crete through artistic and cultural manifestations. Ultimately, drawing from the fields of art history and heritage studies, my purpose is to study and document the integration of war memories in the island's public artistic realm and thus to contribute in the interdisciplinary approach of this volume. The key themes that will be discussed here include the representation, narration and memorialization of the Battle of Crete and of the Resistance through art, the characteristics of the existing artworks, the ideology and the messages that these communicate and their relation to relevant manifestations in the mainland Greece. In this framework, the first section examines the characteristics of the island's war memories, as highlighted by specific works (paintings, sculptures and memorials). The second focuses on one monumental sculptural synthesis, not only in terms of its artistic qualities but also as a source

Figure 15.1 Map of Crete (Copyright Ilias Kitsas).

of visualizing the memories and narratives of war and of highlighting the politics of and approaches to communicating the traumatic past.

VISUALIZING MEMORIES THROUGH ART

The Impact of War-Related Memories on Artistic Expression

Crete is the homeland of important Greek personalities. One of the most prominent, Nikos Kazantzakis (1883–1957), experienced the post-war situation in Crete at first hand, as he served as a member of the Government Committee for the Ascertainment of German Atrocities in the island (1945). Soon afterwards (1946), he entrusted to the Cretan sculptor Nikos Sofialakis (1914–2002)[1] his ideas about two artworks, which are thought to summarize his memories from and his feelings for this experience: *Enslaved Greek Child* and *Execution Pole of Agia*.[2] As a result of this acquaintance, Sofialakis made the *Child of the Occupation*[3] in natural size and in an iconographical type that occupied him at least since 1943.[4] The sculpted child is skeletal and frail, curled up in the foetal position, bent over from hunger, like the 'spectre of war', as Sofialakis remarked in a lyrical account of his work.[5] His child embodies the terror of war and acquires an ecumenical character. The second artistic idea of Kazantzakis referred to the execution pole at a certain place of torture in Crete which attained a memorial character. Polychronakis (2010) described the story of the monument in Agia: in its surrounding area there is a cenotaph and a monument, while the pole is exhibited in the Historical Archive in Chania, Crete. According to the artist's narration, Kazantzakis asked him to represent it as if 'redemption was breaking free' from it (Marmaridis 1983). It seems that the moral courage and heroism that the Cretans expressed while tied to the execution pole associated it in the mind of Kazantzakis not with torment and pain but with redemption and liberation. It is indicative of his feelings that a few months after visiting Crete in 1945 as a member of the Government Committee for the Ascertainment of German Atrocities, he gave a radio speech and made special note of the courage of locals who sang in front of the firing squad and of those who survived the occupation and who, instead of moaning about the disaster, appeared to transcend their tragedy (Andrikakis 2012). Sofialakis, expressing ancient ideals and following Renaissance prototypes, represented the sculpture in the form of a dynamic and proud youth [*kouros*].[6] It is indicative of the power of visual arts to express war memories that Kazantzakis, this notable intellectual, sought to give form to his memories and ideas through sculpture.

VISUALIZING MEMORIES THROUGH PAINTING

The comparison between paintings and sculpture of the Battle of Crete and the Resistance reveals certain iconographical resemblances. Furthermore, it connects the expression of Cretan war narrations not only to that of the mainland but also to that of other countries.

The usual iconography in these works is that of a strong and almost supernatural Cretan that defeats the German enemy. Let us not forget that the outcome of the battle was not in favour of the allies; however, the contribution of Cretans has taken on legendary status. This resulted in artworks that propagandize Cretan strength and bravery. As far as style is concerned, we often find depictions of folk art that combine romantic and Byzantine residues. The traditional elements connect the images with local history and, by extension, with society. Such works appeal to a public that needed to forget the traumatic past and chose to selectively remember only heroic Cretan actions. They are often painted by artists of Cretan origin, expressing the local articulation of war memories.

The work of Alexandros Droudakis (1915–93), for example, represents a traditional aspect of art that is influenced by romantic paintings in general and by those referring to the Greek War of Independence specifically. The *mantinada* (a couplet of the oral Cretan tradition) that he wrote to accompany the work emphasizes that 'no matter if the enemy steps on Cretan bodies, he will not step on Cretan soils'.[7] In this episode, it is the Cretan who almost steps on the dead German. He dominates in space, armed only with his traditional knife and spade, in an iconographical type that implies absolute control over the enemy. The heroic element and the great achievement of romanticism are the prevailing characteristics. Expressing a similar ideology, Petros Vlachakis reverses the roles: in his work, the armed and strong becomes weak, while the unarmed and weak becomes strong. Although the work refers to a real episode and a specific hero, its interpretation is wider: the battle between the Cretan and the German is equated to that of David and Goliath and acquires a holy and moral dimension (Figure 15.2).

In these works, we can often trace Byzantine styles which express a folk conception in depicting war. Dimitris Pothoulakis, for instance, uses an iconographical motif inspired by depictions of the Holy Family.[8] This

Figure 15.2 Petros Vlachakis, *The Battle of Crete* (The Historical Archive of Crete. Copyright The Historical Archive of Crete).

traditional perception of Christian iconography encapsulates symbolically the reasons for the Cretan Resistance. The selection of a traditional work, made by Ioannis Anousakis (1896–1981) as an illustration for a stamp to commemorate the fiftieth anniversary of the battle, indicates the appeal of the folk element to the wider public. In *The Armbands* (2001), Roussetos Panagiotakis (born 1944) visualizes symbolically what distinguishes the motives of the two sides: the German, a deathly ghost, acts under the orders of Nazism; the Cretan, fearless, fights for his family and homeland. Botis Thalassinos ironically depicts Hitler as a man in grief in front of the battlefield. In another portrait he honours Patrick Leigh Fermor for his contribution to the Resistance by depicting him with the characteristic Cretan scarf on his forehead.[9]

Yet war memories do not include only episodes of battle. They also reveal the everyday difficulties of a life that kept going under the worst circumstances. Kanakis I. Geronymakis (born 1926) was only 15 years old at the time of the battle. He expresses his memories both in writing and in painting. Although his depictions have strong personal feeling, they express through their reality not only the individual but also the collective memory. He considers his pictures to be a 'respectful memorial' to 'heroic ancestors'

Figure 15.3 Kanakis I. Geronymakis records a personal memory from September 1943; the Nazis were using people instead of horses to pull the carriages during the construction of a road in Hagia Galini (Copyright Kanakis I. Geronymakis).

(Geronymakis 2006, 153). Geronymakis's work is expressive and direct. It is a chronicle based on real facts. He emphasizes specific and personal war memories. Germans forcing men to dig their own graves before their execution, the inhuman use of people for 'public works' (Figure 15.3) and the great hunger are what he chooses to depict in a plain but narrative language that stems from the tradition of folk painting. In these depictions, one discerns the dignity of the Cretans in contrast to the barbaric instincts of the Nazis.[10]

VISUALIZING MEMORIES THROUGH MONUMENTS

The large war cemeteries in Souda Bay and Maleme are places of remembrance and commemoration which testify to the size of losses for both sides (the Commonwealth servicemen and the Germans, respectively). At the same time, the scattered monuments across the island testify to the number of local casualties and manifest local war memories. It is indicative of the spread of German atrocities and of the need of people to honour and remember those who resisted that virtually every Cretan city and village owns a memorial or public (outdoor) sculpture. My research for this chapter has revealed that a large number of works were commissioned by the authorities, by cultural or resistance societies and by private initiative.

Regarding their morphology, many works are stereotypical in design and often of average quality, similar to cases in the rest of Greece (Mykoniatis 1996, 21). As on the mainland (Papanikolaou 2006, 230), the usual types are columns and headstones, cenotaphs and graves, busts, statues and sculptures, inscriptions or constructions. Some of them are elaborate artworks, others are plain memorial constructions; some are made by notable sculptors, others by local craftspeople.

We often find the form of a plain column or headstone which commemorates the names and sometimes the ages of the dead. The monument in the village of Kallikratis, for instance, is dedicated to the men and women who were executed by the Nazis on 8 October 1943 (Geronymakis 2011, 31). Occasionally, images incorporated into monuments, which are easily understood by the public, describe or remind us of key events. Allegorical and symbolical figures (like Victory, History, Glory or ordinary soldiers) are commonly used and express a message of triumph. The compositions are often quite conventional and employ typical motifs both from neoclassical tradition and from folk art. In this respect, Yiannis Kanakakis (1903–78), in the *Monument of Cretan Resistance* (Heraklion), combines two symbolic and rhetorical personifications (History and Victory) with descriptive medals that narrate the battle under typical iconographical conventions (e.g. the large quantity of parachutists, the giant Cretan that defeats the enemy) and assign a didactic character to the work.

Another motif is that of the eagle, which represents the free spirit that characterizes the Cretans and their unbound soul. Sofialakis' super-scale and proud Cretan eagle, with the wings wide open, was installed in the square of Atsipopoulo, a village in Rethymnon, with a specific aim: to commemorate the battles and sacrifices of the dead heroes.[11]

Monuments referring to or commissioned by the Allies are usually located in places related to their exodus from Crete, such as the example of the Preveli International Memorial for Resistance and Peace and the Memorial in Chora, Sfakia.

Occasionally, remnants of German atrocities acquire a memorial character. The inscriptions that were installed in Kandanos after the destruction of the village and the execution of the locals by the Nazis are an example. But these can also raise debate, as in the case of the 'German Bird' or 'Evil Bird', as it is widely and euphemistically known, carrying the swastika, which was erected by the Nazis in 1941 to commemorate their dead compatriots. Its story may serve as a case study of different meanings and changing attitudes towards war monuments: honour, pride and triumph for the winner—a symbol of oppression, violence and trauma for the loser at the time of its installation; the decision to keep it, made by the victims of Nazism after liberation—the neglect of its preservation. The recent debate—triggered by its destruction—over its restoration or relocation indicates that for a part of the local population it remains an unbearable reminder of the traumatic past.[12]

Similar to Cretan paintings, many monuments are stereotypical and conventional representations of the modern Cretan hero, who is sculptured as a giant, dynamic figure bearing his characteristic clothes and equipment. These are usually idealistic and conservative forms of expression.

However, sometimes this idea is transformed into new and innovative forms. Yiannis P. Markantonakis (born 1952) summarizes the diachronic battles of Crete from the period of the Venetian Rule until WWII by depicting, in a work installed in Lakkoi in 2001, three Cretans of succeeding generations, standing on each others' shoulders. The younger, on the top, points towards the sky—a symbol of liberation, redemption, and heroism. The emblematic character is enhanced by the height of the sculpture (6.2 metres) and by its dynamic upward development.[13] In a second anthropocentric monument (for the Battle of Panagia Kerameion), the same artist incorporates the long Cretan tradition: figures that recall the Minoan frescoes seem to move in the steps of the traditional war dance, the *pentozalis*, over the dead body of a German soldier. The artist represented the latter intentionally larger than in reality to symbolize the 'monster of fascism' (Kakanos 2011). The monumental and dramatic account by Yiannis Parmakelis (born 1932) of the execution of locals in 1943 in his memorial, which stands near Amira (Viannos), appears anthropocentric, abstract and expressionistic all at the same time. Some works honour specific heroes—for example, the statue of one of the leaders of the Cretan Resistance, *Georgios Petrakogiorgis*, a work made by Michalis Kassis (born 1931) in 2008—and not the Cretan man in general.[14]

Similar to war monuments in the rest of Greece from 1950 onwards (Papanikolaou 2006, 230), the works in Crete are installed in central places or in historically charged locations and imply an ethnocentric conception. The locals often regard them as the fulfilment of their debt to the heroic ancestors. At the same time, they act as informative and instructive means of communication to a wider public. What is impressive in the case of Crete is the large number of monuments which narrate the traumatic episodes of the period or emphasize Cretan bravery. The monuments in the small communities along the island express the strong need of the locals to commemorate the events, to honour their heroes, to instruct the younger generations or to inform the public about the island's past. War atrocities were spread all over Greece and Crete alike, but in Crete, memories are strongly expressed in the public sphere, something that is less obvious elsewhere in Greece.

THE *BATTLE OF CRETE* BY NIKOS SOFIALAKIS[15]

The *Battle of Crete* is a monumental composition of 0.9 by 18 metres which commemorates the 'war achievements of the Cretan National Resistance against Nazism and during the war period of the years 1941 to 44'. It was designed as a frieze for the exterior of the Memorial Museum of National Resistance in Patteles, Heraklion. The work was commissioned to Nikos Sofialakis in 1968 by Manolis Bandouvas (1897–1984), the leader of the National Resistance in Crete.[16]

The selection of Sofialakis for the implementation of the work was not random. It is not only his Cretan origin that must have played a role in this decision. By 1968, he was already well known and was considered suitable for public commissions, especially in Crete. As far as the twin subjects of war and occupation are concerned, these topics preoccupied him early in his career, most probably because of his personal experiences. As indicated by Chraniotis, although he was not in Crete at the time of the battle, he served in the Greek–Italian War (1940–1) and never forgot how he returned from the field barefoot.[17] Yet until this point his works were not related to war episodes; they were sensitive depictions of the starving and suffering children of the occupation.

Before coming to a final agreement with Bandouvas, Sofialakis worked on the subject, first on its general conception[18] and afterwards on detailed full-sized drawings. The agreement between Sofialakis and Bandouvas was based on these twelve drawings and on two bas-reliefs that the latter approved with his signature. The terms of the contract and Bandouvas's signature on the drawings indicate the close cooperation between the two men. Bandouvas had a leading role in deciding the subject and perhaps also the order of the plaques. Besides, he was the narrator of the events and the person affirming the historical accuracy of the plaques. It took the artist approximately nine months to complete the majority of the work.[19] At the end of May 1969, the frieze was ready. Before installing the plaques *in situ*, Sofialakis exhibited the composition in the yard of his Athenian studio, together with other works of the previous decade.

The final work comprises eighteen bas-reliefs in Pentelic marble that describe, in chronological order, nine episodes from the battle of Crete to the liberation.[20] Nine descriptive texts that are saved in The Nikos Sofialakis Center of Neoclassical Sculpture reveal the subjects and help us in decoding the hidden meanings and the symbolisms of the various motifs, as will be now described. According to Chraniotis, they accompanied the plaques at the first presentation of the work,[21]and I believe that these might be the ones that Paraskevaidis (1969) reports.

The first plaque (*Fall of the German Paratroopers in Crete, May 1941*) depicts the outset of the battle. The parachutists and the aeroplane emphasize the airborne character of the German attack, revealing simultaneously the main subject and the time. The viewer is positioned in space by the statue of *Liberty* on the right end of the background. This distinctive sculpture of the city of Chania (a work of Thomas Thomopoulos) is not only used as a point of reference for space but also for the aims of the battle and of the resistance. According to the first text, it emphasizes the 'ideological and true target' of the Cretan struggle throughout the centuries. This is the only plaque that refers to the battle (Figure 15.4).

The next scene (*Assembly and Pledge of Warriors at Heraklion*) illustrates the beginning of the Resistance as the gathering of Cretans in Heraklion under the leadership of Manolis Bandouvas on 2 June 1941. They pledge to liberate Greece under the same motto as their heroic ancestors in 1821: 'Liberation or Death'. The composition is divided into two parts, having as a central axis a kneeling Cretan who kisses the hand of the 'captain'. The latter is no other than the commissioner of the work (Bandouvas), who thus becomes the leading figure in the 'pantheon' of Cretan Resistance. The general conception is inspired by scenes that illustrate the heroic events of 1821. The group on horseback with the female figure with a baby in her arms incorporates one of Sofialakis's favourite subjects: *Maternity*. This motif acts as a promise for a better future and creates a religious feeling, as if we were looking at the Madonna and Child. The man who raises his arms to embrace them recalls what Cretans leave behind, what they are ready to sacrifice and what they are fighting for. The surrounding mountains, as symbols of the homeland, suggest the symbols on which the Cretans pledge their loyalty. It is interesting to note that the monk Sofronios Roubakis, who blessed the fighters, is not depicted. The attempt to emphasize the role of Bandouvas is likely to be the reason for this omission.

The third plaque (*Gathering and Transfer of Munitions*) describes the decisive transfer of war supplies to the mountains. The coordination and systematic movements of men and the existence of a woman next to them

Figure 15.4 Nikos Sofialakis, *Fall of the German Paratroopers in Crete (May 1941)*, from the frieze, *The Battle of Crete*, 1969, archives of The Nikos Sofialakis Center of Neoclassical Sculpture (Copyright The Nikos Sofialakis Center of Neoclassical Sculpture).

imply the collective character of resistance. It is worth mentioning that the initial drawing of this scene depicts only men. However, in the final work, the artist replaced the man on the left with a female figure. This attaches a new meaning to the scene, as it underlines the active participation of women. The armed figures emphasize that it is a period of war and death. Yet the surrounding landscape and the woman, who may also be a personification of Spring or Liberty, connect the time of the year to the aim of resistance: as noted in the third text, spring is the time for new life; this will come with liberation, for which people fight.

The next three scenes (*Battle at Vianno*, *Fire and slaughter in Vianno*, *German captives*) describe the Cretan victory at Vianno in September 1943, the resultant German revenge and the reactions in the Cretan observation post. In contrast to the iconography up to this point, which emphasized the dynamic spirit of Cretans, the second scene depicts the evils of war. The Cretans, drawn in the Byzantine and a local style, become martyrs. There is a strong contrast between the executed unarmed civilians and the inhuman, revengeful Germans. However, as the sixth text points out, the captives of the third scene seem desperate as they think of the 'fatal reprisal'. Synthetically, the scene resembles depictions of the Greek military camps during the Revolt of 1821. Furthermore, the 'tree of the knowledge of Good and Evil' in the middle, apart from a synthetic role, has symbolic connotations, as it renders the war between the Cretans and the Nazis into a holy struggle between Good and Evil.

The narrative continues at the historical monastery of Arkadi (*Secret Meeting at the Arkadi*). Abbot Dionysios Psaroudakis blesses the representatives of all Cretan districts, as a second *Palaion Patron Germanos* (the bishop who blessed the warriors and the flag of 1821), continuing the tradition which saw the blessing of warriors at the outbreak of the Greek War of Independence and which attributed a holy aspect to the fight. The gesture of the third man on the row, who touches the shoulder of the second man, underlines the cooperation and support among all representatives, revealing the politics behind the scene, which once again emphasizes the collective character of resistance. According to text seven, which was written to describe this scene, one result of this meeting was the planning of the abduction of General Kreipe, the German commander who operated on Crete and whose abduction by the British and Cretans became legendary. This (*The Abduction of Kreipe*) is described in the next plaque. Various symbols and the mountains locate us in space. Both sides are framed by an armed figure. General Kreipe is surrounded both by the Cretan landscape and the Cretan people: there is no way out!

The story ends in optimism: the last scene (*Time of Peace*) describes life after the war. It takes place in bright daylight. The renaissance is achieved: locals across Crete again take up their everyday tasks with joy. As the descriptive text for this scene underlines, 'life at this point must be won by other means: with work and song'. The powerful scene, a hymn to the fruits of peace, to the fertile Cretan earth and to the hard-working locals, is divided into two plaques. Each figure, as described by the artist, represents a specific Cretan region: in the left is a man with a spade in front of the mill (Lasithi); the young girls who cut raisins (Heraklion); the couple in front of the olive tree (Rethymnon); and the woman gathering oranges, who closes the scene, representing Chania. Thus, the narration begins and ends at the same location. The frieze starts with the aim of resistance (liberty) and ends with the fulfilment of this target (the period of liberation).

In his emblematic work, Sofialakis remains loyal to the anthropocentric character of his production. He also remains faithful to neoclassical art, which he combines with realistic elements but mostly with folk and, by extension, Byzantine art. Tradition becomes a rich source for the Cretan characteristics, clothes and equipment and for a variety of symbols. The inclusion of a strong local element makes the scene comprehensible to the Cretan audience but could also be seen as a tribute to the origins of the artist.

Let us now examine how the specific work communicates war memories. What prevails in this expressive composition is heroism and epical mood. An important observation is that the conception of the iconographic cycle resembles that of the heroic events of the Greek Revolution against the Ottomans. In the latter case, Good (the Greeks) collide with Evil (the Turks), and in the examples discussed here, Good (the Cretans) defend themselves against Evil (the Nazis). Similarly, Cretans are depicted as 'beautiful heroic figures', while their enemies are shown as 'repulsive invaders'.[22] The figures express the catholic, the essential and the moral element of the resistance. There is only one scene in which Cretans are depicted as victims. Their bravery is contrasted to the Nazis' barbarism. The religious aspect is evident. Liberation is presented as God's will, which helps people to believe that they can overcome anything and anyone, including an enemy who is superior to them in the number of both soldiers and weapons. The use of a simple language and of common iconographical conventions is the best way to express meaning easily, which is after all the aim of most public monuments. This iconography captures popular imagination, as people are aware of the meaning of the scenes of 1821. Moreover, this resemblance equates in the mind of the viewer both battles and both sets of Cretan heroes. Even the 'patina' finish acts as a trigger to encourage the imagination to link the older and newer Cretan struggles and, by doing so, adds a diachronic aspect to the narration. Thus, moving beyond the local history, the Cretan resistance acquires an important position in national history. What we face here is the creation of a new national heroic epic and of its iconographical narrative. At the same time, its depiction as the fight of Good against Evil renders it of international and ecumenical importance. It is indicative of this that Paraskevaidis (1969) saw the composition as a protest against violence in general that extends beyond Greek borders.

In this respect, he compared Sofialakis and the *Battle of Crete* to Francisco Goya and his works *The Third of May 1808 in Madrid* and *The Disasters of War*.

In this composition the traumatic past becomes a means of emphasizing the valour of the ancestors and of promoting national elation. The narration sums up the spirit of Cretans, honours and celebrates their courage and nature, regardless of age or gender. The collective Cretan contribution, decisiveness and bravery are underlined in every way. The allies are absent. Events are presented as a series of episodes between Cretans and Germans.

At the same time, the narrative in the *Battle of Crete* promotes the idea that Bandouvas and his men both started and held a leading role in the Resistance. It is indicative of this political use of the work that although the typical features of the figures promote the general stereotype of the brave Cretan rather than of a particular hero (apart from the abbot of Arkadi and General Kreipe), one can identify the commissioner of the work. The narrative is based on and promotes the memories of Bandouvas and his evaluation of the events. The work is a public form of messaging, and in this case Bandouvas used the frieze as a way of legitimizing the role of himself and his group in the Resistance. The tension and emotion in the work suggest the strength of war memories; both the commissioner's and the artist's, who stated that it was an immense satisfaction to work on subject matter such as the resistance of Crete (Lydorikis 1969).

Let us now examine the reception of the composition, which became a point of reference in Sofialakis's career. In presenting the work to the Athenian public, Tomadakis (1969) characterized it as 'the greatest composition' in Sofialakis' life, pointing out that only a Cretan and somebody who experienced the events was able to express them. He summarized the technical difficulties in narrating a four-year period in the limited space of a frieze and claimed that Sofialakis' choices (the small number of figures and the use of readily understood symbols) represented successfully the morale of the struggle. He listed a plethora of symbols and asked the viewers to decode them; the frieze, as he emphasized, 'hides its own truth, the truth of art'.

The response in the press was positive. Paraskevaidis (1969) spoke about 'an artistic event of great interest', a 'grandiose synthesis' that is kept in the 'austere frames of high art'. In another article he stressed that the artist managed, using the ancient Greek *metron* (canon of symmetrical proportions), to illustrate the horror of Hitlerism and the heroic self-sacrifice of the Cretans (Paraskevaidis 1971). Lydorikis (1969) praised the composition not only as 'an evocative reminder of what the heroic ancestors did for the idea of Freedom' but also as an artwork. Panagiotopoulos (1969) understood the scenes as symbols of the legendary and brave Cretan sacrifices for freedom. He claimed that the work revived the glory of a new battle of the giants and characterized it as a 'personal combat' of Sofialakis. Stavroulakis

(1971, 4) considered the 'greatest of his compositional works' as a 'condensing of the diachronic greatness and glory of the last century of Crete'. Another reporter discerned that Sofialakis worked with excitement, power, skillfulness, soul and a 'patriotic pulse' and compared the feelings invoked to those caused by watching an ancient tragedy.[23] The critics generally interpreted the qualities of the work with terms that related to the ancient past. The response to the arrival of Sofialakis in Heraklion for supervising the installation of the work was also positive.[24]

It is worth noting that the *in situ* installation on the museum itself does not match the chronological narration just given. Chraniotis points out that the frieze is divided into two parts with six plaques on the one side (*Fall of the German Paratroopers—Fire and Slaughter in Vianno* and the left part of *Time of Peace*) and four on the other side (*German Captives—the Abduction* and the right-hand part of *Time of Peace).*[25] This arrangement leads to a new reading. Both parts of *Time of Peace* are installed in a central point, contrasting with the other—military—depictions and adding a didactic aspect to the composition (Paraskevaidis 1969). At the same time, they symbolically frame an inscription that bears the oath that Bandouvas and his men gave; this implies that liberation was achieved through this oath. Following the political ideology of its era, the building incorporated into its structure an element of ancient public architecture (in this case, the temple). By positioning the story at the upper part of the exterior walls, the museum (which as a term is connected to the temple) symbolically became the temple of Resistance and of its leaders. The thirteen busts surrounding the building are a contemporary addition dating to 2008 and represent Bandouvas and his group.[26]

CONCLUSION

Paintings, sculptures and monuments are only one form of expressing and visualizing war-related memories. Although not always historically accurate or free of political implications, they provide us with an idea of the impact on and the perception of events by communities. In the case of Crete, they indicate that war memories are long-lasting.

The historical importance and the nature of the Battle of Crete have had a strong impact on the maintenance of memory. This airborne attack was of a large scale, and the enemy suffered heavy casualties. The works discussed in this chapter show that both factors impressed locals, who later drew from these to build their narrative. The geographical morphology of the island and the local mentality has contributed to this preservation of memories.

To a large extent, Crete is composed of small, isolated, closed and conservative communities, where family, homeland, tradition and heritage are highly valued. Bravery, pride, independence and the resistance to foreign domination are aspects that Cretans emphasize as elements of their character, and together they form a strong and unique identity. The war episodes of the period and its local heroes are part of Cretan heritage. Together with those war memories from previous eras, they act as a means to emphasize Cretan bravery and the struggle of locals both for the island and for the mainland. While survivors of WWII and their families still live, there is a strong and locally felt need to express respect towards them.

The case of Crete helps to trace and examine differences and similarities in the expression of war memories in comparison to the mainland and thus enriches our knowledge about how small islands react to war. At the same time, it visualizes how the Cretans publicly expressed their war memories and contributes to the wider picture and mode of expressing war memories, in this case through art.

Art becomes a vehicle to connect to the past and to validate the present. Similarly to the mainland and to the heritagization of the 1821 narrative (the Greek War of Independence from Ottoman rule), Cretans draw their prototypes from ancient and Byzantine art. Their heroes resemble those of the Greek Revolution; they are depicted under the classical canon, hallowed in an idealistic manner. The connection to antiquity and to previously well known motifs is not only a way to ensure comprehension. It is also a way to validate the scenes—the real war episodes—and to render them diachronic. Thus, the 'new' heritage and narrative become acceptable as part and extension of the 'old' and 'common' heritage.

The commemoration of the battle that led to the occupation by emphasizing the 'brave contribution of locals' mirrors how Cretans have come to terms with their traumatic past. Moreover, the example of Crete on the one hand encapsulates the influence of the mainland on islands, and on the other hand it indicates the need of islands to connect and integrate their memories to the ones of the nation. Cretans, who have great self-belief, offer to the wider Greek nation 'memories which become everlasting presences', in the words of Panagiotopoulos (1959).

Today the commemoration of the battle of Crete remains vigorous and is highlighted in annual anniversaries and events. It is interesting to note that it is also commemorated by Australians, New Zealanders and Greek/Pancretan communities who have settled in these countries and who adopt similar commemoration practices but emphasize the joint struggle. The building of the Australian Hellenic Memorial in Melbourne, the Greek–New Zealand Memorial in Wellington and the New Zealand Memorial in Galatas, Crete, for example, is not only a medium of honouring heroes but also of underlining the bonds and the friendship between these countries, while anniversary commemorations are a chance for official representatives to express their honour to heroes and to the ideal of peace.

These conclusions reflect that although this chapter has focused on the expression of war memories in Crete, its narratives extend beyond the island's and the country's borders.

ACKNOWLEDGEMENTS

I owe special thanks to Kanakis I. Geronymakis (Vouvas, Sfakia), Georgios Chraniotis and Franjeska-Nicole Chraniotis (The Nikos Sofialakis Center of Neoclassical Sculpture, Athens), and Eirini Bandouva-Mela (Union of the Fighters for the Cretan Resistance, Heraklion) for contributing valuable material.

NOTES

1. For Sofialakis, see The Nikos Sofialakis Center of Neoclassical Sculpture (n.d.-a and 2004); Alevizou (2010, 246); Koundouridis (1991, 567–70); Mathiopoulos (1997–2000, 194); Stavroulakis (1971, 1, 4). Archival and photographic material is available online through www.sofialakis.org.
2. Sofialakis described this acquaintance to Marmaridis (1983). Stavroulakis (1971, 4) also mentions Kazantzakis and the two works.
3. Bas-relief, Pentelic marble, 1948, The Nikos Sofialakis Center of Neoclassical Sculpture, Athens (Nikos Sofialakis Center of Neoclassical Sculpture n.d.-b).
4. This is indicated by a drawing dated to 1943 (The Nikos Sofialakis Center of Neoclassical Sculpture, Athens). We cannot exclude the possibility that the author saw this or other relevant sketches during his visit at the artist's studio.
5. See The Nikos Sofialakis Center of Neoclassical Scupture (n.d.-b). The Director of Research at The Nikos Sofialakis Center of Neoclassical Sculpture and granddaughter of the sculptor, Mrs Franjeska-Nicole Chraniotis pointed out, in a personal communication of 15 March 2013 that the theme is inspired by the Iliad's 'spectre of war' (rhapsody L, 3–4), which was translated by Kazantzakis at the same period. She also noted that in a letter to Kazantzakis of 25 October 1954, Sofialakis referred to the *Occupation,* as the *Enslaved Young Greek*, folded in a foetal position into the mothers' belly, in order to return to the 'big mother', the earth.
6. *The Defender*, study on plaster, 1948, The Nikos Sofialakis Center of Neoclassical Sculpture, Athens.
7. Droudakis was not only injured in the battle but was also active in the Resistance. A brief biography is available online: www.kritesegaleo.gr/alexandros-droudakes (accessed 30 September 2013). An image of the work and the *mantinada* are available online: https://www.facebook.com/photo.php?fbid=363083240400281&set=a.363077667067505.78477.360727903969148&type=1&theater (accessed 16 September 2013).
8. Cover illustration of the book by M. Polioudakis (2005).
9. Images are online. Available HTTP: http://sadentrepese.blogspot.gr/2009/05/blog-post_4084.html; www.roussetos-art.gr and https://www.facebook.com/pages/ΜΠΟΤΗΣ-ΘΑΛΑΣΣΙΝΟΣ/104726542903846 respectively (accessed 30 September 2013).
10. For these illustrations, see Geronymakis 1998, 189–90; Geronymakis 2006, 22.
11. Agreement regarding the Atsipopoulo memorial, 14 December 1963, archives of The Nikos Sofialakis Center of Neoclassical Sculpture, Athens.
12. For a photographic record of the memorial, see Terniotis 2000.
13. 'Για τους αγωνιστές της λευτεριάς [For the freedom fighters]', *Rizospastis*, 28 August 2001, 22.

14. Photographs of the works are online. Available HTTP: www.lakkoi.gr/photos/lakkoi; www.haniotika-nea.gr/82607-Μνήμηηρώων.html and http://en.wikipedia.org/wiki/File:Amiras_Memorial_R02.jpg, respectively (accessed 16 September 2013).
15. See Pavlopoulos and Antonopoulou (2013).
16. Private agreement regarding the composition *Battle of Crete*, 15 June 1968, Archives of The Nikos Sofialakis Center of Neoclassical Sculpture, Athens. Designed by the architect Erifyli Mathioudaki-Lyberi, the museum was erected by the Union of the Fighters for the Cretan Resistance, which was established in 1948 for those who served the Cretan Resistance from 1941 until 28 May 1945 and for their adult ancestors. The commission was made one year after its erection. In 1998, the building was donated to the National Defence Fund (personal communication with Eirini Bandouva-Mela, letter correspondence, 20 March 2013).
17. Personal communication with Franjeska-Nicole Chraniotis, email, 19 March 2013.
18. There are nine sketches in ink dated in 1968 and bearing the artist's signature in The Nikos Sofialakis Center of Neoclassical Sculpture.
19. Anonymous, *Ethnos,* 28 March 1969.
20. A comparison between the early drafts and the final work reveals that two subjects (two and eight) appear at the stage of the life-size drawings, while two do not exist at the final scenes and their meaning needs clarification. The rest of the drawings were transferred to marble with minor alterations. Besides, according to the contract, the artist was not allowed to make any changes to the approved sketches. It is interesting that Tomadakis (1969) and Stavroulakis (1971, 4) mention the existence of 12 *metopes.*
21. Personal communication with Franjeska-Nicole Chraniotis, email, 15 and 19 March 2013. I will refer to them as text 1, text 2, etc.
22. Anonymous, *Ethnos*, 13 June 1969.
23. Anonymous, *Ethnos*, 13 June 1969.
24. 'Ο γλύπτης Νίκος Σοφιαλάκης εις την πόλιν μας [The sculptor Nikos Sofialakis in our town]', *Patris,* 1 July 1969.
25. Personal communication with Franjeska-Nicole Chraniotis, email, 18 March 2013.
26. Personal communication with Eirini Bandouva-Mela, letter correspondence, 20 March 2013.

BIBLIOGRAPHY

Alevizou, Nt.-Ch. (2010) *Η Κρήτη των Καλλιτεχνών, 19ος–20ός αιώνας. Αγιογραφία-Ζωγραφική-Γλυπτική* [*Crete of the Artists, 19th–20th Century. Hagiography-Painting-Sculpture*], Heraklion: Dokimakis.

Andrikakis, A. A. (2012) 'Μία ραδιοφωνική ομιλία του Ν. Καζαντζάκη, τον Δεκέμβριο του 1945, για την κατεστρεμμένη Κρήτη και τους μαυροφορεμένους αλλά αλύγιστους Κρητικούς, τον πρώτο καιρό μετά από την απελευθέρωση από τον Γερμανό [A radio speech by N. Kazantzakis, in December 1945, about the destroyed Crete and the black dressed but implacable Cretans, in the period after the liberation from the German]', *Patris*, 2 July. Online. Available HTTP: www.patris.gr/articles/224856/151385 (accessed 19 September 2013).

Anonymous. *Ethnos*, 28 March 1969.

Anonymous. *Ethnos*, 13 June 1969.

Geronymakis, K. I. (1998) *Η Κρήτη στο πρόσφατο παρελθόν* [*Crete in the Recent Past*], Athens: Vivlioepilogi.

Geronymakis, K. I. (2006) *Ταξίδι στην Παράδοση* [*Journey in Tradition*], Chania: Prefecture of Chania.

Geronymakis, K. I. (2011) 'Το ολοκαύτωμα του Καλλικράτη το 1943 από τους Γερμανούς [The holocaust of Kallikratis in 1943 by the Germans]', *Kritika Nea,* December, 31.

Kakanos, Y. (2011) 'Στην Παναγιά των Κεραμειών. Μνήμη Ηρώων [At Panagia of Kerameion. Memory of heroes]', *Haniotika Nea,* 14 November. Online. Available HTTP: www.haniotika-nea.gr (accessed 19 September 2013).

Koliopoulos, I. S. (1978) 'Η μάχη της Κρήτης [The Battle of Crete]', in *Ιστορία του Ελληνικού Έθνους* [*History of the Greek Nation*], vol. IE, Athens: Ekdotike Athinon, 452–4.

Koundouridis, V. (1991) *90 ζωγράφοι, χαράκτες και γλύπτες* [*90 Painters, Engravers and Sculptors*], Athens: Pylarinos.

Lydorikis, A. (1969) 'Η αντίστασις της ηρωϊκής Κρήτης σε μια έκθεσι στο χαγιάτι του Νίκου Σοφιαλάκη [The resistance of the heroic Crete in an exhibition at the hayati of Nikos Sofialakis]', *Apogevmatini,* 7 June.

Marmaridis, G. (1973) 'Νίκος Σοφιαλάκης: μια "καθαρή αναπνοή" στην Τέχνη [Nikos Sofialakis: a "clear breath" in art]', *Ergatiki Epitheorisis,* August, 31–4.

Marmaridis, G. (1983) 'Νίκος Σοφιαλάκης' ['Nikos Sofialakis'], *Akropolis tis Kyriakis,* 14 August, 7.

Mathiopoulos, E. (ed). (1997–2000) 'Σοφιαλάκης Νίκος [Sofialakis Nikos]', in *Λεξικό Ελλήνων Καλλιτεχνών. Ζωγράφοι-Γλύπτες-Χαράκτες, 16ος–20ός αιώνας* [*Dictionary of Greek Artists. Painters-Sculptors-Engravers, 16th–20th century*], Vol. 4, Athens: Melissa.

Mykoniatis, H. (1996) *Νεοελληνική Γλυπτική* [*Neohellenic Sculpture*], Athens: Ekdotike Athinon.

The Nikos Sofialakis Center of Neoclassical Sculpture. (n.d.-a) *The Artist. A Short Biography.* Online. Available HTTP: www.sofialakis.org/#!the-artist/c1a7h (accessed 19 September 2013).

The Nikos Sofialakis Center of Neoclassical Sculpture. (n.d.-b) *Child of the Occupation.* Online. Available HTTP: www.sofialakis.org/#!untitled/zoom/c1a7h/i017c2 (accessed 19 September 2013).

The Nikos Sofialakis Center of Neoclassical Sculpture. (2004) *The Official Informative Brochure,* Athens: The Nikos Sofialakis Center of Neoclassical Sculpture. Online. Available HTTP: www.sofialakis.org/#!information-brochures/csy5.

Panagiotopoulos, I. M. (1959) 'Επί τη επετείω της Μάχης της Κρήτης. Το νησί που λέει πάντοτε "όχι" στους όποιας φυλής επιδρομείς. Εκεί που λατρεύεται με πάθος η Ελευθερία [On the occasion of the anniversary of the Battle of Crete. The island that always says "no" to the invaders of any race. There, where Freedom is passionately adored]', *Taxydromos (Egypt),* 128 (31 May), 1.

Panagiotopoulos, V. (1969) 'Μνημείον Εθνικής Αντιστάσεως του Νίκου Σοφιαλάκη στην Κρήτη [National Resistance Memorial in Crete by Nikos Sofialakis]', *H Vradyni* (11 June).

Papanikolaou, M. (2006) 'The two aspects of nationalistic art in Greece, 1950–1960', *The Journal of Art Theory & Practice,* Seoul, 219–39.

Paraskevaidis, M. (1969) 'Η Εθνική Αντίστασις της Κρήτης στα ανάγλυφα του Σοφιαλάκη [The National Resistance of Crete in the reliefs of Sofialakis]', *Eleutheros Kosmos,* 5 June, 9.

Paraskevaidis, M. (1971) 'Ο νεοκλασικισμός του Ν. Σοφιαλάκη. Η έκθεσις του στην "Ροτόντα" [The neoclassicism of N. Sofialakis. His exhibition in "Rotonda"]', *Eleutheros Kosmos,* 21 October.

Pavlopoulos, D. and Antonopoulou, Z. (2013) *The Battle of Crete by the Sculptor Nikos Sofialakis*, Athens: The Nikos Sofialakis Center of Neoclassical Sculpture.
Polioudakis, M. (2005), *Μαρτυρίες Αγωνιστών. Για τη Μάχη της Κρήτης και την Εθνική Αντίσταση σ΄ αυτήν 20.5.1941–30.6.1945* [*Testimonies of Fighters. About the Battle of Crete and the National Resistance in it 20.05.1941–30.06.1945*], Rethymnon: Prefecture of Rethymnon. Online. Available HTTP: www.libret.gr/2005%20.html (accessed 30 September 2013).
Polychronakis, P. (2010) 'Η ιστορία του μνημείου του Γολγοθά της Κρήτης [The history of the memorial of Golgothas of Crete]', *Haniotika Nea,* 9 July. Online. Available HTTP: http://haniotika-nea.gr/48973.html (accessed 20 March 2013).
Stavroulakis, A. (1971) 'Ρεθυμνιώτες Καλλιτέχνες. Ο γλύπτης Νίκος Σοφιαλάκης τιμή για το Ρέθυμνο [Rethymnian artists. The sculptor Nikos Sofialakis honor for Rethymnon]', *Kritiki Epitheorisis,* 13 February, 1, 4.
Terniotis, Y. (2000) 'Το Γερμανικό πουλί στα Χανιά [The German bird in Chania]', *Stratiotiki Historia,* 51, November, 74–5.
Tomadakis, N. B. (1969) *Λόγος εις τα εγκαίνια εκθέσεως των έργων του Κρητός Γλύπτου Νικόλαου Σοφιαλάκη* [Speech at the opening of the exhibition with the works of the Cretan sculptor Nikolaos Sofialakis], Athens, 4 June, archives of 'The Nikos Sofialakis Center of Neoclassical Sculpture', Athens.

16 Remembering War and Occupation in Post-Independence Timor-Leste

Michael Leach

INTRODUCTION

This chapter examines the way difficult sites of conflict, imprisonment, trauma and resistance are being remembered in the now independent nation of Timor-Leste (Figure 16.1). While the difficult challenge of memorializing massacre sites, as well as places of political imprisonment, torture and human rights abuses, confronts many post-conflict societies, few represent as profound a loss as Timor-Leste, having suffered an estimated minimum 102,000 casualties during the Indonesian occupation from 1975 to 1999, along with forced population displacements and extensive nonfatal human rights violations through arbitrary detention, torture and rape (CAVR 2005, 43).[1] A generation earlier, an estimated 60,000 East Timorese were killed in the course of the Japanese occupation of Portuguese Timor from 1942 to 1945.

In Timor-Leste, these difficult legacies have been complicated by the distinct cultural and linguistic affiliations promoted by successive colonial regimes, political schisms within the former independence movement, a lack of justice for the victims of human rights abuses during the Indonesian occupation and, in 2006–7, the brief but violent rise of regional tensions. These fissures have complicated the process of nation building and the articulation of a unifying postcolonial national identity since independence was restored in 2002. As such, they are critical to understanding the cultural heritage of the independence struggle and its conservation in Timor-Leste, which is itself an exercise in articulating cultural nationalism.

In examining East Timorese responses to these difficult issues since independence in 2002, this chapter discusses some of the challenging contexts of cultural heritage management in Timor-Leste and surveys the colonial and postcolonial 'layers' of the cultural heritage landscape, examining their competing visions of East Timorese identity. It then focuses on East Timorese nationalist conservation of difficult sites in the struggle for independence, including key jails and interrogation centres, massacre sites, and recent monuments to the armed resistance movement FALINTIL (Armed Forces for the National Liberation of Timor-Leste) and to civilian actors in the independence movement.

Figure 16.1 Map of Timor-Leste (Creative Commons).

Despite limited resources in the early years after independence, Timor-Leste has had notable success in the conservation of key sites and events in the liberation struggle, at both the national and the local levels. Nonetheless, it is argued that the cultural heritage landscape reflects a major 'fault line' in post-independence politics, in that the contribution of younger East Timorese nationalists in the struggle for independence still remains relatively neglected.

CONTEXTS

Timor-Leste[2] is a half-island state at the eastern end of the Indonesian archipelago, with an area of 15,000 square kilometres and a population of some 1.1 million. As a result of colonial division of the island of Timor between the Portuguese and Dutch, the western half of Timor island has been part of Indonesia since independence from the Netherlands in 1949. Timor-Leste also includes the small exclave of Oecusse located within west Timor: a legacy of the first Portuguese settlement on the island in the early sixteenth century before the capital was moved to Dili in 1769. Despite a short-lived period of independence from Portugal in late 1975, a truly national sense of East Timorese identity and community was consolidated most clearly in the struggle for independence from Indonesia, from the collective experiences of suffering under the occupation, which had a unifying effect among various ethnic and language groups under the 'colonial gaze' (Anderson 1993). As a half-island community sharing a land border only with its occupier, the domestic wings of the resistance were isolated for twenty-four years. Since the restoration of independence in May 2002, maintaining this unified sense of a common national identity has proven a more challenging task. Timor-Leste has witnessed intergenerational disputes over national identity and official languages (Leach 2003, 2008, 2012), a major political-military crisis in 2006, and wider 'history wars' within the former independence movement over the symbolic 'ownership' of the independence struggle and its core historical narratives (Leach 2006, 2007).

One widely endorsed and popular narrative of East Timorese nationalism is that of *funu*: a 450-year 'national' resistance struggle against consecutive foreign occupiers (Ramos-Horta 1987). However, beneath this broadly unifying historical narrative, a more complex and ongoing struggle over postcolonial cultural affiliations and national identity is evident. Put simply, the distinct experiences and educational backgrounds of two generations of nationalists, respectively encountering Portuguese and Indonesian colonialism, have complicated the task of articulating a simple, unifying postcolonial national identity. Older nationalists have politically dominated the post-independence state, and it is clear that significant numbers of young people have felt misrecognized by some 'official' articulations of national identity embedded in the East Timorese constitution and in the policies of the first government. As such, while the process of articulating the cultural components of nationalism may be metaphorically understood as one of 'imagining' a nation (Anderson 1983), in practice it may involve the universalization of cultural and political values of a dominant nationalist grouping (Leach 2002, 45).

One key issue is the choice of official languages and the official cultural affiliations of the independent state. While the indigenous *lingua franca*, Tetum, is accorded a high degree of cross-generational endorsement, the co-official language, Portuguese, has been less popular with younger generations educated in Indonesian, especially in the early years after independence. The co-official use of the language made sense for an older generation of the political élite, literate in Portuguese and instrumental in the rise of East Timorese nationalism in the early 1970s. For this generation, Portuguese was important as a unifying language across the élite of local language groups (before the spread of Tetum in the 1980s) and as a language of the armed resistance. Its choice as an official language also acknowledged the critical diplomatic support of lusophone[3] countries during the occupation. For a younger generation educated in Indonesian and with little knowledge of the old colonial language, however, the choice of Portuguese raised fears of their exclusion from symbolic sources of power and cultural identity in an independent Timor-Leste. There is some evidence that this tension is moderating over time (Leach 2012) and in flux as a new generation less fluent in Indonesian emerges, though it remains a more controversial element of post-independence political settlements.

Another example of this 'fault line' is evident in intergenerational debate over nationalist historiography and the nature of an 'authentic' postcolonial national identity. For an older generation of nationalists, emphasizing the long history of *funu*, the Portuguese presence is critical to East Timorese nationalism: unifying different regions against a common colonial occupier, bringing Catholicism, and marking the nation as a distinctive grouping not only in relation to Indonesia as a whole but, equally, to the indigenous peoples of Dutch-colonized, Protestant-influenced west Timor.[4] Many nationalists in a younger generation look for what they see as a more authentic postcolonial identity, looking primarily to its precolonial indigenous roots and more interested in the pan-island commonalties with west Timor. As one interviewee put it (Leach 2006, 232), 'I would prefer to study indigenous history, not related to colonialism. Timor-Leste's history itself; the local things.'

With some 65 per cent of the population under thirty years of age—but the political élite still dominated by an older generation—these differences are important political fault lines. Indeed, two key background factors behind the political crisis in Timor-Leste in 2006 were political divisions within the former independence movement and these well documented intergenerational tensions. These wider tensions over national identity and history are an important context to understanding issues of cultural heritage management in the independent nation.

'LAYERS' OF CULTURAL HERITAGE

Timor-Leste joined the international community in 2002, following 450 years of Portuguese colonial rule ending in 1975, a brief interregnum of independence and a twenty-four-year struggle against Indonesian occupation between 1975 and 1999. As the future president and Nobel Peace Prize recipient Jose Ramos-Horta (1996) noted, 'East Timor is at the crossroads of three major cultures: Melanesian, which binds us to our brothers and sisters of the South Pacific region; Malay-Polynesian, binding us to South East Asia; and the Latin Catholic influence, a legacy of almost 500 years of Portuguese colonization'. These influences have offered disparate resources for competing colonial and nationalist accounts of East Timorese history and identity (Leach 2006, 224). During the twenty-four-year Indonesian occupation, these tensions became a site of symbolic struggle. Indonesian neocolonial historiography tended to emphasize historical Malay connections, while East Timorese nationalists highlighted Melanesian affinities and the 450-year impact of Portuguese colonialism, by which East Timor could be identified as a political community distinct from Dutch-colonized west Timor. In this way, competing historical links with distinct island cultures to the west and east entered debates over national identity.

Similarly, the various colonial and postcolonial layers of cultural heritage in Timor-Leste represent contested attempts to reinterpret the past in ways which suit, respectively, Portuguese, Indonesian and nationalist ideas of East Timorese identity.

The built landscape of cultural heritage and monuments charts a changing history from 'heroes of the Portuguese empire', through neocolonial 'integration' monuments of Indonesia, to the monuments of an independent state. As Wiley (1994, 145) notes, collective identities commonly interpret the past as the linear 'origin' of the present political self or of a future self it is in the process of 'becoming' (see also Anderson 1983, 22–36). Each layer of the memorial landscape of Timor-Leste exhibits this process of remembering the past in ways which ideologically buttress 'contemporary' political projects of collective identity construction. Timor-Leste still has many Portuguese colonial monuments, including memorials and statues to early navigators or to various 'heroes of the empire'. These are primarily dedicated to metropolitan Portuguese but also prominently include monuments to loyal Timorese *Liurai* (Kings) who fought or died '*por Portugal*', either in helping to suppress indigenous rebellions or, as in the case Dom Aleixo Corte-Real of Ainaro, in fighting the Japanese occupation. Portuguese colonial historiography emphasized positive relations with the mother country and the progress of 'Portugalization'—a colonial metaphor for a 'civilizing' mission involving the spread of Catholicism and the 'pacification' of periodic rebellions (Gunn 1999, 22–4).

Of particular interest here are monuments remembering victims of the Japanese occupation, such as the prominent monument to the 1942 massacre of Portuguese troops in the town of Aileu. With typical colonial focus, the largest of these commemorate fallen Portuguese soldiers and officials, not the estimated 60,000 Timorese who died during the occupation. In various parts of Timor-Leste, smaller post-WWII Portuguese-era monuments can be found at crossroads, dedicated in a particular colonial logic 'to the victims of the foreign occupation'. Also in the landscape are a small number of memorials to the Australian so-called Sparrow Force (comprising 282 members primarily of 2/40th Infantry Battalion)—which harassed the Japanese occupiers and received essential support from Timorese civilians from 1942 to 1943. The most notable of these memorials is at Dare in the hills above Dili, though other remnants of the Japanese and Allied presence exist in the landscape, including cave networks, pillboxes, plane wreckage and abandoned artillery. These few reminders conceal elements of deeper historical controversy. In 1942, Japan was under considerable pressure from its Axis ally Germany to respect Portuguese neutrality, as Germany feared the Allies would be granted access to Portugal's naval bases in the Azores. Fearing it would be used as a base for attacks on Australia, Allied Australian and Dutch forces pre-emptively occupied Portuguese Timor. This was followed by a Japanese invasion of the neutral territory, an outcome which many argue might otherwise have been avoided, as it was in the

Portuguese colony of Macau. At one of several recent low points in the East Timorese–Australian relationship, Prime Minister Gusmao raised this issue publicly in 2010, accusing Australia of sacrificing the lives of 60,000 Timorese (Murdoch 2010).[5] Towards the end of the war, as Germany had feared, Portugal granted the Allies access to the bases in the Azores. This saw the Allies support the return to Portugal of its colonial possessions after WWII.

Indonesian-era monuments also reflect on the Portuguese colonial past. These typically seek to depict the forced integration of East Timor into Indonesia as a 'return to the fatherland' and portray elements of East Timorese nationalism against the Portuguese as consonant with Indonesia's own anticolonial struggle against the Dutch (Gunn 2001, 10). For example, most major towns in Timor-Leste have Indonesian integration monuments, some of which depict Dom Boaventura, an anti-Portuguese rebel *Liurai* (King) of the early twentieth century, in traditional dress, breaking free from the chains of Portuguese colonialism (Figure 16.2).

In this way, the Indonesian regime appropriated a key image of then nascent East Timorese nationalism and adapted it to an integrationist purpose, celebrating the forced integration as a triumph of Timorese anticolonialism. Where East Timorese traditional houses have four pillars, the integration monuments often have five-sided platforms, reflecting the five Indonesian

Figure 16.2 Indonesian integration monument, Dili, depicting a traditional Timorese warrior breaking loose from the chains of Portuguese colonialism (Copyright Leach).

citizenship principles of *Panca Sila*.[6] Importantly, too, the Indonesian regime took pains to conserve certain sites as *warisan nasional* (national heritage), such as the nineteenth-century Portuguese jail at Aipelo, preserved as a monument to the brutality of colonial era. Here, it is possible to witness the serial connections between the ideology of the successive colonial regimes and the heritage conservation practices in each era (Logan 2003).

EAST TIMORESE NATIONALIST CULTURAL HERITAGE

The final layer of cultural heritage consists of East Timorese memorials reflecting on the Indonesian era, as well as the pain and trauma of the liberation struggle. Prominent among these post-independence sites are the Comarca Balide (Balide Jail), a former jail and interrogation centre; the 'heroes monument' to Falintil resistance fighters at Metinaro; and memorials remembering the victims of massacres in the lead-up to and aftermath of the 1999 referendum on independence, such as those in the Suai and Liquica churches. More recently, this has included a monument to the victims of the Santa Cruz massacre, erected in Motael in 2012 (Figure 16.3).

Figure 16.3 Santa Cruz massacre monument, Motael, Dili (Copyright Leach).

These newer sites are central to the process of forging a postcolonial national identity. Some, like the 'heroes monument' at Metinaro, are designated as sacred spaces of the nation, to be under permanent honour guard. Others, like the monuments to the victims of the Suai and Liquica massacres, recall traumatic events that took place in already sacralized spaces, honouring the memory of victims who died sheltering from Tentara Nasional Indonesia (TNI) army forces or their proxy militias in churches or cemeteries. These difficult sites of cultural heritage are especially important in this process of articulating a nationalist view of East Timorese history and identity, as they are intimately tied in with wider processes of national and international reconciliation. It is no accident that one of the key sites, the Comarca Balide, was home to the Commission for Reception, Truth and Reconciliation (CAVR) until 2005, and now contains its records and the office of the Post-CAVR Secretariat.

COMARCA BALIDE: THE COMMISSION FOR RECEPTION, TRUTH AND RECONCILIATION

The most important repository of national memories of the Indonesian occupation is the Comarca Balide. A former Portuguese colonial jail built in 1963, the site was used as an incarceration facility by several regimes, including the short-lived unilaterally declared Democratic Republic of Timor-Leste in late 1975. Employed briefly by the pro-independence party FRETILIN[7] to house political prisoners after the civil war in 1975, the jail was then an Indonesian interrogation centre run by the notorious military police and intelligence organizations Kodim and Morem. The Comarca's importance as a site of pain and suffering under successive regimes made it a symbolically compelling choice for the headquarters of the CAVR after independence. Once the CAVR's primary testimony collection activities were wound down in late 2005, the site was designated as a permanent memorial and archive for CAVR documents, including the thousands of victim testimonies.

The initial proposal to rehabilitate the Comarca came from the association of ex-political prisoners (ASSEPOL) in 2000 and was adopted by the then nascent CAVR as an appropriate site to house the 'human rights history' of Timor-Leste. In 2002, a memorandum of understanding determined that the Comarca would become the Dili office of the CAVR for its mandate period, then stand as an archive under a long-term objective to 'preserve the former Balide prison for future generations as a memorial to repression and as a centre for the promotion of human rights and reconciliation'. These objectives were endorsed by the government of Timor-Leste.

Restored with the support of the Japanese and Irish governments, the Comarca building was ready for the formal opening of the CAVR hearings in February 2003. The original ASSEPOL inspiration was not forgotten, with the publicity leaflet featuring a quote from one of the founding committee members: 'We will show that flowers can grow in a prison'. The restoration process was participatory and inclusive, with ex-prisoners strongly involved in landscaping and other features of the restoration as a form of rehabilitative therapy. One woman who had spent her childhood there while her mother was imprisoned transformed the inner courtyards into gardens. Former political prisoners also built the furniture and lecterns in the courtyard meeting space area. Before the opening, another former prisoner conducted a traditional cleansing ceremony. In this way, the Comarca restoration project was actively conceived both as a site for personal recovery and as a national historical repository.

On 20 December 2005, at the end of the CAVR mandate period, the Comarca became a permanent memorial for the victims of human rights abuses in Timor-Leste and home to the post-CAVR technical secretariat, charged with disseminating the CAVR report and maintaining its permanent archives. The process of converting the Comarca site into a memorial and historical repository was conducted in consultation with UNESCO and other relevant international museums.[8]

THE COMARCA AS A MEMORIAL

The Comarca houses both standing memorials to victims of human rights abuses and other less conventional memorials, literally embedded in the architecture of the site. In the former category, the Santa Cruz room is the archive for the thousands of CAVR records collected during its mandate, its name recalling the site of the 1991 massacre of students which put the occupation of East Timor firmly back on the world stage. Another room houses the Suai Circle, a memorial to the victims of the Suai massacre during militia rampages in 1999, with photographs, traditional *tais* (woven cloth) and votive painted stones recalling individual victims.

One innovative method of preserving the memory of human rights abuses is through the conservation of prisoner graffiti. In total, sixty-five graffiti are preserved in whole or part, including one from a future CAVR commissioner imprisoned during the Indonesian era. Most graffiti are in Portuguese, and many express the simple remembrances of prisoners such as 'Here lay Zeca'. Others mark extended periods of arbitrary imprisonment in the early years of the Indonesian occupation, such as one scratching in which a prisoner laments: 'I spent my past in this cell'. Some show a sense of humour under adversity, such as one which declares, '[S]pecial cell for world leadership candidates'. Yet others are more disturbing, such as those found in the isolation cells under the main jail: 'You tortured my body in the fetters of your empire'—a line from the East Timorese nationalist poet Borja da Costa who was killed by Indonesian forces in 1975. Some graffiti in the isolation cells record perhaps the last testament of political prisoners, such as one dated 10 August 1976, nine months after the Indonesian invasion, in which a list of names follows an etching on a wall: 'In this cell of death were . . .'. The power of these graffiti lies in the fact that they are intensely personalized artefacts of a lived present of suffering rather than abstract, general or reconstructed memories of the past. As the CAVR (2003) notes, these conserved graffiti remind the visitor of the wider function of the building as a memorial and historical repository of narratives.

The graffiti from the jail and the un-renovated isolation cells serve as stark testimony to the plight of political prisoners, and the CAVR archives hold many written, audio and visual records to supplement the memory of this time and the story of what happened in the building.

While the internal rooms of the Comarca have been renovated, the prisoner graffiti is conserved under plastic frames, deeper under the modern paint layers, where many were inadvertently preserved in later Indonesian times. By contrast, the isolation cells have been left as they are, aside from the installation of lighting "so that visitors can see for themselves conditions in these 'cells of death'" (CAVR 2003). The Comarca also contains the legacies of other occupants, including graffiti and coarse artworks by TNI officers, such as one depicting a large number of Indonesian soldiers with a woman stripped of her clothes. The site also conserves the more recent graffiti of rampaging Aitarak and Mahidi militias in 1999, declaring Xanana Gusmão 'a mongrel'.

THE HEROES MONUMENT, METINARO

At the apex of East Timorese nationalist memorial sites, the impressive Heroes Monument at Metinaro, dedicated to the fallen FALINTIL soldiers, is designed as a sacralized national site, to be under permanent honour guard. Built close to the main Timor-Leste Defence Force (FDTL) barracks, the centrepiece of the site is an open platform monument, designed to accommodate official ceremonies, with three flagpoles, facing the open sea (Figure 16.4). The site also comprises a national memorial garden and natural reserve—to honour all the victims of the struggle for independence—a chapel, and ossuary houses, which initially contained the remains of several hundred FALINTIL fighters, at which visitors paid their respects. In more recent years, the Metinaro site has seen the establishment of the graves of hundreds of FALINTIL fighters, transferred from the ossuary houses or other graves in the country. The most significant addition in recent years was the grave of Xavier do Amaral, the first president of Timor-Leste during its short-lived independence in 1975, who died in 2012.

Built with the support of the United Nation Development Program, as well as USAID and other donors, the site was a major project of the

Figure 16.4 Heroes Monument, Metinaro (Copyright Leach).

Recovery, Employment and Stability Program for Ex-Combatants and Communities in Timor-Leste (RESPECT). Despite its success in establishing the Metinaro site, the program has been a controversial one, tarred by the first government's initial failure to adequately recognize and compensate former combatants for their sacrifices during the twenty-four-year struggle for independence—an agenda which was vigorously prosecuted by subsequent governments after the 2006 political-military crisis. Aside from those who joined the FDTL, a number of former veterans have since formed the backbone of various 'anti-system' political groups who reject the government and the 2002 constitution.

The difficulties of nation building are reflected in one long-running controversy over the Heroes Monument site, with one of these veteran-dominated groups, the CPD-RDTL (Committee for the Popular Defence of the Democratic Republic of Timor-Leste), arguing that their fallen comrades should be buried in their own districts. In recent years, such practices have been evident in Lautem, with FALINTIL monuments built by locals that differ from the national monuments, often including FRETILIN party symbols and with traditional objects and symbols integrated into the memorial construction (Feijo and Viegas 2014). This subsequently led to the Government funding smaller ossuary houses and memorials in each district capital. Once again, this critique highlights a perennial tension in Timor-Leste between nation-building projects and the ongoing strength of local and regional identities based on language and ethnic groups. These tensions, so evident in the crisis of 2006, touch the cultural heritage landscape just as they influence broader debates over national identity and history.

NATIONAL MUSEUM OF THE RESISTANCE

Housed in the former Portuguese-era courthouse in Dili, the Archive and Museum of the East Timorese Resistance opened in early 2006 and was expanded considerably in 2012 and 2013. In a more conventional museum style but with innovative exhibits, the museum differs from the CAVR in that its archives preserve contemporary documents, artefacts and photos of the resistance movement between 1975 and 1999, rather than subsequent victim testimonies. Supported by the Mario Soares Foundation and the Association of Resistance Veterans, the archives include the document collections of key resistance leaders and FALINTIL brigades, as well as those of the clandestine front operating in towns, along with exhibits such as FALINTIL leader Konis Santana's typewriter.

The key exhibition is a series of fifty-two life-size panels on the history of the resistance, with photos and interpretative text, ordered chronologically and thematically, with an accompanying catalogue in Tetum, Portuguese and English. As one of the curators, José Mattoso, argues, in a country in which 43 per cent of the inhabitants are under 15 years of age, the collective history of the resistance in Timor-Leste 'will remain a fact expressed by a fragile memory' unless preserved and recorded quickly for future generations (Mattoso 2004). While the exhibition is impressive, beautifully housed in the renovated court of justice, only three panels focus specifically on the key role of the youth- and student-dominated civilian resistance. As I will argue further, this lower level of recognition accorded to the youth contribution to the independence struggle is a feature of the cultural heritage landscape, and one that reflects the broader intergenerational 'fault line' in post-independence politics.

MASSACRE SITES: SUAI AND LIQUICA

Some 2,600 East Timorese are estimated to have been killed by the TNI or their proxy militias in the violence leading up to and following the independence referendum in September 1999 (CAVR 2005, 44). The most notorious massacres of civilians took place in or around church sites, while the victims were seeking shelter. Some of them have been honoured at the district level, normally with some outside assistance. For example, the monument to the 200 victims of the Suai massacre was built by people of the Covalima district with support from the East Timorese government, the governments of Ireland and the UK, and the UN Serious Crimes Unit.

In examining the post-independence cultural heritage landscape of Timor-Leste, it is important to acknowledge different levels of memorialization and to look beyond the realm of formal monuments. In the grounds of the Suai church itself, local memorials, often family-made, offer simple remembrances of particular individuals killed on that day, including three priests. As these sites were already sacralized spaces, the church and local parishioners have exercised much of the responsibility for the less formal memorials of these tragic events. Similarly, the small but moving monument to the sixty victims of the Liquica church massacre was clearly designed and constructed at the local community and parish level by those closest to the victims. All over Timor-Leste, local memorials for those who died in the violence in 1999 may be found, organized at the village or community level. Indeed, there is scarcely a town above village size without one.

OTHER MEMORIALS AND 'IMMANENT' CULTURAL HERITAGE

These more informal memorials include locally made monuments in hiding places for FALINTIL resistance fighters, built by the proud community members who assisted them. The site of the Santa Cruz massacre in 1991 is itself commemorated informally with votive candles on the front gate. Though the informal memorial is very moving, to some of the younger generation the lack of a more formal memorial was considered a strange omission for the first ten years of independence, given the critical nature of the Santa Cruz massacre in garnering world attention after 1991.

Other more difficult sites are partly remembered, recalled by one side of the brief but bloody civil war in 1975 between FRETILIN and UDT (Timorese Democratic Union). For example, sites such as the *Armagem* (storehouse) in Aileu—used as a FRETILIN jail during the early and difficult years of Indonesian occupation when FRETILIN still controlled the hinterland—stir difficult and divisive memories in the town even today. Similarly, a small local memorial on the road to Same in central Timor, commemorating two victims of the civil war 'barbarically assassinated by FRETILIN', shows small signs of local dissent from official narratives of the cultural heritage landscape. This 'whispered' remembering of a difficult era highlights the unresolved legacies of the civil war. Despite some genuine efforts at reconciliation, including the CAVR process which covered the civil war period, these sites highlight aspects of East Timorese history considered inimical to the nation-building task, if not to the wider process of reconciliation.

The final level—and in some ways the most important—is the unintentional or 'immanent' cultural heritage landscape, consisting of the unrestored wreckage of houses and buildings burned or damaged by departing TNI and their militia in 1999. Alongside these ubiquitous sites, invariably festooned with militia graffiti, are other unforgotten but abandoned sites, such as the Indonesian interrogation centre in Baucau, which lies boarded up immediately behind one of the key tourist attractions of the area, the colonial era Baucau Pousada hotel.

CULTURAL HERITAGE AS RECOGNITION

Memorials and repositories of difficult national memories, like the Comarca, seek to make sense of the collective experiences of a people in ways that foster a sense of national unity and valorize the pain and trauma of all those who suffered in the struggle for liberation. An essential element of 'nation building', more broadly, is the cultural production of *unifying* narratives of collective identity and history. For these reasons, a certain level of popular legitimacy must support 'official' (constitutional or state-endorsed) narratives of cultural nationalism. However, as I have argued, the 'national' values and culture of the 'imagined community' may in fact privilege those of a dominant nationalist grouping and contribute to cultural and political conflict after independence. In this section, I employ a 'recognition approach' to examine the way in which the nationalist cultural heritage landscape in Timor-Leste has valorized some contributions to the independence struggle more than others.

Broadly speaking, a recognition approach examines the way distorted or inadequate forms of recognition may become important sources of motivation for political mobilization and resistance (Honneth 1995, 138–9). Perceived 'disrespect' to a group's sense of self, to its traditions and values, or a perceived 'misrecognition' of its contribution to shared and valued social goals, such as national independence, may create the conditions for political conflict (Honneth 1995, 121–43). As previously noted, young people clearly feel misrecognized by some aspects of East Timorese cultural nationalism after independence.

With these intergenerational tensions in mind, it is useful to review Timor-Leste's cultural heritage landscape on a 'recognition' basis. Broadly speaking, the success of Timor-Leste's independence struggle was due to a unique combination of three forces: the armed resistance of FALINTIL (the armed wing of FRETILIN, and later of CNRT (National Council of the Timorese Resistance), both of which were dominated by first-generation nationalists); the underground civilian resistance (dominated by youth and student groups); and the diplomatic front and international solidarity networks.

While the political leadership of FRETILIN and CNRT fared well in the post-independence government, the armed resistance of FALINTIL initially had some serious problems with recognition, with many former veterans feeling inadequately reintegrated into society and inadequately compensated for their sacrifices. There has been substantial progress since 2005 in the form of veteran registration processes, pensions and retraining programs, and in cultural heritage terms, the Heroes Monument at Metinaro and the National Museum of the Resistance strongly highlight this first key element of the resistance in the memorial landscape.[9]

By contrast, the youth-dominated civilian resistance still remains comparatively neglected in the nationalist cultural heritage landscape. The 12 November anniversary of the Santa Cruz massacre is a public holiday, dedicated to the victims of the massacre (National Youth Day), and there is a system of medals for veterans of the of clandestine resistance. Yet it is surprising that there was no formal monument to the victims of the internationally infamous massacre until 2012, and there is still, at the time of writing, no formal monument on or near the Santa Cruz site itself.[10] There is some further controversy attached to the new Santa Cruz monument at Motael, which was installed without consulting the 12 November committee led by Gregorio Saldanha, the leader of the original demonstration who now presides over the committee to maintain awareness of their sacrifices (da Silva 2012). It has also met criticism for not listing the names of the estimated 250 victims. The significance of this site to the East Timorese independence struggle cannot be overstated, as it was the footage of the massacre in 1991 that put East Timor's plight firmly back on the world stage. Indeed, as early as 1999, some commentators noted that the FALINTIL leader Xanana Gusmao 'took the time to console the widows of fallen FALINTIL heroes . . . but did not take the time to join the young people in the first Santa Cruz massacre commemoration after independence' (Aditjondro 1999). More broadly, the extent of the contribution of the *juventude* (youth) tends to be publicly neglected in favour of military veterans' issues and histories in post-independence politics. In a parallel feature, as Fernandes (2011, 125) notes, the civilian clandestine resistance has as yet been relatively neglected by historians. As he argues, the history of the clandestine front movements and their 'vital yet often unacknowledged role' in the independence struggle is yet to be fully documented. In sum, there is to date comparatively little recognition in the nationalist memorial landscape of the youth contribution to the liberation struggle. This misrecognition forms part of a broader set of intergenerational tensions over postcolonial political settlements in Timor-Leste, which together represented key background factors contributing to the crisis of 2006.

The third force—that of the diplomatic front and international solidarity campaigns, including thousands of diaspora Timorese—has also been relatively neglected in nationalist memorials. In 2002, an International People's Park commemorating the role of international solidarity in securing the referendum for independence was created in a prominent beachside park at Lecidere in Dili. By 2005, the original memorial had been mysteriously sidelined by a new and larger monument commemorating the UN's and various international governments' peacekeeping contributions since 1999. This resulted in a sail-sculpture in the original park being removed and a small original monument and plaque being physically sidelined by a new structure and large signs emphasizing the corporate and governmental sponsors of the monument. Rather than finding a new site to acknowledge the important role of state-based internationalism since 1999, the prior site—recognizing broader notions of the international 'people-to-people' solidarity so critical to the independence struggle—was simply displaced.

By the 2007 elections, some of these gaps in the heritage landscape were being acknowledged by key opposition figures from the older generation, signalling a growing awareness of intergenerational tensions as a potent issue in Timor-Leste's politics. Perhaps responding to the former FRETILIN government's neglect of these issues, President Gusmão, then running for prime minister with a new opposition party, awarded 'clandestine resistance' medals in the presidential election campaign period, honouring many former activists, including victims and survivors of the Santa Cruz massacre. Heritage issues were also spoken of as part of the agenda of the then newly elected president, Jose Ramos-Horta, including the need for a memorial to the victims of the massacre at Santa Cruz, and—again reflecting the links between ideology and cultural heritage, and a shift in power to a new government closer to the Catholic Church—a new memorial on the site of Pope John Paul II's visit in 1989.[11] Despite this promising start to a new government in 2007 and substantial progress in recognition of veterans in the years since, the youth and clandestine groups still do not feature prominently in the heritage landscape. For victims groups, seeking justice for the crimes of 1974–99, the lack of recognition is even more profound.

CONCLUSION

A patchwork of cultural heritage sites, authored by several generations of colonial and nationalist elites, serve as a reminder of Timor-Leste's long and difficult history of occupation, resistance and ultimately of national liberation. These different historical 'layers' of cultural heritage offer competing visions of the past in Timor-Leste—visions which still echo in the memorial landscape, recalling the distinct political projects of successive regimes. For the Portuguese colonial era, with a visual narrative of 'Portugalization' evident in monuments and grand churches, a typical *mission civilitrice*, colonial discourse of progress from animist 'backwardness' to 'civilization' was emphasized. For the Indonesian era, with its integration monuments seeking to highlight imagined precolonial unities, the visual narrative is one of 'reunification' and of Asian resistance to European colonialism, subsuming the East Timorese struggle under the aegis of its own nationalist narrative. And for East Timorese nationalists, the broad narrative is one of *funu*, or the struggle of a united people against consecutive colonial occupations. Each offers competing visions of East Timorese collective identity, its origins and its history.

Behind the visual competition between layers, each layer has its own contradictions—and the East Timorese nationalist project is no exception.

It has been argued that while key sites in the nationalist heritage landscape have succeeded in preserving difficult national memories for future generations, more could yet be done to valorize all participants in the struggle for independence equally and, in particular, the youth- and student-dominated civilian resistance. Secondly, it has been argued that this pattern of misrecognition relates strongly to other cleavages in post-independence politics, which contributed directly to the political-military crisis of 2006. These include broad intergenerational tensions but also the 'east-west' conflict, itself linked to unresolved legacies of the occupation and 'recognition' disputes over contributions to the resistance struggle. Finally, as ongoing uncertainty over the Comarca site demonstrates, cultural heritage policy is also linked with other political problems of the independent state: of reconciling good relations with neighbours with the continuing need for post-conflict justice. Perhaps inevitably, then, the difficult issues of cultural heritage management mirror the larger challenges of nation building in Timor-Leste: of recognizing different generations' experiences in the liberation struggle and accommodating the ongoing power of local and regional identities in presenting a unified national story.

NOTES

This chapter is an updated and expanded version of Leach (2009).

1. CAVR's estimate of the minimum total number of conflict-related deaths is 102,800 (±12,000). This figure includes both killings and deaths due to privation. The often cited figure of 180,000 is CAVR's upper estimate of total conflict-related mortality.
2. This chapter uses the official name *Timor-Leste* to refer to the post-2002 nation state and *East Timor* when referring to the pre-2002 territory. *East Timorese* is employed as the demonym throughout.
3. That is, Portuguese-speaking.
4. As Jose Ramos-Horta put it, 'If you take away Portuguese language and religion, there is no such thing as East Timor' (cited in Chesterman 2001).
5. Prime Minster Gusmao was quoted as saying that 'this suffering could have been prevented if the Australian forces had not come to Timor-Leste in order to wage war here so as to prevent the Japanese from invading Australia'.
6. *Panca Sila* (Five Principles) is the official Indonesian 'state philosophy', comprising monotheism, justice, unity, democracy and social justice.
7. The Revolutionary Front for an Independent East Timor.
8. Including the Famine Museum in Ireland, Hiroshima Peace Memorial Museum and the Port Arthur Museum in Tasmania.
9. Likewise, government pensions and infrastructure contracts to veterans now constitute a substantial feature of the budget of government, to the point where many believe it should be reduced to become more sustainable.
10. This is despite a government-announced design competition, co-sponsored by the November 12 committee in 2010 (RDTL 2010), and parliamentary mention of the idea as early as 2004.
11. The Catholic Church had been critical of the former FRETILIN government and protested its strong opposition to compulsory religious education in government schools.

BIBLIOGRAPHY

Aditjondro, G. (1999) 'Self-determination under globalisation: Timor Lorosae's transformation from Jakarta colony to a global capitalist outpost', paper presented to Protesting Globalisation: Prospects for Transnational Solidarity conference, Sydney, University of Technology, Sydney, 10–11 October.

Anderson, B. (1983) *Imagined Communities: Reflections on the Origin and Spread of Nationalism*, London: Verso.

Anderson, B. (1993) 'Imagining East Timor', *Arena Magazine*, 4 (April–May), 23–7.

CAVR. (2003) 'Comarca: from colonial prison to centre for reconciliation and human rights', Dili: CAVR.

CAVR (2005) *Chega! The Report of the Commission for Reception, Truth and Reconciliation in Timor-Leste*, Executive Summary, Dili: CAVR.

Chesterman, S. (2001) 'East Timor in transition: from conflict prevention to state-building', *International Peace Academy Reports* (May).

da Silva, M. (2012) 'Harii Monumentu 12 Novembru St. Cruz Sai Fali Estatua [Monument to 12 November Santa Cruz is instead a statue]', *Jornal Independente*, 14 November.

Feijo, R. and Viegas. S. (2014) 'Mártires e Antepassados: Notas de uma pesquisa em curso [Martyrs and Ancestors: Notes on a study in progress]', paper presented to the A Produção do Conhecimento Científico em Timor-Leste conference, Universidade Nacional de Timor Lorosae, Dili 13–15 August.

Fernandes, C. (2011) *The Independence of East Timor: Multi-Dimensional Perspectives—Occupation, Resistance and International Political Activism*, Brighton: Sussex Academic Press.

Gunn, G. C. (1999) *Timor Loro Sae: 500 Years*, Macau: Livros do Oriente.

Gunn, G. C. (2001) 'The five-hundred year Timorese *Funu*', in R. Tanter, M. Selden and S. Shalom (eds), *Bitter Flowers, Sweet Flowers: East Timor, Indonesia and the World Community*, Sydney: Rowman and Littlefield, 3–14.

Honneth, A. (1995) *The Struggle for Recognition: The Moral Grammar of Social Conflicts*, Cambridge: Polity.

Leach, M. (2002) 'Valorising the resistance: national identity and collective memory in East Timor's constitution', *Social Alternatives*, 21 (3), 43–7.

Leach, M. (2003) 'Privileged ties: young people debating language, heritage and national identity in East Timor', *Portuguese Studies Review*, 11 (1), 137–50.

Leach, M. (2006) 'East Timorese history after independence', *History Workshop Journal*, 61 (1), 222–37.

Leach, M. (2007) 'History teaching in East Timor: challenges and alternatives', in D. Kingsbury and M. Leach (eds), *East Timor: Beyond Independence*, Clayton: Monash Asia Institute Press, 193–207.

Leach, M. (2008) 'Surveying East Timorese tertiary student attitudes to national identity: 2002–2007', *South East Asia Research*, 16 (3), 405–31.

Leach, M. (2009) 'Difficult memories: the independence struggle as cultural heritage in East Timor', in W. Logan and K. Reeves (eds), *Places of Pain and Shame: Dealing with 'Difficult Heritage'*, London: Routledge, 144–61.

Leach, M. (2012) 'Longitudinal change in East Timorese tertiary student attitudes to national identity and nation building, 2002–2010', *Journal of the Humanities and Social Sciences of Southeast Asia and Oceania*, 168 (2–3), 219–52.

Logan, W. (2003) 'Hoa Lo: a Vietnamese approach to preserving places of pain and injustice', *Historic Environment*, 17 (1), 27–31.

Mattoso, J. (2004) 'The resistance archives and national identity', *Timorese Resistance in Documents*, Dili: Mario Soares Foundation.

Murdoch, L. (2010) 'East Timor leader accuses Australia over war', *The Age*, 9 April. Online. Available HTTP: www.theage.com.au/national/east-timor-lea

der-accuses-australia-over-war-20100408-rv43.html#ixzz3CzikbZqU (accessed 15 August 2014).

Ramos-Horta, J. (1987) *Funu: The Unfinished Saga of East Timor*, Trenton, NJ: Red Sea Press.

Ramos-Horta, J. (1996) *Nobel Lecture*, Stockholm: Nobel Foundation. Online. Available HTTP: www.nobelprize.org/nobel_prizes/peace/laureates/1996/ramos-horta-lecture.html (accessed 11 August 2014).

RDTL. (2010) 'Open competition for the best drawing of the monument dedicated to the 12th of November 1991 Massacre. 25 October.' Online. Available HTTP: http://timor-leste.gov.tl/?p=4153&lang=en&n=1(accessed 18 August 2014).

Wiley, N. (1994) 'The politics of identity in American history', in C. Calhoun (ed), *Social Theory and the Politics of Identity*, Cambridge: Blackwell, 130–49.

Contributors

Gilly Carr is a Senior Lecturer and Academic Director in Archaeology at the University of Cambridge's Institute of Continuing Education. She is also a Fellow and Director of Studies in Archaeology and Anthropology St Catharine's College. Her fieldwork is currently based in the Channel Islands, where she has worked and published extensively on the archaeology, heritage and memory of the German occupation. Her latest monograph is *Legacies of Occupation: Archaeology, Heritage and Memory in the Channel Islands* (Springer, 2014). She also works with former deportees who were sent to German internment camps during WWII. Her museum exhibition, *Occupied Behind Barbed Wire*, was on display in Guernsey and Jersey Museums in 2010 and 2012, respectively. Other publications include *Prisoners of War: Archaeology, Memory and Heritage of 19th and 20th Century Mass Internment*' (Springer, 2012) and *Cultural Heritage and Prisoners of War: Creativity Behind Barbed Wire*' (Routledge, 2012), both co-edited with Harold Mytum, and *Protest, Defiance and Resistance in the Channel Islands, 1940–1945*, coauthored with Paul Sanders and Louise Willmot (Bloomsbury Academic, 2014).

Joseph Cheer is currently Lecturer and Associate Director of the Australia & International Tourism Research Unit (AITRU) at the National Centre for Australia Studies (NCAS), Monash University. His recently completed PhD (anthropology) *Tourism and the Neotraditional: An Ethnography of Tourism and Traditional Culture in Vanuatu* builds on his professional experience in the related fields of international development and tourism in the Pacific Islands. His research interests are focused on development, livelihoods, tradition, transformation, cultural change, postcolonial legacies and, in particular, the impacts of tourism expansion. Joseph's research has been published internationally including in *Annals of Tourism Research*, *Journal of Heritage Tourism*, *Pacific Economic Bulletin*, and *Tourism Analysis and Tourism Planning and Development.*

Sandro Debono studied art history at the University of Malta and is currently Senior Curator at the National Museum of Fine Arts (Malta) and leading the new national museum of art project (MUŻA). Research interests include cultural policy, collections management and the socio-historical relativity of art and artistic expression. He has also researched core-periphery issues and the migration of aesthetic ideas, iconographies and stylistic trends. Sandro is currently concluding his doctorate on heritage policy and collections development at University College (London).

Rob van Ginkel is a Senior Lecturer in Cultural Anthropology at the University of Amsterdam's Department of Anthropology. He has done extensive field research on the island of Texel in 1989–1991 and 2005–2006, while also being a resident there 2006–2011. His most recent research project concerns the commemoration of WWII in the Netherlands, which resulted in a book entitled *Rondom de stilte. Herdenkingscultuur in Nederland* [*Around the Silence. Commemorative Culture in the Netherlands*] (2011).

Maria Kagiadaki is an Art Historian at the Hellenic Ministry of Culture, Education and Religious Affairs and a tutor at the Hellenic Open University. She was Visiting Lecturer at the Department of Primary Level Education of the Democritus University of Thrace (2008–2013) and at the Department of Applied and Visual Arts of the University of Western Macedonia (2008–2011). She did her undergraduate degree in History and Archaeology and her PhD in Art History at Aristotle University of Thessaloniki. She completed her MPhil in Heritage Management and Museums at the University of Cambridge.

Rick Knecht is Senior Lecturer in Archaeology at the University of Aberdeen, Scotland. He specializes in the archaeology of high-latitude North America, especially Alaska and the Aleutians, where he works closely with indigenous communities and has established several museums. Rick was previously ethnographer and director of the Oral History Program for the Bureau of Arts and Culture in Palau. In addition to further research in the Pacific, he is currently directing excavations at a number of Yup'ik sites in Alaska.

Irene Lagani is a Professor of European History (twentieth century) at the University of Political and Social Sciences in Athens. She studied Political Sciences in Athens (*Kapodistriako Panepistimio*) and History in Paris (*Universite de Paris I-Sorbonne*), where she completed her thesis on 'Greece and Its Balkan Neighbours from 1941 to 1949'. Her research interests are in the field of Cold War, the history textbooks and the refugees of the Greek Civil War. She is author of 'The 'Abduction' of Children and Greek Yugoslav Relations 1948–52' and co-editor of 'The Children of the Greek Civil War in Central and Eastern Europe' (2013). Her publications also include, inter alia, '*Nostos (nostalgia)* and Cold War Political Considerations: The Greek Civil War Refugees' (*Journal of the Hellenic Diaspora*, NY, 2013), *The Presentation of Wars in History Textbooks of Greece* (Georg Eckert

Institute, Hannover, 1996), *The Resistance Movement Through the Greek and French history Textbooks* (Salonica, 2008).

Michael Leach is an Associate Professor in Politics and Public Policy at Swinburne University of Technology in Melbourne, Australia. His has researched and published widely on nation building and national identity in Timor-Leste and Melanesia. He is co-editor (with D. Kingsbury) of *The Politics of Timor-Leste: Democratic Consolidation After Intervention* (Cornell Southeast Asia Program Publications, 2013) and coauthor of *Attitudes to National Identity in Melanesia and Timor-Leste: A Survey of Future Leaders in Papua New Guinea, Solomon Islands, Vanuatu and Timor-Leste* (Peter Lang, 2013).

Jacqueline Leckie is Associate Professor and Head of the Department of Anthropology and Archaeology at the University of Otago. She has undertaken extensive research and produced several publications on the South Pacific. She is currently working on a history of madness and colonization in Fiji. Her books include *Indian Settlers: The Story of a New Zealand South Asian Community* (Otago University Press, 2007), *To Labour with the State* (University of Otago Press, 1997), editing *Asians and the New Multiculturalism in Aotearoa New Zealand* (Otago University Press, 2015), *Development in an Insecure and Gendered World* (Ashgate, 2009), *Localizing Asia in Aotearoa* (Dunmore, 2011), *Recentring Asia: Histories, Encounters, Identities* (Brill, 2011) and *Labour in the South Pacific* (James Cook University, 1990). She is President of the Pacific History Association.

Gavin Lindsay is a PhD student in Archaeology at the University of Aberdeen, Scotland. His research has included wartime landscape characterization and public interpretation work for the Island of Hoy Development Trust in Orkney and the Peleliu War Historical Society in Palau. A specialist in island conflicts and landscapes of war, Gavin has many years of fieldwork experience in archaeology and on community engagement projects involving recent conflict and memory.

Lamont Lindstrom, Kendall Professor and Chair of Anthropology at the University of Tulsa, has long-term research interests in Vanuatu and other Melanesian countries including WWII ethnohistory, local social movements, contemporary chiefs, the politics of tradition, cultural policy and sociolinguistics. His current research focuses on the careers of early cinematographers Martin and Osa Johnson and on urban migration and personhood in Vanuatu.

Elena Mamoulaki is currently lecturing as visiting Assistant Professor at the Programme in Hellenic Studies at Columbia University. She holds a PhD in Social and Cultural Anthropology from the University of Barcelona

(2013). She received a diploma in Architecture from Aristotle University of Thessaloniki, an MSc in Design-Space-Culture from the National Technical University of Athens (2005) and an MA in Anthropology from the University of Barcelona (2008). She has taught as Lecturer at the Department of Architecture of the Aristotle University of Thessaloniki (2007–2009), at the Arcadia University in Athens (2012–2014) and at the National and Kapodistrian University of Athens for the Programme Plato's Academy (2012–2015). Her research focuses on the narratives, practices and materialities of the social memory of the 1940's in Greece—especially political exile and its transgenerational legacy.

Nota Pantzou is currently a Lecturer of Heritage Management at the University of Patras, Greece. Her research interests include archaeological heritage and museum management, with a focus on local communities, contemporary archaeology, traumatic heritage and socio-politics of the past. She received her undergraduate degree in Archaeology and History of Art from Aristotle University of Thessaloniki and her postgraduate degree in Archaeological Heritage Management and Museums from Cambridge University. In 2009 she completed her doctorate in Archaeology at University of Southampton.

Hazal Papuccular has an undergraduate degree in Political Science and International Relations, and an MA degree in Modern Turkish History. She is a PhD candidate at Bogazici University in Istanbul, Turkey, in the Department of Modern Turkish History. She is now writing her dissertation on the Dodecanese Islands with the title of 'War or Peace? Dodecanese Islands in Turkish Foreign and Security Policy, 1923–1947'. Her research interests are in Turkish foreign policy, twentieth-century European history, and Aegean Sea studies. Her publications include *Turkish–Italian Relations in the Interwar Period: Italian Mare Nostrum Policy* and the *Formulation of Turkish Foreign Policy in Response* (Lambert, 2010).

Tony Pollard is Director of the Centre for Battlefield Archaeology at the University of Glasgow, where he is a Senior Lecturer in History and Archaeology. In addition to teaching military history, he is Convenor of the Centre's master's programme in battlefield and conflict archaeology and a supervisor of PhD students. He has written widely for both popular and academic audiences and in 2005 cofounded the *Journal of Conflict Archaeology*, which he co-edits with Dr Iain Banks.

Neil Price is Professor of Archaeology at the University of Uppsala, Sweden. A leading specialist on the Vikings and ancient religion, he has global interests in a wide range of archaeological subjects. Alongside further work on WWII in the Pacific, he is currently researching Viking burials, historical piracy, and the archaeology of the nineteenth-century opium trade.

Keir Reeves is Professor of Australian History and is the founding director of the Collaborative Research Centre for Australian History (CRCAH) at Federation University Australia. Previously he was a Senior Research Fellow in the School of Journalism, Australian and Indigenous Studies at Monash University. His current research concentrates on Asia and the Pacific cultural heritage and history. He is currently involved in two major Australian Research Council projects that interrogate war and memory. In 2013, he was a Visiting Fellow at Clare Hall Cambridge and a visiting researcher at the McDonald Institute for Archaeological Research, University of Cambridge, where he worked with the Cambridge Heritage Research Group in the Department of Archaeology and Anthropology. He is also a Visiting Researcher at Ghent University. Keir's key publications include co-editing (contributing to) *Places of Pain and Shame: Dealing with 'Difficult Heritage'* (Routledge, 2009) with Bill Logan, and more recently he has contributed to *Anzac Journeys: Walking the Battlefields of the Second World War* (Cambridge University Press, 2013). He has published on heritage, history and travel in *Annals of Tourism Research*, *International Journal of Heritage Studies*, *Tourism Analysis*, *Tourism Management*, *Landscape Research*, *Australian Historical Studies* and *Critical Asian Studies*. He is the current historian member of the Heritage Council of Victoria and is interested in cultural heritage in Asia, Australia and the Pacific.

Daniel Travers is currently Adjunct Professor of History at Laurentian University in Canada. Specializing in how the identity of island societies can be shaped in relation to historical and cultural association with Britain, Daniel was awarded his PhD with Vice-Chancellor's Distinction from the University of Huddersfield in May 2012. He is the author of published articles on the memory of war in Jersey, Orkney and the Isle of Man and was co-editor (with Jodie Matthews) of *Islands and Britishness: A Global Perspective* (Cambridge Scholars, 2012).

Geoffrey White is a Professor of Anthropology at the University of Hawaii. From the 1980s to the present, he has been working on indigenous memories of WWII in the Pacific, American memories of Pearl Harbor at the national memorial and, more recently, American tourism of D-Day memorial sites. Relevant publications include *The Pacific Theater: Island Memories of World War II* (co-edited, University of Hawai'i Press, 1989), *Black and White Memories of the Pacific War* (coauthored, Smithsonian, 1990), and *Perilous Memories: The Asia Pacific War(s)* (co-edited, Duke University Press, 2001).

Index

For Product Safety Concerns and Information please contact our EU representative GPSR@taylorandfrancis.com
Taylor & Francis Verlag GmbH, Kaufingerstraße 24, 80331 München, Germany

www.ingramcontent.com/pod-product-compliance
Ingram Content Group UK Ltd.
Pitfield, Milton Keynes, MK11 3LW, UK
UKHW022115190926
475702UK00013B/1465
* 9 7 8 0 8 1 5 3 4 6 9 6 8 *